Edward John Chapman

Blowpipe Practice

An outline of blowpipe manipulation and analysis, with original tables for the determination of all known minerals

Edward John Chapman

Blowpipe Practice

An outline of blowpipe manipulation and analysis, with original tables for the determination of all known minerals

ISBN/EAN: 9783337219819

Printed in Europe, USA, Canada, Australia, Japan

Cover: Foto ©berggeist007 / pixelio.de

More available books at **www.hansebooks.com**

CHAPMAN'S

BLOWPIPE PRACTICE

AND

MINERAL TABLES.

BY THE SAME AUTHOR.

AN OUTLINE OF THE GEOLOGY OF CANADA,

BASED ON A SUBDIVISION OF THE PROVINCES
INTO NATURAL AREAS.

With six sketch-maps and 86 figures of characteristic fossils.

By E. J. CHAPMAN, Ph.D., LL.D.

This work presents a synoptical view of the geology of the entire Dominion. It is used as a book of reference in University College, Toronto ; in Queen's College and University, Kingston ; and in the University of Halifax and Science Department of Dalhousie College, Nova Scotia.

COPP, CLARK & CO., 1877.

BLOWPIPE PRACTICE.

AN OUTLINE

OF

BLOWPIPE MANIPULATION AND ANALYSIS,

WITH ORIGINAL TABLES

FOR THE

DETERMINATION OF ALL KNOWN MINERALS.

BY

E. J. CHAPMAN,

Ph.D., LL.D.

PROFESSOR OF MINERALOGY AND GEOLOGY IN UNIVERSITY COLLEGE AND SCHOOL OF PRACTICAL SCIENCE, TORONTO.

TORONTO:
COPP, CLARK & CO., 47 FRONT STREET EAST.
1880.

INTRODUCTORY NOTICE.

The title-page to this little work indicates succinctly the scope and character of the book. The work comprises two distinct parts: an introductory sketch of the use of the Blowpipe in qualitative mineral examinations; and a series of Tables, with chemical and crystallographic notes, for the practical determination of minerals, generally. In the first portion of the work, the writer's aim has been to systematise and condense as far as possible: but, although confessedly a mere outline of the subject, this introductory portion will not be found altogether devoid of original matter. The sixth section, more especially, contains a new and greatly simplified plan of BLOWPIPE ANALYSIS, by which the general composition of an unknown substance may be determined in most cases very rapidly and with comparatively little trouble. As a rule, the methods of Blowpipe Analysis, hitherto published, are little more than Tables of Reactions. They attempt no separation of electro-negative bodies from bases, but mix up the two, very illogically; and they exact the performance of many unnecessary experiments, by which certain components become detected over and over again, whilst others escape detection, altogether, or are recognized only after much unnecessary delay.* These defects are remedied very materially, it is thought, in the method now proposed. The Determinative Tables, which occupy the second and principal portion of the work, are also original. In their arrangement, an attempt is made to place bodies of related composition, only, under the same subdivision: so as to avoid, wherever possible, the unnatural collocations so commonly seen in Tables of this character. It will be evident, however, that without greatly increasing the number of the Tables, complete success in this respect is not always attainable. The Tables include, practically, all

---

* After the first part of this work was in type and entirely struck off, the author received from HERR LANDAUER, of Brunswick, a copy of his "*Systematischer Gang der Löthrohr-Analyse.*" Herr Landauer's method entirely meets the above objections, and is without doubt the most satisfactory plan of Blowpipe Analysis hitherto published. It has been subsequently incorporated by its author into a little work on the Blowpipe, an English translation of which, under the title of "Blowpipe Analysis," has recently appeared.

known minerals; but as many of these are rarely met with, or are comparatively of little importance, an Explanatory Note, referring only to species of ordinary occurrence, is attached to each Table. In these Notes, more especially in those which relate to the concluding Tables of the series, additional information is given respecting the crystallization, spectroscopic reactions, and other distinctive characters of leading species. The spectroscope recommended for use, in these investigations, is a simple, direct-vision pocket-spectroscope, such as can be carried very conveniently, with accompanying Bunsen-burner (the foot unscrewed), in a spare corner of the blowpipe case.

SCHOOL OF PRACTICAL SCIENCE, TORONTO:
August 12th, 1880.

BRIEF SKETCH OF THE HISTORY OF THE BLOWPIPE.

The use of the Blowpipe, in the arts, dates from a very distant period—a simple form of the instrument having been long employed, in the process of soldering, by jewellers and other workers in gold and silver. This employment must naturally have suggested its use to the alchemists; and in the curious collection of woodcuts known as the *Liber mutus*, in which an alchemist, assisted by his wife, is depicted in the performance of various chemical operations, the use of the blowpipe is clearly indicated. The *Liber mutus* is of very uncertain date, but it belongs, in all probability, to the beginning of the seventeenth century. The alchemist is here employed, it is true, not in the actual examination of a substance by his blowpipe, but in the construction or sealing up of a glass vessel. Nevertheless, the use of the instrument in the conversion of calc spar into lime is pointed out by ERASMUS BARTHOLIN in his treatise on Iceland Spar, written in 1670; and in the *Ars vitraria experimentalis* of KUNCKEL, published in 1679, the blowpipe is recommended for use in the reduction, on charcoal, of metal-holding bodies, the requisite blast being produced by a pair of air-tight bags. In 1702, the celebrated alchemist JOHANN GEORG STAHL distinctly refers to the reduction of lead and antimony, by the fusion of what are now known as the oxides of these metals, on a piece of charcoal, by means of a "soldering pipe" or *tubulo cæmentorio aurifabrorum*. JOHANN ANDREAS CRAMER, in his *Elementia docimasticæ* (1739) describes the use of the instrument in the examination of small particles of metallic bodies, and suggests the use of borax (long previously employed in soldering, and also by the alchemists in crucible operations) for this purpose. He gives also a description of a mouth blowpipe provided at its lower end with a cylindrical reservoir for the retention of the moisture which condenses from the operator's breath.

In Sweden, a few years later (1746), SWEN RINMAN published some details on the examination of ferruginous tin-ore, and other minerals, by the blowpipe; and, in 1748, ANTON VON SWAB—usually, but erroneously, cited as the first person by whom the blowpipe was used in its scientific applications—referred to the use of the instrument in a paper on the occurrence of native antimony. BERGMAN states that VON SWAB employed the blowpipe in 1738, but the date of his first publication in which reference is made to its use is ten years later, as pointed out by Dr. HERMANN KOPP in his valuable *Geschichte der Chemie*: 1844.

Up to this time, however, no general or systematic use of the blowpipe appears to have been attempted; but in 1758, AXEL FREDERIC CRONSTEDT, who had previously employed the blowpipe in his researches on nickel (1751), published anonymously at Stockholm his celebrated treatise on Mineralogy, in which a chemical classification of minerals was first definitely essayed. In this work, the pyrognostic characters of minerals, as determined by the blowpipe, are brought prominently into notice; and in addition to borax, the two general reagents still in use, bicarbonate of soda ("*sal sodæ*") and microcosmic salt or phosphor-salt ("*sal fusibile microcosmicum*") are employed as blowpipe fluxes. To the English translation of Cronstedt's work published in 1770, GUSTAV VON ENGESTROM appended a short but complete sketch of the use of the Blowpipe, as then known; and JOHN HYACINTH DE MAGELLAN added somewhat to this sketch in the second (English) edition of the work, published in London in 1788. The plate which accompanies VON ENGESTROM's essay, exhibits a portable case of blowpipe apparatus, comprising, in addition to the blowpipe as devised by Cronstedt, a hammer, anvil, magnet, silver spoon and other articles (but none, of course, of platinum), with candle, charcoal, and three small bottles for fluxes. This essay of VON ENGESTROM, attached to his translation of Cronstedt's work, was translated into Swedish by RETZIUS in 1773; and in the same year the Swedish chemist TORBERN BERGMAN published a memoir on the blowpipe reactions of lime, magnesia, alumina, and silica; whilst, in 1774, SCHEELE described the action of the blowpipe on manganese ores, molybdenite, and other minerals. A few years later (1777) a complete treatise in Latin on the use of the Blowpipe was drawn up by Bergman, and published, soon after, under the editorship of Baron VON BORN, the metallurgist, at Vienna (*Commentatio de tubo ferruminatorio, etc.: Vindobonæ*, 1779). A Swedish translation, by HJELM, was issued at Stockholm in 1781.

In the preparation of this work, BERGMAN was very materially assisted by JOHANN GOTTLIEB GAHN. The latter chemist subsequently carried out an extended series of experiments with the blowpipe, and discovered various new methods of research. BERZELIUS, to whom at an after period he communicated personally his mode of operating, states that GAHN always carried his blowpipe with him, even on his shortest journeys, and submitted to its action every new or unknown substance that came in his way. In this manner he acquired great skill in the use of the instrument. He published nothing, however, on the subject; but, finally, drew up at the instigation of BERZELIUS the short sketch of the blowpipe and its applications contained in the latter's *Lärbok i Kemie* first issued in 1812. GAHN then undertook, in conjunction with

BERZELIUS, a complete blowpipe examination of all known minerals; but his death, in 1818, occurred almost at the commencement of this undertaking. BERZELIUS therefore carried on the investigation alone; and the results, together with all the improvements and new processes introduced by Gahn and by himself, were published at Stockholm under the title of *Afhandling om Blasrorets användende i Chemien*, in 1820. This work has formed the basis of almost all that has subsequently been published on the use of the Blowpipe in qualitative researches, although many new tests and methods of investigation have been discovered since its date. At the death of its distinguished author in 1853 it had entered its fourth edition, and had been translated into all the leading European languages. An English translation (taken however from a French version) by CHILDREN, appeared in 1821; and another by Whitney (from the fourth German edition by HEINRICH ROSE) was published at Boston, United States, in 1845.

A new era of blowpipe investigation commenced in 1827, when EDUARD HARKORT, of Freiberg in Saxony, applied the instrument to the assaying or quantitative examination of silver ores. HARKORT left Germany for Mexico, and died there, soon after the publication of his essay on this subject (*Probirkunst mit dem Löthrohre, Freiberg, 1827*); but CARL FRIEDRICH PLATTNER, to whom he had shewn his method of working, carried on this important application of the blowpipe, and published elaborate memoirs on the assaying, by this method, of gold, lead, copper, tin, nickel, and other metallic ores and furnace products. His great work on the Blowpipe, bearing a similar title to HARKORT'S earlier publication, appeared in 1835. It reached a third edition in 1853; and since Plattner's death in 1858, two other editions (the last in 1878) have been issued under the editorship of DR. THEODOR RICHTER, Plattner's successor in the Freiberg Mining Academy. This work has been translated into various languages. An American edition, by PROF. H. B. CORNWALL, appeared in 1875.

Of late years, the use of the Blowpipe has been greatly extended; and numerous original memoirs on points relating to Blowpipe Practice and Analysis have appeared from time to time in scientific journals. But the discussion of these more modern investigations belongs properly to a future time. The principal works published since the date of Plattner's treatise are mentioned at page 21 of the present volume. To these must be added the *Systematic Course of Analysis* of J. LANDAUER, referred to in the preceding note.

CONTENTS.

PART I.

§ 1.—THE BLOWPIPE: ITS STRUCTURE AND GENERAL USE	1
§ 2.—ACCESSORY APPLIANCES AND REAGENTS	4
§ 3.—STRUCTURAL PARTS AND CHEMICAL PROPERTIES OF FLAME	5
§ 4.—BLOWPIPE OPERATIONS:	
I. The Fusion Trial	8
II. Treatment in closed Tube:	
(i) Treatment in Flask or Bulb-Tube	9
(ii) Treatment in Closed Tube, proper	10
III. Roasting, and Treatment in Open Tube:	
(i) Roasting on Charcoal, Porcelain, and other supports	11
(ii) Roasting and Sublimation in Open Tubes	11
IV. Treatment with Nitrate of Cobalt	12
V. Formation of Glasses on Platinum Wire, or on Charcoal:	
(i) Details of Process; Flaming, &c.	12
(ii) Table of Borax Glasses	12
(iii) Phosphor-Salt Glasses	15
(iv) Glasses formed with Carb. Soda	15
VI. Reduction	15
VII. Cupellation	18
VIII. Fusion with Reagents in Platinum Spoon	20
§ 5.—BLOWPIPE REACTIONS:	
(i) *Non-metallic Bodies*	22
1, Oxygen; 2, Hydrogen; 3, Sulphur; 4, Selenium; 5, Nitrogen; 6, Chlorine; 7, Bromine; 8, Iodine; 9, Fluorine; 10, Phosphorus; 11, Boron; 12, Carbon; 13, Silicon.	
(ii) *Unoxidizable Metals*	30
14, Platinum; 15, Gold; 16, Silver.	
(iii) *Volatilizable Metals*	32
17, Tellurium; 18, Antimony; 19, Arsenic; 20. Osmium; 21, Mercury; 22, Bismuth; 23, Lead; 24, Thallium; 25, Cadmium; 26, Zinc; 27, Tin.	

CONTENTS.

§ 5.—BLOWPIPE REACTIONS—(Continued).

   (iv) *Flux-colouring Metals* 40
      28, Copper; 29, Nickel; 30, Cobalt; 31, Iron; 32, Tungstenum; 33, Molybdenum; 34, Manganese; 35, Chromium; 37, Uranium; 38, Cerium; 39, Titanium.

   (v) *Earth Metals* 51
      40, Tantalum (?); 41, Aluminium; 42, Glucinum; 43, Zirconium; 44, Yttrium.

   (vi) *Alkaline-Earth Metals* 54
      45, Magnesium; 46, Calcium; 47, Strontium; 48, Barium.

   (vii) *Alkali Metals* 57
      49, Lithium; 50, Sodium; 51, Potassium; 52, Ammonium.

§ 6.—PLAN OF ANALYSIS.

   (i) Determination of the Chemical Group to which a mineral substance belongs 60
   (ii) Determination of the Base or Bases 63

APPENDIX—ORIGINAL CONTRIBUTIONS TO BLOWPIPE ANALYSIS.

   1. Reaction of Manganese Salts on Baryta.. 71
   2. Detection of Baryta in the presence of Strontia 71
   3. Detection of Alkalies in the presence of Magnesia 72
   4. Method of Distinguishing the red flame of Lithium from that of Strontium 72
   5. Method of Distinguishing FeO from Fe^2O^3 in Silicates and other compounds 73
   6. Detection of Lead in presence of Bismuth.. 74
   7. Detection of Lithia in presence of Soda 74
   8. Action of Baryta on Titanic Acid 75
   9. Detection of Manganese when present in minute quantity in mineral bodies 75
   10. The Coal Assay 76
   11. Phosphorus in Iron Wire 81
   12. Detection of minute traces of Copper in Iron Pyrites and other bodies 82
   13. Detection of Antimony in tube sublimates 83
   14. Blowpipe reactions of Thallium 84
   15. Opalescence of Silicates in Phosphor-salt 86
   16. Reactions of Chromium and Manganese with Carbonate of Soda 87
   17. Detection of Cadmium in presence of Zinc in blowpipe experiments 88
   18. Solubility of Bismuth Oxide in Carbonate of Soda before the blowpipe 88
   19. Detection of Carbonates in Blowpipe Practice 89
   20. Detection of Bromine in Blowpipe Experiments 90
   21. Blowpipe reactions of Metallic Alloys 91

PART II.

ORIGINAL TABLES FOR THE DETERMINATION OF MINERALS.

    Introduction: Explanation of Crystal Symbols, &c. 95

    Analytical Index to the Tables 99

        Table I., 101; T. II., 103; T. III., 105; T. IV., 110 (N.B.—Cinnabar to be erased from this Table); T. V., 113; T. VI., 115; T. VII., 116; T. VIII., 117; T. IX., 121; T. X., 124; T. XI., 130; T. XII., 132; T. XIII., 135; T. XIV., 143; T. XV., 149; T. XVI., 151; T. XVII., 163; T. XVIII., 171; T. XIX., 174; T. XX., 178; T. XXI., 181; T. XXII., 182; T. XXIII., 186; T. XXIV., 195; T. XXV., 213; T. XXVI., 227; T. XXVII., 256.

    Index to Minerals described in Part II. 279

ADDITIONS AND CORRECTIONS.

Page 21.—The following works should be added to the list given in the foot-note on this page:—Blowpipe Analysis by J. LANDAUER (English edition), 1880; "Clavis der Silicate" by Dr. LEOP. H. FISCHER, 1864.

P. 24, line 8:—*for* "sulphates" *read* "most sulphates." See exceptions, in Note to Table XVI., page 162.

P. 28, bottom line: *for* BO^3, *read* B^2O^3.

P. 33, line 9:—*after* "the solution has a distinct reddish-purple colour," *add*, "and imparts a dark stain to metallic silver or lead test-paper, in the manner of a sulphur or selenium compound."

P. 58, line 3:—*erase* the comma after the word "various."

P. 59, at close of Potassium reactions, *add*, "If a piece of deep-blue glass, however, be held between the spectroscope and the flame, the potassium line will alone be visible."

P. 59, Foot-note. In reference to the statement in this note it may be observed that the ash of tobacco shews the red K-line, in the spectroscope, by simple immersion in the flame, but the Ca-lines only appear when the ash is moistened with hydrochloric acid. If lithium be present (as in the Perique tobacco, &c.) the crimson Li-line also comes out *per se*.

P. 61, under "Substances Indicated" (Expt. 1), *add*, "(2), Antimony, Tellurium."

P. 66. Molybdenum, placed under Group 2 on this page, should be placed, strictly, by itself, apart—as yielding infusible metallic grains and forming under certain conditions a slight sublimate. But its true place (as stated in the text) is in the Electro-Negative Table, and no error is likely to arise from the arrangement adopted.

P. 105. To description of MARCASITE, *add*, "but sp. gr. slightly lower, viz., 4·7–4·9;" and to description of PYRRHOTINE, *add*, "decomposed by hydrochloric acid, with separation of sulphur and emission of sulphuretted hydrogen odour.

P 110. *Cancel* CINNABAR. [The paragraph relating to this mineral slipped in, here, by some oversight during the printing of the work.] See page 121, its proper place.

P. 112, line 5 from bottom:—*for* "$\frac{1}{3}R$," *read* "—$\frac{1}{2}R$."

P. 119, foot-note:—*for* "HC acid," *read* "HCl acid."

P. 163, first and second lines under APATITE, *for* "CaO," *read* "3 CaO."

P. 164, line 14:—*for* D *read* C^3.

P. 174:—*Erase* the heading "A^1.—NO WATER IN BULB-TUBE;" or, otherwise, *add* "A^2.—HYDROUS SPECIES," above line 8 from bottom.

P. 185, line 10:—*for* 130° 33′, *read* 113° 52′. The latter angle is that of the more commonly occurring pyramid of Scheelite, over a middle edge.

In the Note to Table XIV., page 143, for Olivine **read** Olivenite.

AN OUTLINE

OF

BLOWPIPE PRACTICE,

AS APPLIED TO THE

QUALITATIVE EXAMINATION OF MINERAL BODIES.

§ 1.

THE BLOWPIPE—ITS STRUCTURE AND GENERAL USE.

The blowpipe, in its simplest form, is merely a narrow tube of brass or other metal, bent round at one extremity, and terminating, at that end, in a point with a very fine orifice, Fig. 1. If we place the pointed end of this instrument just within the flame of a lamp, common candle, or gas-jet with narrow aperture, and then blow gently down the tube, the flame will be deflected to one side in the form of a long narrow cone, and its heating power will be greatly increased. Many minerals, when held in the form of a thin splinter at the point of a flame thus acted upon, may be melted with the greatest ease; and some are either wholly or partially volatilized. Other minerals, on the contrary, remain unaltered. Two or more substances, therefore, of similar appearance, may often be separated and distinguished in a moment, by the aid of the blowpipe.

FIG. 1.

The blowpipe (in its scientific use) has, strictly, a three-fold application. It may be employed, as just pointed out, to distinguish minerals from one another: some of these being fusible, whilst others are infusible; some attracting the magnet after exposure to the blowpipe, whilst others do not exhibit that reaction; some imparting a colour to the flame, others volatilizing, and so forth. Secondly, the blowpipe may be employed to ascertain the general composition of a mineral; or to prove the presence or absence, in a given body, of some particular substance, as silver, copper, lead, iron, cobalt, manganese, sulphur, arsenic, antimony, and the like. Thirdly, it may be used to determine, in certain special cases, the actual amount of a metallic or other ingredient previously ascertained to be present in the substance under examination.

In using the blowpipe, the mouth is filled with air, and this is forced gently but continuously down the tube by the compression of the muscles of the cheeks and lips, breathing being carried on simultaneously by the nostrils. By a little practice, this operation becomes exceedingly easy, especially in ordinary experiments, in which the blast is rarely required to be kept up for more than twenty or thirty seconds at a time. The beginner will find it advisable to restrict himself at first to the production of a steady continuous flame, without seeking to direct this on any object. Holding the blowpipe in his right hand (with thumb and two outside fingers below, and the index and middle finger above the tube), near the lower extremity, he should let the inner part of his arm, between the wrist and the elbow, rest against the edge of the table at which he operates. The jet or point of the blowpipe is turned to the left, and inserted either into or against the edge of the flame, according to the nature of the operation, as explained below. After a few trials, when sufficient skill to keep up a steady flame has been acquired, the point of the flame may be directed upon a small splinter of some easily fusible material, such as natrolite or lepidolite, held in a pair of forceps with platinum tips.* Some little difficulty will probably be experienced at first in keeping the test-fragment exactly at the flame's point; but this, arising partly from irregular blowing, and partly from the beginner feeling constrained to look at the jet of the blowpipe and the object simultaneously, is easily overcome by half-an-hour's practice. A small cutting of metallic tin or copper supported on a piece of well-burnt soft-wood charcoal can be examined in a similar manner.

---

* If forceps of this kind cannot be procured, a pair of steel forceps with fine points, such as watchmakers use, may serve as a substitute. It will be advisable to twist some silk thread or fine twine round the lower part of these, in order to protect the fingers. The points must be kept clean by a file.

In these experiments, the beginner must be careful not to operate on fragments of too large a bulk. The smaller the object submitted to the flame, the more certain will be the results of the experiment.

In out-of-the-way places, the common form of blowpipe described above is frequently the only kind that can be obtained. It answers well enough for ordinary operations, but the moisture which collects in it, by condensation from the vapour of the breath, is apt to be blown into the flame. This inconvenience is remedied by the form of construction shewn in the annexed figures, in which the instrument consists of two principal portions, a main stem closed at one end, and a short tube fitting into this, at right angles, near the closed extremity. The short tube is also commonly provided with a separate jet or nozzle of platinum. In this case, the jet can be cleaned by simple ignition before the blowpipe-flame, or over the flame of the spirit-lamp. In

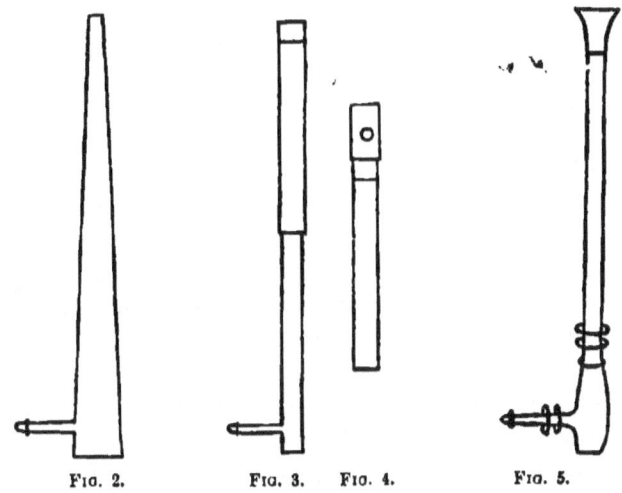

Fig. 2.   Fig. 3.   Fig. 4.   Fig. 5.

the variety of blowpipe known as "Black's Blowpipe," Fig. 2, the main tube is usually constructed of japanned tin-plate, and the instrument is thus sold at a cheap rate. Mitscherlich's Blowpipe, Fig. 3, consists of three separate pieces which fit together, when not in use, as shewn in Fig. 4. This renders it as portable as an ordinary pencil-case. Fig. 5 represents Gahn's or Berzelius's Blowpipe, with a trumpet-shaped mouth-piece of horn or ivory as devised by Plattner. This mouth-piece is placed, of course, on the outside of the lips. It is preferable to the ordinary mouth-piece, but is not readily used by the beginner. In length, the blowpipe varies from about seven-and-a-half to nine inches, according to the eyesight of the operator.

§ 2.
ACCESSORY APPLIANCES AND REAGENTS.

In addition to the blowpipe itself, and the forceps described above, a few other instruments and appliances are required in blowpipe operations.* The principal of these comprise: Some well-burnt, softwood charcoal, and a thin narrow saw-blade to saw the charcoal into rectangular blocks for convenient use; a few pieces of platinum wire, three or four inches in length, of about the thickness of thin twine, to serve as a support in fusions with borax, &c. (see below); some pieces of open glass-tubing of narrow diameter, and two or three small glass flasks, or, in default, a narrow test-tube or two—the latter used chiefly for the detection of water in minerals (see below); a small hammer and anvil, or piece of hard steel, half-an-inch thick, polished on one of its faces; a triangular file; a bar or horse-shoe magnet; a pen-knife or small steel spatula; a small agate pestle and mortar; a small spirit-lamp; a platinum spoon; a small porcelain capsule with handle; and eight or ten turned wooden boxes or small stoppered bottles to hold the blowpipe reagents. These latter are employed for the greater part in the solid state, a condition which adds much to their portability, and renders a small quantity sufficient for a great number of experiments. The principal comprise: Carbonate of soda (abbreviated into *carb. soda*, in the following pages), used largely for the reduction of metallic oxides and detection of sulphides and sulphates, manganese, &c., as explained below; biborate of soda, or borax, used principally for fusions on the platinum wire, many substances communicating peculiar colours to the glass thus formed; and phosphate of soda and ammonia, commonly known as microcosmic salt or phosphor-salt, used for the same purposes as borax, and also for the detection of silicates and chlorides, as explained further on. Reagents of less common use comprise: nitrate of cobalt (in solution); bisulphate of potash; black oxide of copper; chloride of barium; metallic tin; bone ash; strips of yellow turmeric paper, and blue and red litmus paper; with a few other substances of special employment, mentioned under § 5, below.

---

* Only the more necessary operations, instruments, &c., are here alluded to.

§ 3.
STRUCTURAL PARTS AND CHEMICAL PROPERTIES OF FLAME.

The effects produced by the blowpipe cannot be properly understood without a preliminary knowledge of the general composition and structural parts of Flame. If the flame of a lamp or candle, standing in a place free from draughts, be carefully examined, it will be seen to consist of four more or less distinct parts, as shown in the annexed diagram, Fig. 6. A dark cone, a, will be seen in the centre of the flame.

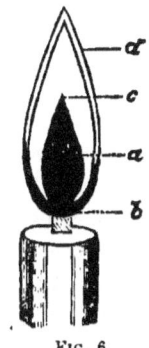

Fig. 6.

This consists of gases, compounds of carbon and hydrogen, which issue from the wick, but which cannot burn, as they are cut off from contact with the atmosphere. A bright luminous cone surrounds this dark central portion, except at its extreme base. In this bright cone the carbon, or a portion of it, separates from the hydrogen of the gaseous compounds pumped up by the wick. The carbon becomes ignited in the form of minute particles, and these, with the liberated hydrogen and undecomposed gas, are driven partly outwards, and partly downwards, or into the blue cup-shaped portion which lies at the base of the flame. At this latter spot, the carbon, meeting with a certain supply of oxygen, is converted into carbonic oxide, a compound of equal combining-weights of carbon and oxygen. Finally, in the flame-border or outer envelope, of a pale pinkish colour, only discernible on close inspection, complete combustion, i.e., union with oxygen, of both gases, carbon and hydrogen, takes place. The carbon burns into carbonic acid, a compound of two combining weights of oxygen with one of carbon; and the hydrogen, uniting with oxygen, forms aqueous vapour. If a cold and polished body, for example, be brought in contact with the edge of a flame of any kind, its surface will exhibit a streak or line of moisture.

These different parts of flame, possess, to some extent, different properties. The dark inner cone is entirely neutral or inert. Bodies placed in it become covered with soot or unburnt carbon. The luminous or yellow cone possesses *reducing powers*. Its component gases, requiring oxygen for their combustion, are ready to take this from oxidized bodies placed in contact with them. This luminous cone, however, in its normal state, has not a sufficiently high temperature to decompose oxidized bodies, except in a few special cases; but its temperature, and consequently its decomposing or deoxidizing power, becomes much increased by the action of the blowpipe, as shewn below. The blue portion of flame possesses also reducing powers, but of comparatively feeble intensity, as the carbon is there able to obtain from the atmosphere a partial supply of oxygen. Finally, in the outer or feebly luminous envelope, in which complete combustion takes place, the flame attains its highest temperature; and, having all the oxygen it requires from the surrounding atmosphere, it exerts an oxidizing influence on bodies placed in contact with it, since most bodies absorb oxygen when ignited in the free air.

In subjecting a body to the action of the blowpipe, we seek: (1) to raise its temperature to as high a degree as possible, so as to test the relative fusibility of the substance; or (2) to oxidize it, or cause it, if an oxide, to combine with a larger amount of oxygen; or (3) to reduce it,* either to the metallic state, or to a lower degree of oxidation. The first and second of these effects may be produced by the same kind of flame, known as an oxidating flame (or O. F.), the position of the substance being slightly different; whilst the third effect is obtained by a reducing flame (or R. F.), in which the yellow portion is developed as much as possible, and the substance kept within it, so as to be cut off from contact with the atmosphere.

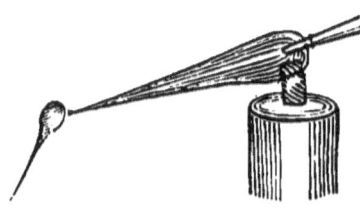

Fig. 7.

An oxidating and fusion flame is thus produced. The point of the blowpipe is inserted well into the flame of the gas-jet, lamp or candle under use, so as almost to touch the surface of the gas-burner or wick. The deflected flame is thus well supplied with oxygen, and its reducing or yellow portion becomes obliterated. It forms a long narrow blue cone, surrounded by its feebly luminous mantle. The body to be oxidized should be held a short distance beyond the point of the cone, as in Fig. 7; but to test its fusion, it must be held in contact with this, or even a little within the flame. In this position, many substances, as those which contain lithia, strontia, baryta,

---

* A substance in metallurgical language is said to be "reduced," or to undergo "reduction," when, from the condition of an oxidized (or other) compound, it becomes converted into metal.

copper, &c., impart a crimson, green, or other colour to the outer or feebly luminous cone.

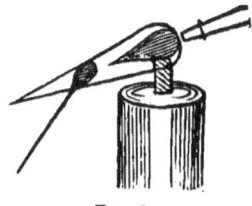

Fig. 8.

For the production of a reducing flame the orifice of the blowpipe must not be too large. The point is held just on the outside of the flame, a little above the level of the burner or wick, as shewn in Fig. 8. The flame, in its deflected state, then retains the whole or a large portion of its yellow cone. The substance under treatment must be held within this (although towards its pointed extremity), so as to be entirely excluded from the atmosphere; whilst, at the same time, the temperature is raised sufficiently high to promote reduction. As a general rule, bodies subjected to a reducing treatment should be supported on charcoal.

For ordinary experiments, such as testing the relative fusibility, &c., of minerals, the blowpipe may be used with the flame of a common candle. The wick of the candle should be kept rather short (but not so as to weaken the flame), and it should be turned slightly to the left, or away from the point of the blowpipe, the stream of air being blown along its surface. A lamp flame, or that of coal gas, however, gives a higher temperature, and is in many respects preferable. The upper part of the wick-holder (or jet, if gas be used) should be of a rectangular or flattened oblong form, with its surface sloping towards the left at a slight angle.* Either good oil, or, better, a mixture of about 1 part of spirit of turpentine, or benzine, with 6 parts of strong alcohol, may be used with the lamp. If the latter mixture be used, equal volumes of the two ingredients must be first well shaken up together, and then the rest of the alcohol added. If the wick crust rapidly, the turpentine will be in excess, in which case another volume of alcohol may be added to the mixture.

§ 4.

BLOWPIPE OPERATIONS.

The following are some of the more general operations required in

---

* The most convenient flame for blowpipe use is that of a small Bunsen burner, into which is dropped a narrow tube (somewhat longer than the tube of the burner, and with sloped and flattened upper surface), to cut off the supply of air and produce a luminous flame. This accessory tube is of course to be removed when bulb-tubes or solutions are heated, or when a substance is ignited without the aid of the blowpipe.

blowpipe practice. The student should master them thoroughly, before attempting to employ the blowpipe in the examination or analysis of minerals. A few additional operations of special employment are referred to in a subsequent section.

(1) *The Fusion Trial.*—In order to ascertain the relative fusibility of a substance, we chip off a small particle, by the hammer or cutting pliers, and expose it, either in the platinum-tipped forceps or on charcoal, to the point of the blue flame (Fig. 7, above). If the substance be easily reduced to metal, or if it contain arsenic, it must be supported on charcoal (in a small cavity made by the knife-point for its reception), as substances of this kind attack platinum.* In other cases, a thin and sharply-pointed splinter may be taken up by the forceps, and exposed for about half-a-minute to the action of the flame. It ought not to exceed, in any case, the size of a small carraway seed—and if smaller than this, so much the better. If fusible, its point or edge (or on charcoal, the entire mass) will become rounded into a bead or globule in the course of ten or twenty seconds. Difficultly fusible substances become vitrified only on the surface, or rounded on the extreme edges; whilst infusible bodies, though often changing colour, or exhibiting other reactions, preserve the sharpness of their point and edges intact.

The more characteristic phenomena exhibited by mineral bodies when exposed to this treatment, are enumerated in the following table: †

(*a*) The test-fragment may "decrepitate" or fly to pieces. Example, most specimens of galena. In this case, a larger fragment must be heated in a test-tube over a small spirit-lamp, and after decrepitation has taken place, one of the resulting fragments can be exposed to the blowpipe-flame as directed above. Decrepitation may sometimes be prevented if the operator expose the test-fragment cautiously and gradually to the full action of the flame.

(*b*) The test-fragment may change colour (with or without fusing) and become attractable by a magnet. Example, carbonate of iron. This becomes first red, then black, and attracts the magnet, but does not fuse. Iron pyrites, on the other hand, becomes black and magnetic, but fuses also.

---

* In order to prevent any risk of injury to the platinum forceps, it is advisable (even if not strictly necessary in all cases) to use charcoal as a support for bodies of a metallic aspect, as well as for those which exhibit a distinctly coloured streak or high specific gravity.

† Blowpipe operations, as described in this section, are not intended to serve as a course of analysis. Merely a few examples, therefore, are given in illustration of their effects. For Plan of Analysis, see § 6.

(c) The test-fragment may colour the flame. Thus, most copper and all thallium compounds impart a rich green colour to the flame; compounds in which tellurium or antimony is present, also those containing baryta, and many phosphates and borates, with molybdates and the mineral molybdenite, colour the flame pale green; sulphur, selenium, lead, arsenic, and chloride of copper colour the flame blue of different degrees of intensity; compounds containing strontia and lithia impart a crimson colour to the flame; some lime compounds impart to it a pale red colour; soda compounds, a deep yellow colour; and potash compounds, a violet tint.

(d) The test-fragment may become caustic. Example, carbonate of lime. The carbonic acid is burned off, and caustic lime remains. This restores the blue colour of reddened litmus paper.

(e) The test-fragment may take fire and burn. Example, native sulphur, cinnabar, common bituminous coal, &c.

(f) The test-fragment may be volatilized or dissipated in fumes, either wholly or partially, and with or without an accompanying odour. Thus, gray antimony ore volatilizes with dense white fumes; arsenical pyrites volatilizes in part, with a strong odour of garlic; common iron pyrites yields an odour of brimstone; and so forth. In many cases the volatilized matter becomes in great part deposited in an oxidised condition on the charcoal. Antimonial minerals form a white deposit or incrustation of this kind. Zinc compounds, a deposit which is lemon-yellow whilst hot, and white when cold. Lead and bismuth are indicated by sulphur-yellow or orange-yellow deposits. Cadmium by a reddish brown incrustation.

(g) The test-fragment may fuse, either wholly, or only at the point and edges, and the fusion may take place quietly, or with bubbling, and with or without a previous "intumescence" or expansion of the fragment. Most of the so-called zeolites, for example (minerals abundant in trap rocks), swell or curl up on exposure to the blowpipe, and then fuse quietly; but some, as prehnite, melt with more or less bubbling.

(h) The test-fragment may remain unchanged. Example, quartz, and various other infusible minerals.

(2) *Treatment in the Flask or Bulb-Tube (The Water Test).*—Minerals are frequently subjected to a kind of distillatory process by ignition in small glass tubes closed at one end. These tubes are of two general kinds. One kind has the form of a small flask, and is commonly known as a "bulb-tube." Where it cannot be procured, a small-sized test-tube may supply its place. It is used principally in testing minerals for water. Many minerals contain a considerable amount of water, or the elements of water, in some unknown physical condition. Gypsum, for example, yields nearly 21 per cent. of water. As the presence of this substance is very easily ascertained, the water test is frequently resorted to, in practice, for the formation of determinative groups, or

separation of hydrous from anhydrous minerals. The operation is thus performed. The glass is first warmed gently over the flame of a small spirit-lamp to ensure the absence of moisture, and is then set aside for a few moments to cool. This effected, a piece of the substance under examination, of about the size of a small pea, is placed

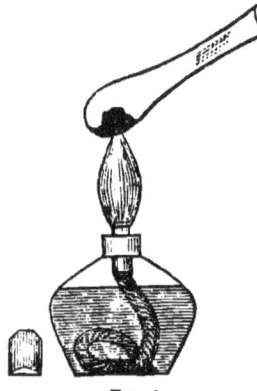

Fig. 9.

in it, and ignited over the spirit-lamp—as shewn in the annexed figure—the tube being held in a slightly inclined position. If water be present in the mineral, a thin film, condensing rapidly into little drops, will be deposited on the neck or upper part of the tube. As soon as the moisture begins to shew itself, the tube must be brought into a more or less horizontal position, otherwise a fracture may be occasioned by the water flowing down and coming in contact with the hot part of the glass. The neutral, acid, or alkaline condition of the water, can be determined by slips of blue and red litmus paper. A mineral may also be examined for water, though less conveniently, by ignition before the blowpipe-flame in a piece of open

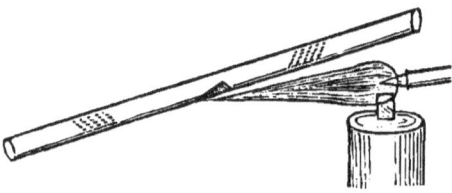

Fig. 10.

tubing, as shewn in Fig. 10. To prevent the tube softening or melting, a strip of platinum foil may be folded around it where the test-fragment rests. The latter is pushed into its place by a thin iron wire. The moisture condenses on each side of the test-matter.

(3) *Treatment in Closed Tubes, proper.*—In addition to the flask or bulb-tube, small pieces of narrow glass tubing—closed, and sometimes drawn out to a point, at one extremity—are frequently used in the examination of mineral bodies. The substance is ignited (either alone, or mixed with thoroughly dry carb. soda or other flux) at the closed end of the tube. After the insertion of the test-substance, the upper part of the tube must be cleaned by a piece of soft paper twisted round an iron wire, or by the feather end of a quill pen, &c.; but this will not be necessary if the substance be inserted by means of a narrow

slip of glazed paper, folded lengthwise. A characteristic sublimate is produced in many cases by ignition of bodies in tubes of this kind. The operation serves especially for the detection of mercury and arsenic. (*See* § 5.)

(4) *Roasting.*—The principal object of this process is the elimination of sulphur, arsenic, and certain other volatile bodies, from the mineral under examination, as these bodies prevent the reduction of many substances to the metallic state, and also mask, to some extent, their other characteristic reactions. By roasting, the substance is not only deprived of sulphur, &c., but is also converted in the majority of instances into an oxidized condition. The operation is most readily performed as follows. A small fragment of the mineral is reduced to powder. Some of this, made into a paste by moistening with a drop of water, is spread over the surface of a block of charcoal, or, better, over a small piece of porcelain, resulting, for example, from a broken evaporating dish or thin crucible. It is then ignited before the point of an oxidating flame (Fig. 7), the heat being kept low, at first, to prevent fusion. It is sometimes necessary to remove the ignited paste to the mortar, and to grind it up again and renew the operation. When the roasting is terminated, the powder will present a dull earthy aspect, and cease to omit fumes or odour. It is then ready for Operations 7 and 8, described below. By reducing the substance to powder before roasting, the risk of decrepitation and fusion is prevented, and the process itself is more efficiently performed.

(5) *Treatment in Open Tube.*—Roasting is sometimes effected in a piece of open glass tubing, as in Fig. 11, the test object being placed near one end of the tube, whilst the tube itself is held in an inclined position. For the better retention of the substance, the tube may be softened at this end in the flame of the spirit-lamp, and then bent into a slight elbow. Sulphur eliminated from bodies by this treatment, is converted into sulphurous acid (a compound of sulphur and oxygen, the latter taken up from the atmosphere); and arsenic forms arsenious acid, which deposits itself in the shape of numerous microscopic octahedrons on the cool sides of the glass near the upper part of the tube. Sulphurous

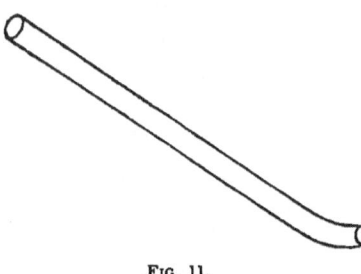

FIG. 11.

acid in escaping from the open end of the tube is easily recognized by its odour (identical with that emitted by an ignited match), as well as by its property of changing the blue colour of a slip of moistened litmus paper to red. Antimonial compounds form a dense white uncrystalline sublimate.

(6) *Treatment with Nitrate of Cobalt.*—This operation is required in special cases only. It serves for the detection of alumina, magnesia, oxide of zinc, and some few other substances; but it is not applicable to deeply coloured or easily fusible bodies, nor to such as possess a metallic lustre or coloured streak. A fragment of the substance, under treatment, is reduced by the hammer and anvil, and afterwards by the use of the agate mortar, to a fine powder. This is moistened with a drop of the cobalt solution (nitrate of cobalt dissolved in water), and the resulting paste is strongly ignited on charcoal by being held about an inch before the point of the flame, fusion being carefully avoided. Thus treated, alumina assumes on cooling a fine blue colour; magnesia (and the comparatively rare tantalic acid), a flesh-red tint; baryta, a dull brownish-red colour; oxide of zinc, bin-oxide of tin, antimony oxides, a green colour. With other substances a gray, bluish-grey, brownish-black, or other indefinite coloration is produced, unless fusion takes place, in which case a glass may be obtained, coloured blue by the dissolved oxide of cobalt.

(7) *Formation of Glasses on Platinum Wire or Charcoal.*—This operation is one of constant utility in the determination of the constituents of minerals. The glasses in question are formed by the fusion of small portions of borax, phosphor-salt, or carbonate of soda: the latter reagent, however, being only occasionally used. Most substances dissolve in one or the other of these fluxes before the blowpipe, and many communicate peculiar colours to the glass, by which the nature of the test-matter is made known. If the matter to be tested contain sulphur or arsenic, it should be roasted before being subjected to the action of these fluxes. Metals and metallic alloys, as well as metallic oxides, chlorides, &c., of very easy reduction, must be examined on charcoal, but in other cases it is more convenient to employ a piece of platinum wire as a support. One end of the wire may be inserted into a cork or special handle, or, if the wire be from $2\frac{1}{2}$ to 3 inches in length, it may be held in the naked fingers, as platinum conducts heat very slowly. The other end is bent into a small loop or ear. This, when borax or phosphor-salt is used, is ignited by the blowpipe-flame, and

plunged into the flux, the adhering portion of the latter being then fused into a glass. If a sufficient portion to fill the loop be not taken up at first, the process must be repeated. With beginners, the fused glass is often brownish or discoloured by smoke, but it may be rendered clear and transparent by being kept in ignition for a few moments before the extreme point of the flame, the carbonaceous matter becoming oxidized and expelled by this treatment. When carbonate of soda is used, a small portion of the flux must be moistened and kneaded in the palm of the left hand, by a knife-point or a small spatula, into a slightly cohering paste, which is placed on the loop of the wire, and fused into a bead. Whilst hot, the soda bead is transparent, but it becomes opaque on cooling. The portion of test-matter added to a glass or bead, formed by these reagents, must be exceedingly small, otherwise the glass may become so deeply coloured as to appear quite black. In this case, the colour may be observed by pinching the bead flat between a pair of forceps, before it has time to cool. It is always advisable, however, in the first instance, to take up merely a minute particle or two of the test-substance, and then to add more if no characteristic reaction be obtained. The glass, in all cases, must be examined first before an oxidating flame, and its colour observed both whilst the flux is hot and when it has become cold; and, secondly, it must be kept for a somewhat longer interval in a good reducing flame (Fig. 8), and its appearance noted as before.* With certain substances (lime, magnesia, &c.) the borax and phosphor-salt glasses become milky and opaque when saturated, or when subjected to the intermittent action of the flame—the latter being urged upon them in short puffs, or the glass being moved slowly in and out of the flame—a process technically known as *Flaming*.

The colours, &c., communicated to these glasses by the more commonly occurring constituent bodies, are shewn in the annexed tabular view.

BORAX.

Colour of Bead after exposure to an Oxidating Flame.	Compounds of:	Colour of Bead after exposure to a Reducing Flame.
Violet or amethystine	Manganese	Colourless, if quickly cooled. Violet-red, if slowly cooled.
Violet-brown (whilst hot) Clear-brown (when cold)	Nickel	Gray and opaque.
Blue (very intense)	Cobalt	Blue (very deep).

---

* The colour of the glass ought not, of course, to be examined by the *transmitted* light of the lamp or candle flame. Strictly, it should be observed by daylight.

Colour of Bead after exposure to an Oxidating Flame.	Compounds of:	Colour of Bead after exposure to a Reducing Flame.
Green (whilst hot) Blue or greenish-blue (cold)	Copper	More or less colourless whilst hot; brownish-red & opaque on cooling.
Green or bluish-green	Cobalt + Iron	Green or bluish-green.
Green (dark)	Copper + Nickel Copper + Iron	Brownish-red, opaque, on cooling.
Yellowish or reddish (hot) Yellowish-green (when cold)	Chromium	Emerald-green.
Yellow (whilst hot) Greenish-yellow (cold)	Vanadium	Brownish (whilst hot). Emerald-green (when cold).
Yellowish or reddish	Iron	Bottle-green.
Yellowish or reddish Enamelled by flaming	Uranium	Green (black by flaming).
Yellow (whilst hot) Pale yellowish (cold) Enamelled by flaming	Cerium	Colourless or yellowish. Opaque-white, if saturated.
Yellow (hot) Colourless (cold) Enamelled by flaming	Titanium	Yellow or yellowish-brown. Enamelled light-blue by flaming. See under Phosp.-salt, below.
Yellow (hot) Colourless (cold) Enamelled by flaming	Tungstenum	Yellow or yellowish-brown. Enamelled by flaming. See under Phosphor-salt, below
Yellow (hot) Colourless or yellowish (cold) Grayish and opaque by flaming	Molybdenum	Brown or gray, semi-opaque, often with separation of black specks. See under Phosphor-salt, below
Yellow or yellowish-red (hot) Yellowish or colourless, and often opaline, when cold	Lead Bismuth Silver Antimony	Gray and opaque on cooling; but after continued subjection to the flame, the glass becomes clear: the reduced metallic particles either collecting together or volatilizing.
Yellowish (hot) Colourless (cold) Opaque-white when saturated	Cadmium	Colourless — the reduced metal being volatilized.
Colourless (permanently clear) Slowly dissolved See under Phosphor-salt, below	Aluminium Silicon Tin	Colourless: permanently clear. (Tin compounds dissolve in small quantity only. On charcoal, they become reduced to metal, especially if a little carb. soda be added to the glass).
Colourless. When saturated, opaque-white on cooling or by flaming	Tantalum Zirconium Glucinum Yttrium, &c. Thorium Magnesium Calcium Strontium Barium Lithium Natrium Kalium	Colourless. When saturated, opaque-white on cooling or by flaming. See REACTIONS, § 5.

PHOSPHOR-SALT.

The glasses produced by the fusion of constituent bodies with this reagent are for the greater part identical with those obtained by the use of borax, although somewhat less deeply coloured as a general rule. The principal exceptions are the glasses formed in a reducing flame with compounds of molybdenum, tungstenum, and titanium, respectively. The molybdenum glass presents, when cold, a fine green colour, and the tungstenum glass becomes greenish-blue. If the latter contain iron, the colour of the glass is changed to blood-red or brownish-red. Titanium in the presence of iron gives a similar reaction; but when free from iron, the glass is yellow whilst hot, and violet-coloured when cold. Phosphor-salt is an important reagent for the detection of silica in silicates, as the silica remains for the greater part undissolved in the glass, in the form of a translucent flocculent mass, technically known as a "silica skeleton," the associated constituents being gradually taken up by the flux. A small amount of silica is also generally dissolved, but this is precipitated as the bead cools, rendering it semi-transparent or opaline. Phosphor-salt is likewise employed for the detection of chlorides, &c. (*See* under REACTIONS, § 5.) In other respects, it is especially adapted for fusions on charcoal, as it does not spread out like borax, but forms a globule on the support.

CARBONATE OF SODA.

This reagent is principally used to promote the reduction of oxidized and other bodies to the metallic state, as explained below, under that process. It is also of very frequent employment as a test for sulphur in sulphides and oxidized bodies. (*See* under REACTIONS, § 5.) It is rarely used, on the other hand, for the formation of glasses on platinum wire, except as a test for the presence of manganese; although, when employed in this manner, it serves to distinguish salts of the alkalies, and those of strontia and baryta, from all other salts: the alkalies, with baryta and strontia, dissolving completely and rapidly in the bead, whereas lime, magnesia, alumina, and other bases, remain unattacked. Manganese compounds form by oxidizing fusion with this reagent a green glass, which becomes blue or bluish-green and opaque on cooling. A very minute amount of manganese may be thus detected. The delicacy of the test is increased by the addition of a small quantity of nitre, as this promotes oxidation; and if the substance contain much lime, magnesia, iron oxides, or other bodies more or less insoluble in carb. soda, it is advisable to add a little borax to the test-mixture. The blue or bluish-green bead thus produced, is technically known as a "turquoise enamel." Chromium compounds produce a somewhat similar reaction; but if the bead be saturated with silica or boracic acid, it will remain green in the latter case; while if the green colour result from the presence of manganese, a violet or amethystine glass will be obtained. Some other applications of carbonate of soda as a blowpipe reagent will be found under the head of REACTIONS, § 5.

(8) *Reduction.*—This term denotes the process by which an oxidized or other compound is converted into the metallic state. Some com-

pounds become reduced by simple ignition; others require for their reduction the addition of certain reagents; and some, again, resist reduction altogether. The reduced metal is in some cases so highly volatile that it cannot be obtained except by a kind of distillatory process. In other cases, one or more fusible globules, or a number of minute infusible grains, are obtained in blowpipe operations. Reducible metals may be thus distributed into three groups, as shown (with omission of a few metals of rare occurrence) in the annexed table:

A. *Yielding metallic globules.*—Gold, silver, copper, tin, lead, bismuth, antimony.
B. *Yielding infusible metallic grains.*—Platinum, iron, nickel, cobalt, molybdenum, tungstenum.
C. *Yielding metallic vapours only, when treated on charcoal.*—Mercury, arsenic, cadmium, zinc.

A metal of the first group may be obtained, unless present in very small quantity, by a simple fusion of the previously roasted test-substance, with some carbonate of soda, on charcoal, in a good reducing flame (Fig. 8, above). In ordinary cases, metallic globules are rapidly produced by this treatment. By a little management the globules may be brought together, so as to form a single large globule. This must be tested on the anvil as regards its relative malleability,* &c. Gold, silver, copper, tin and lead are malleable; bismuth and antimony, more or less brittle. Gold and silver (if pure) retain a bright surface after subjection to an oxidating flame. Copper becomes covered with a black film, and tin with a white crust. Lead and bismuth volatilize more or less readily, and deposit on the charcoal a yellow coating of oxide. Antimony is rapidly volatilized with deposition of a dense white incrustation on the charcoal. It is not, of course, always necessary to subject the test-substance to a previous roasting (Operation 4, above), but it is always safer to do so. Sulphur in most, and arsenic in all cases, must be driven off by this preliminary treatment before the actual process of reduction is attempted.

When the metal to be reduced belongs to the second group, or if

---

* To test the relative malleability of a metallic globule as obtained by the blowpipe, the globule must be placed on a small steel anvil, and a strip of thin paper (held down by the forefinger and thumb of the left hand) being placed over it to prevent dispersion, it is struck once or twice by a light hammer. Thus treated, malleable globules become flattened into discs, whilst brittle globules break into powder

the amount of a fusible metal in the test-substance be less than 4 or 5 *per cent.*, the operation is performed as follows: A small portion of the substance in powder—subjected previously to the roasting process, if it contain sulphur or arsenic—is mixed with 3 or 4 volumes of carbonate of soda (or neutral oxalate of potash, or a mixture of about equal parts of carb. soda and cyanide of potassium—the latter, it must be remembered, a highly poisonous substance), and the mixture is exposed on charcoal to a good reducing flame, until all the alkaline salt has become absorbed. More flux is then added, and the operation is repeated until the whole or the greater part of the test-matter is also absorbed. This effected, the charcoal, where the assay rested, is removed by a sharp knife-point, and carefully ground to powder in a small agate mortar or porcelain capsule, whilst a fine stream of water is projected upon it from time to time, until all the carbonaceous and other non-metallic particles are gradually washed away. For this purpose, the mortar or capsule may be placed in the centre of an ordinary plate; and if the operator be not provided with a chemical washing-bottle, he may use a small syringe, or, in place of this, a simple piece of glass tubing, five or six inches in length and about the fourth of an inch in diameter, drawn out at one end to a point. This is filled by suction, and the water is expelled, with the necessary force, by blowing down the tube. The metallic grains or spangles obtained by this process must be examined by the magnet. Those of iron, nickel and cobalt are magnetic. Sometimes, however, when but a trace or very small percentage of reducible metal is contained in the test-substance, its presence is only indicated by a few metallic streaks on the sides and bottom of the mortar. Metallic markings of this kind can be removed by a piece of pumice.

Metallic compounds referable to the third group, yield no metal on charcoal, or by other treatment in open contact with the atmosphere. The presence of arsenic, however, is easily made known by the garlic-like odour evolved during fusion with reducing agents (or alone) on charcoal. Cadmium and zinc may also be recognized by the oxidized sublimates which they deposit on the charcoal. The cadmium sublimate is reddish-brown; the zinc sublimate, lemon-yellow and phosphorescent whilst hot, and white when cold. Mercury forms no incrustation on charcoal; but its presence in any compound may be determined by reduction with carbonate of soda or iron-filings in a glass tube of narrow diameter. A small test-tube or piece of glass tubing closed at one end before the blowpipe, may be used for the experiment. The test-substance, in powder, mixed with 3 or 4 vols. of perfectly dry carb. soda, is inserted into the tube by means of a narrow strip of glazed writing-paper bent into the form of a trough, so as to prevent the sides of the glass from being soiled, and the mixture is strongly ignited by the spirit-lamp or by the blowpipe-flame. If mercury be present, a gray metallic sublimate will be formed near the upper part of the tube. By friction with an iron wire, or the narrow end of a quill-pen, &c., the sublimate may be brought into the form of fluid globules, which can be poured out of the tube, and are then easily recognized as metallic mercury.

(9) *Cupellation.*—Gold and silver are separated by this process from other metals. The test-metal is fused with several times its weight of pure lead. The button, thus obtained, is exposed to an oxidating fusion on a porous support of bone ash, known as a cupel. The lead and other so-called base metals become oxidized by this treatment, and are partly volatilized, and partly absorbed by the bone ash, a globule of gold or silver (or the two combined) being finally left on the surface of the cupel. For blowpipe operations, cupels are generally made by pressing a small quantity of dry bone ash into a circular iron mould, the latter being fixed, when presented to the flame, in a special support, consisting essentially of a wooden foot and pillar supporting a wire stem, with three or four short cross-wires at the top, between which the cupel-mould rests. Instruments of this kind cannot be obtained in remote places, but the process may be performed equally well by pressing some dry bone ash into a suitable cavity fashioned at the extremity of a cylindrical piece of pumice or well-baked clay, or even charcoal. The smooth end of the agate pestle, or a glass button cemented to a cork, or the rounded end of a glass stopper, may be used for this purpose. The cupel, thus formed, must then be exposed for a few moments to the point of the blowpipe-flame, so as to render the bone ash perfectly dry; and if its surface become blistered or be in any way affected by this drying process, it must be rendered smooth again by pressure with the pestle. The substance to be cupelled must be in the metallic state; if not in this condition, therefore, it must first be subjected to the reducing operation described above. The piece of test-metal, which may weigh about a couple of

OPERATIONS. 19

grains (or from 100 to 150 milligrammes) is wrapped in a piece of pure lead-foil of at least four times its weight, and the whole is exposed, on the surface of the cupel, to the extreme point of a clear oxidating flame. If the substance consist of argentiferous lead, as obtained from galena, &c., the addition of the lead-foil is of course unnecessary.* As soon as fusion takes place, the cupel must be moved somewhat farther from the flame, so as to allow merely the outer envelope of the latter, or the warm air which surrounds this, to play over the surface of the globule. By this treatment, the lead will become gradually converted into a fusible and crystalline slag. When this collects in large quantity, the position of the cupel must be slightly altered, so as to cause the globule to flow towards its edge, the surface of the lead being thus kept free for continued oxidation. When the globule becomes reduced to about a fourth or fifth of its original bulk, the process is discontinued, and the cupel is set aside to cool. This is the first or concentration stage of the process. Another cupel is then prepared and dried; and the concentrated globule—after careful separation from the slag in which it is imbedded—is placed on this new cupel, and again subjected to the oxidizing influence of the flame. During this second part of the process, the flame is made rather to play on the surface of the cupel around the lead button than on the button itself, a complete absorption of the oxidized lead being thus effected. The flame should be sharp and finely-pointed, and urged down on the cupel at an angle of forty or forty-five degrees. Finally, if the test-metal contain gold or silver, a sudden flash or gleam will be emitted at the close of the operation, and a minute globule of one (or both) of these metals will be left on the surface of the bone ash. By concentrating several portions of a test-substance, melting the concentrated globules together, again concentrating, and finally completing the cupellation, as small an amount as half an ounce

---

* In reducing galena, with a view to test the lead for silver by cupellation, the reduction may be conveniently performed as follows: A small portion of the galena, crushed to powder, is mixed with about twice its volume of carb. soda, to which a little borax has been added. This is made into a paste by the moistened knife-blade, and a short piece of thin iron wire is stuck through it, and the whole is then placed in a charcoal cavity, and exposed for a couple of minutes to the action of a reducing flame. By a little management, the minute globules of lead which first result can easily be made to run into a single globule. The iron serves to take up the sulphur from the galena. When the fused mass is sufficiently cool, it is cut out by a sharp knife-point, and flattened (under a strip of paper) on the anvil. The disc of reduced lead, thus separated, is then ready for cupellation. See also, under silver, § 5.

of gold or silver in a ton of ore—or in round numbers, about one part in sixty thousand—may be readily detected by the blowpipe.*

During cupellation, the process sometimes becomes suddenly arrested. This may arise from the temperature being too low, in which case the point of the blue flame must be brought for an instant on the surface of the globule, until complete fusion again ensue. Or, the hindrance may arise from the bone ash becoming saturated, when a fresh cupel must be taken. Or, it may be occasioned, especially if much copper or nickel be present, by an insufficient quantity of lead. In this latter case, a piece of pure lead must be placed in contact with the globule, and the two fused together; the cupel being then moved backward from the flame, and the oxidating process again established.

(10) *Fusion with Reagents in Platinum Spoon.*—This operation is only required in certain special cases, as in the examination of a substance suspected to be a tungstate or molybdate, or in searching for the presence of titanic acid, &c. The substance, in fine powder, is mixed with three or four parts of the reagent (carb. soda, or bisulphate of potash, &c.), and the mixture, in successive portions, is fused in a small platinum spoon. As a rule, the flame may be made to impinge upon the bottom of the spoon; and the operation is terminated when bubbles cease to be given off and the mixture enters into

Fig. 12.

quiet fusion. During the operation the spoon is held in the spring-forceps (Fig. 12), the points of which remain in close contact when the sides are not subjected to pressure. The fusion accomplished, the

Fig. 13.

spoon is dropped, bottom upwards, into a small porcelain capsule (Fig. 13) provided with a handle. Some distilled water is then added and brought to the boiling point over the spirit-lamp. The fused mass quickly separates from the spoon, and it can then be crushed to powder and again warmed until

---

* A cupellation bead may appear from its pure white colour to consist of silver only, and may yet contain a notable amount of gold. A white bead, therefore, should be flattened into a disc, and fused with some bisulphate of potash in a small platinum spoon. By this treatment the silver is removed from the surface of the disc, and the latter, if gold be present, assumes a yellow colour. If the metal be again fused into a globule, the white colour is restored.

solution, or partial solution, takes place. When the undissolved matters have settled, the clear supernatant liquid is decanted carefully into another capsule (or into a test-tube) for further treatment. (*See* Experiments 7 and 8, in § 6, beyond.)

§ 5.
BLOWPIPE REACTIONS.

In this section, the leading reactions of the more important elementary bodies and chemical groups are passed rapidly under review. Bodies of exceptional occurrence as mineral components—or such of these, at least, as cannot be properly detected by the blowpipe—are omitted from consideration.* The other elementary substances are taken in the order shewn in the following index:

I. *Non-metallic Bodies.*—1, Oxygen; 2, Hydrogen; 3, Sulphur; 4, Selenium; 5, Nitrogen; 6, Chlorine; 7, Bromine; 8, Iodine; 9, Fluorine; 10, Phosphorus; 11, Boron; 12, Carbon; 13, Silicon.

II. *Unoxidizable Metals.*—14, Platinum; 15, Gold; 16, Silver.

III. *Volatilizable Metals.*—17, Tellurium; 18, Antimony; 19, Arsenic; 20, Osmium; 21, Mercury; 22, Bismuth; 23, Lead; 24, Thallium; 25, Cadmium; 26, Zinc; 27, Tin.

IV. *Flux-colouring Metals.*—28, Copper; 29, Nickel; 30, Cobalt; 31, Iron: 32, Tungstenum; 33, Molybdenum; 34, Manganese; 35, Chromium; 36, Vanadium; 37, Uranium; 38, Cerium; 39, Titanium.

V. "*Earth*" *Metals.*—(40, Tantalum?); 41, Aluminium; 42, Glucinum; 43, Zirconium; 44, Yttrium.

---

* For full details respecting the blowpipe reactions of inorganic bodies generally, the following works may be especially consulted: 1. The old work by Berzelius, "Die Anwendung des Löthrohrs," etc.; translation of the 4th edition, by J. D. Whitney: Boston, 1845. 2. "Handbuch der Analytischen Chemie," von Heinrich Rose, 6th edition, by R. Finkener: Leipsig, 1871. 3. Plattner's "Probirkunst mit dem Löthrohr," 5th edition, by Richter, 1878. American translation of 4th edition, by H. B. Cornwall: New York, 1875. 4. "Untersuchungen mit dem Löthrohr," by Dr. H. Hartmann: Leipsig, 1862. 5. "Löthrohr-Tabellen," by Dr. J. Hirschwäld: Leipsig und Heidelberg, 1875. 6. "Manual of Determinative Mineralogy and Blowpipe Analysis," by George J. Brush: 2nd edition, 1878. 7. "Leitfaden bei qual. und quan. Löthrohr-Untersuchungen," von Bruno Kerl, 2nd edition, Clausthal, 1877. For the determination of minerals, &c., the far-renowned "Tabellen" of Von Kobell (in addition to the work of Prof. Brush, essentially constructed on that of Von Kobell, although with much amplification and addition of new matter) may be especially consulted. The "Anleitung zum Bestimmen der Mineralien," of Dr. Fuchs, is also a very serviceable little book; and some useful tables will be found at the end of E. S. Dana's excellent "Text Book of Mineralogy."

VI. *Alkaline-Earth Metals.*—45, Magnesium; 46, Calcium: 47, Strontium; 48, Barium.

VII. *Alkali Metals.*—49, Lithium; 50, Sodium; 51, Kalium; 52, Ammonium.

I.—NON-METALLIC BODIES.

(1) *Oxygen.*—Although this element occurs so abundantly as a constituent of mineral bodies, its presence, as a rule, can only be inferred by negative evidence. If a substance be neither one of the few known simple bodies of natural occurrence, as gold, carbon, &c., nor a sulphide, selenide, arsenide, chloride, &c., it may be regarded with tolerable certainty as an oxidized body. And if, farther, its examination shew that it is not an oxygen-salt, *i.e.*, a sulphate, carbonate, silicate, or the like, we can then only infer that it must be a simple oxide, either electro-negative or basic in its characters.

All non-oxidized bodies attackable by nitric acid, decompose the latter in taking oxygen from it, and thus cause the evolution of ruddy nitrous fumes; but this decomposition is also effected by certain oxides in passing into a higher state of oxidation, as by Cu^2O, for example.

Some few bodies, as binoxide of manganese, nitrates, chlorates, bichromates, &c., give off oxygen on strong ignition. If these be ignited (in not too small a quantity) in a test-tube containing at its upper part a charred and feebly glowing match-stem, the latter, as the evolved oxygen reaches it, will glow more vividly. These bodies, also, if fused with borax or phosphor-salt, dissolve with strong ebullition; but carbonates produce the same reaction.

(2) *Hydrogen.*—This element, apart from its occurrence in bitumen and other hydro-carbonaceous substances, is only present in oxidized minerals. From these, it is evolved, with oxygen, in the form of water, during the ignition of the substance. (*See* Operation 2, page 9.)

(3) *Sulphur.*—Occurs in the free state, as "native sulphur;" also combined with metals in sulphides and sulphur-salts; and in combination with oxygen, as SO^3, in the large group of sulphates. Native sulphur is readily inflammable, burning with blue flame, and volatilizing (with the well known odour of burning brimstone) in the

form of sulphurous acid SO^2. Metallic sulphides and sulphur-salts (especially if previously reduced to powder and moistened into a paste), when roasted in an open tube of not too narrow diameter, give off the same compound (SO^2), easily recognized by its odour, and by its action on a slip of moistened litmus paper placed at the top of the tube, the paper becoming reddened by the acid fumes. In very narrow (as in closed) tubes, part of the evolved sulphur may escape oxidation, and may deposit itself on the inside of the tube near the test-substance. The sublimate, thus formed, is distinctly red whilst hot, and yellow on cooling. From many arsenical and antimonial sulphides also, a coloured sublimate of this kind, but consisting of As^2S^3, or $2Sb^2S^3 + Sb^2O^3$, &c., may be deposited in narrow tubes, especially if the tube be held more or less horizontally.

Sulphides of all kinds, if fused on charcoal with carb. soda, (or better, with carb. soda mixed with a little borax) readily form an alkaline sulphide or "hepar." This smells, when moistened, more or less strongly of sulphuretted hydrogen, and imparts a dark stain to silver, or to paper previously steeped in a solution of lead acetate. A glazed visiting card may be used as a substitute for the latter. The stain is removed from the silver surface by friction with moistened boneash.

Sulphates fused with carb. soda and a little borax (the borax in the case of earthy sulphates greatly assisting the solvent power of the flux) produce the same reaction. This reaction is of course produced also by sulphites (which do not occur, however, as minerals), and by bodies which contain selenium in any form. Sulphites, treated with hydrochloric acid, evolve sulphurous acid, easily recognized by its smell and its action on litmus paper; and, in acid solutions, they yield no precipitate with chloride of barium. Sulphates, on the other hand, emit no odour of SO^2 when treated with hydrochloric acid; and chloride of barium produces an insoluble precipitate in their acid or other solutions. Bodies containing selenium, are distinguished from sulphur compounds by the strong odour, resembling that of "cabbage-water," which they evolve on ignition.

The efficacy of the sulphur-test is imperilled however by two causes: (1), the difficulty, in many places, of procuring carbonate of soda perfectly free from traces of sulphates; and (2), the very frequent presence of sulphur in the flame, where gas is used in blowpipe operations. The first defect may be remedied (if the carb. soda, employed alone, produce the reaction) by substituting, as proposed by Plattner, oxalate of potash for the test, as that salt is generally pure and free from sulphates; and the flame of a candle, or an oil or spirit-flame, may be used in this experiment, when the gas flame is found by trial with pure soda or oxalate of potash to give the reaction.

Sulphides of natural occurrence are distinguished from sulphates, by emitting sulphurous acid (or, strictly, by emitting sulphur vapour which combines with atmospheric oxygen and forms sulphurous acid) on ignition; although in the case of certain sulphides (blende, molybdenite, &c.) a strong reaction is only produced by the ignition of the substance in powder. Most natural sulphides, also, present a metallic aspect, or otherwise are highly inflammable (orpiment, cinnabar, &c.), or yield a strongly-coloured streak. Light-coloured varieties of zinc blende are the only exception. On the other hand, no sulphate possesses a metallic aspect; and, in all, the streak is either colourless or very lightly tinted.

(4) *Selenium.*—Met with only in a few minerals of very rare occurrence. In these, its presence is revealed by the formation of a "hepar" with carb. soda, and simultaneous emission of strongly-smelling fumes, the odour resembling that of decaying vegetable matters or "cabbage water." In volatilizing, selenium, like sulphur, burns with a blue flame.

(5) *Nitrogen.*—Found only, as regards minerals proper, in an oxidized condition (Ni^2O^5) in nitrates. These are soluble or (as regards certain metallic nitrates) sub-soluble in water; and they deflagrate when ignited on charcoal or in contact with other carbonaceous bodies. Heated with a few drops of sulphuric acid (or fused with bisulphate of potash) in a test-tube, nitrates evolve, also, ruddy fumes of nitrous acid; and many nitrates, moistened with sulphuric acid, impart a dull green coloration to the flame-border.

(6) *Chlorine.*—Occurs, among minerals, in combination with various bases, forming the group of chlorides. In these, its presence is very easily recognized by the bright azure-blue coloration of the

flame-border which originates during the fusion of a chloride with a bead of phosphor-salt coloured by oxide of copper. The fusion may be performed on a loop of platinum wire, the phosphor-salt being first fused with some black oxide of copper into a somewhat deeply coloured glass, and the test-substance, in the form of powder, being then added. Or the fusion may be made on a thin copper-wire with phosphor-salt alone, the end of the wire being cut off after each experiment. By this treatment, chlorides become decomposed, and chloride of copper is formed. The latter compound rapidly volatilizes, and imparts a remarkably vivid bright-blue colour to the flame. The coloration soon passes, but can, of course, be renewed by the addition of fresh test-matter to the bead. Care must be taken to use pure phosphor-salt, as that reagent, unless carefully made, is frequently found to contain traces of chloride of sodium.

Oxidized chlorine-compounds do not occur as minerals, but it may be stated that chlorates produce the same flame-reaction as chlorides, when fused with phosphor-salt and copper oxide. All chlorates, however, detonate like nitrates, only more violently, when ignited in contact with carbonaceous bodies; and they turn yellow, decrepitate, and emit greenish fumes when warmed with a few drops of sulphuric acid (or fused with bisulphate of potash) in a test-tube. The fumes smell strongly of chlorine, and bleach moistened litmus paper. Chlorides, when thus treated with sulphuric acid, effervesce and give off white fumes of hydrochloric acid.

(7) *Bromine.*—Only known, among minerals, in some rare silver bromides. Its blowpipe reactions closely resemble those of chlorine, but the flame-coloration of bromide of copper is a bright blue with green streaks and edges. A small sharply-pointed flame is required to shew the reaction properly; and care must be taken not to add the test-matter to the cupreous phosphor-salt bead until all traces of the green coloration, arising from the oxide of copper, have disappeared. Heated in a test-tube with sulphuric acid (or fused with large excess of bisulphate of potash) bromides yield brownish or yellowish-red, strongly smelling vapours of bromine. Bromates produce the same reaction, but this is accompanied by sharp decrepitation; and when fused on charcoal they detonate more or less violently. (*See* Appendix, No. 20).

(8) *Iodine.*—In nature, occurs only in one or two rare minerals, compounds of iodine and silver, or iodine, silver, and mercury. In these, as well as in all artificial iodides, its presence may be recognized by the vivid green coloration imparted to the flame during fusion with a cupreous phosphor-salt bead. The test-matter must not be added to the bead until the copper oxide is completely dissolved in the latter, and all traces of green (communicated by the CuO) have disappeared from the flame. Iodides, also, when warmed with a few drops of sulphuric acid (or fused with excess of bisulphate of potash) in a test-tube, evolve strongly smelling violet-coloured vapours, which impart a deep blue stain to matters containing starch. A strip of moistened tape or starched cotton may be held at the top of the tube. Iodates exhibit the same reactions, but deflagrate when ignited with carbonaceous bodies. (*See* Appendix, No. 20).

(9) *Fluorine.*—This element, as an essential component of minerals, occurs in combination with calcium and other bases, forming the various fluorides. It is also largely present in topaz, probably in combination with silicon and aluminium; and it occurs, though in smaller proportion, in chondrodite, and as an accidental or inessential component in many other silicates. Its presence is revealed most readily, by warming the substance, in powder, with a few drops of sulphuric acid (or fusing it with bisulphate of potash) in a test-tube, when stifling fumes, which strongly corrode the inside of the glass, are given off. Or, the trial may be made in a platinum crucible covered with a glass plate: on washing the test-tube or glass, and drying it, the corrosion is rendered visible. When fluorine is present in very small quantity in a substance, it is generally driven off the more readily, often by the mere ignition of the substance (either alone, or with previously fused phosphor-salt) at one end of an open narrow tube—the flame being directed into the tube, so as to decompose the test-matter and drive the expelled gases before it. A slip of moistened Brazil-wood paper, placed at the mouth of the tube, is rendered yellow. Many silicates which contain only traces of fluorine lose their polish when strongly ignited, in the form of a small splinter, *per se.*

(10) *Phosphorus.*—Occurs, in minerals, in an oxidized condition only, *i.e.*, as phosphoric acid (or anhydride) in the group of phosphates.* As first pointed out by Fuchs, these bodies, when moistened with sulphuric acid, impart a green coloration to the flame-border, and many produce this reaction *per se*. A closely similar coloration, however, is communicated to the flame by borates (when moistened with sulphuric acid), as well as by bodies containing barium, copper, &c. It only serves, therefore, as a probable indication of the presence of phosphoric acid. The readiest and most certain method of detect-

---

* It is assumed to be in this condition simply because phosphates give the known reactions of phosphoric acid or phosphoric anhydride, although these reactions may, of course, be modified to some extent by the presence of other bodies. In like manner, when iron is present in an oxidized body, we assume that it is present in the condition of FeO if the substance give the known reactions of that compound, and increase in weight on ignition; and that it is present as Fe^2O^3 if the reactions of sesquioxide of iron be given by the substance. As to the *actual conditions*, either physical or chemical, of bodies in combination, we know absolutely nothing, but we *have* a certain knowledge of the secondary components of most bodies. We are able to examine these components apart, and to form more complex bodies by their union. Thus, from a piece of limestone or calcite we can obtain two well known compounds, lime and carbonic acid (or carbonic anhydride); and with these compounds we can readily produce limestone or its equivalent. Hence, the simplest and most practically useful way of stating, either verbally or by symbols, the composition of limestone and other mineral bodies, is surely that which makes known to us at once the components into which the body readily splits up or decomposes, or which characterize it directly by their reactions. This method, therefore, is adhered to in the present handbook. It may be urged that a formula of the kind represented by CaO, CO^2 asserts too much, and that consequently the more modern $Ca CO_3$ is preferable. But rightly considered, the old formulæ need not be assumed to make any assertions regarding the actual condition of bodies in combination, but only to indicate clearly the well known simple compounds into which (in the great majority of cases) substances may be more or less readily decomposed, and the reactions which substances exhibit. As a strict matter of fact, moreover, the new formulæ are not free from assertion. They carry upon their face, at least, a seeming assertion that the elementary bodies in compounds are present in an absolutely free, separate and independent state; or that unknown problematical compounds, as CO_3, SiO_4, SiO_5, SiO_6, etc., etc., are present in the substances to which these formulæ refer. To take another illustration. A student has two minerals before him : one he finds to be the well known mineral, corundum, and consequently Al^2O^3 (alumina); and the second he finds to be ordinary quartz, and consequently SiO^2 (silica), according to the commonly received formula. He has also before him a third mineral, one that gives the reactions of alumina and silica, and yields these separate bodies on analysis. Naturally, therefore, he writes the formula (assuming the two components to be in equal atomic proportions) Al^2O^3, SiO^2. But, to his bewilderment, he finds it given in modern books as Al_2SiO_5. Practically, we do not want to know how much aluminium, silicon and oxygen, are present in a body of this kind, but how much alumina and silica; and the first formula shews us this, or enables us to determine it at once. Were only simple elements and their complex combinations known to us, the new views, carried out properly to their full conception, might pass without opposition; but the question becomes entirely altered by the occurrence of simple binary compounds so abundantly in the free state. In mineral analysis, and in the practical study of minerals, it is not possible to ignore these binary formulæ without great inconsistency. Among other works, they are retained essentially, we are glad to find, in the standard and very copious "Handwörterbuch der Chemie," now being published under the editorship of Dr. Von Fehling of Stuttgart. See also Von Kobell's remarks on this subject in the 5th edition of his "Mineralogie :" 1878.

ing the latter, is to boil or warm the powdered substance in a test-tube with a few drops of nitric acid, and after half-filling the tube with distilled water, to drop into it a small fragment of molybdate of ammonia. In the presence of phosphoric acid, this will turn yellow immediately, especially if the solution be warmed, and a canary-yellow precipitate (soluble in ammonia) will rapidly form. All natural phosphates, with the exception of the rare phosphate of yttria, xenotime, are dissolved or readily attacked by nitric acid; and xenotime, if in fine powder, is generally attacked sufficiently to yield the reaction. Phosphates may also be decomposed by fusion, in fine powder, with three or four parts of carbonate of soda in a platinum spoon or loop of platinum wire. An alkaline phosphate, soluble in water, is formed by this treatment—with xenotime as with other phosphates—and the solution, rendered acid, may then be tested by molybdate of ammonia. Or it may be rendered neutral by a drop of acetic or very dilute nitric acid, and tested with a fragment of nitrate of silver, in which case a canary-yellow precipitate will also be produced. Or it may be tested by adding to it a small fragment or two of acetate of lead, and fusing the resulting precipitate on charcoal. On cooling, the surface of the fused bead shoots into crystalline facets.

(11) *Boron.*—Present in nature in an oxidized condition only, as boracic acid. This occurs: (1), in the hydrated state; (2), in combination with bases, in the group of borates; and (3), in certain so-called boro-silicates. Boracic acid (or anhydride) and many borates and boro-silicates impart *per se* a green coloration to the flame-border, and all produce this coloration if previously saturated with sulphuric acid. In some few silicates, however, in which little more than traces of BO^3 are present, the reaction is scarcely or only very feebly developed unless the test-substance, in fine powder, after treatment with sulphuric acid, and partial desiccation, be moistened with glycerine, according to a process first made known by Iles. But a similar flame-coloration is produced by phosphates and certain other bodies. For the proper detection of borates, therefore, the following long-known method should be resorted to. The test-matter, in fine powder, is saturated with sulphuric acid, and allowed to stand for a minute or two; a small quantity of alcohol is then added, and the mixture is stirred and inflamed. The presence of BO^3—unless in

very minute or accidental quantity—communicates to the point and edges of the flame a peculiar green or yellowish-green colour. Phosphates do not colour the flame under this treatment.

(12) *Carbon.*—Occurs in the simple state in the diamond and graphite, and practically so in the purer kinds of anthracite; also combined with hydrogen, &c., in ordinary coals and bituminous substances; and in an oxidized condition, as carbonic acid (or anhydride) in the group of carbonates. Free (mineral) carbon is infusible and very slowly combustible in the blowpipe-flame, a long continued ignition being necessary to effect the complete combustion of even minute splinters. Ignited with nitre, it deflagrates and is dissolved, carbonate of potash resulting. With other blowpipe reagents it exhibits no characteristic reactions. The presence of carbonic acid in carbonates is readily detected by the effervescence which ensues during the fusion of a small particle of the test-substance with a previously-fused bead of borax or phosphor-salt on platinum wire, CO^2 being expelled. All carbonates, even in comparatively large fragments, dissolve readily under continued effervescence in these fluxes. A mixture of carbonate of lime in silicates, sulphates, and other bodies, may thus be easily recognized. (*See* Appendix, No. 19). It should be remembered, however, that bodies which evolve oxygen on ignition, produce also a strong effervescence by fusion with borax; but, with the exception of binoxide of manganese, very few of these bodies are of natural occurrence.

(13) *Silicon.*—This element occurs in nature only in an oxidized condition, as Silica, SiO^2. The latter compound, in the form of quartz and its varieties, is the most widely distributed of all minerals. In the various opals, it occurs combined with water, and in combination with bases (especially with Al^2O^3, Fe^2O^3, CaO, MgO, FeO, Na^2O, and K^2O), it forms the large group of silicates. In the simple state, silica is quite infusible in the ordinary blowpipe-flame. With carb. soda, it dissolves with effervescence (due to the expulsion of CO^2 from the flux), and it forms with that reagent, in proper proportions, a permanently clear glass—*i.e.*, a glass that remains clear on cooling. To obtain this, the flux should be added little by little, until perfect fusion ensue: with too much soda, the bead is opaque.

Borax attacks silica very slowly, and in phosphor-salt it is still more slowly attacked. A portion may be taken up by the hot glass, but this is precipitated on cooling, and the glass becomes opalescent. (*See* Appendix, No. 15). Silicates vary greatly in their comportment before the blowpipe, the variation depending chiefly on the relative proportions of silica and base, and on the nature of the base. Many silicates are infusible; others become vitrified on the thin edges; and others, again, melt more or less readily,—most of the so-called zeolites (hydrated silicates of alumina, lime, soda, &c., especially characteristic of trap rocks) exhibiting the phenomenon of intumescence. Silicates, as a rule, are very readily detected by their comportment with phosphor-salt: the bases are gradually taken up, whilst the silica remains for the greater part undissolved, forming a "silica-skeleton." This is seen as a diaphanous, flocculent mass (of the shape and size of the test-fragment) in the centre of the hot bead. A small portion of the silica, or in one or two exceptional cases the greater part of it, may be dissolved with the bases, but this precipitates as the glass cools, and renders it semi-translucent or opalescent. Practically, silicates are readily distinguished from phosphates, carbonates, sulphates, &c., by these reactions with phosphor-salt: namely, very slow or partial solution, and formation in most cases of a silica skeleton or opalescent glass. The trial is best made on platinum wire, and the test-substance should be added, if possible, in the form of a thin scale or splinter. (*See* Appendix, No. 15).

II.—UNOXIDIZABLE METALS.

As regards their blowpipe reactions, the metals of this group fall into two series: *Infusible metals*, comprising platinum (with palladium, &c.); and *Fusible metals*, comprising gold and silver. Strictly, silver absorbs a small amount of oxygen when fused in contact with the atmosphere, but the oxygen is evolved as the metal solidifies. It is this which causes cupelled silver to "spit" or throw out excrescences, if the button be allowed to cool too quickly. All the metals of this group (palladium slightly excepted) retain a bright surface when exposed to the action of an oxidating flame.

(14) *Platinum.*—Occurs in the metallic state, alloyed with iridium, and commonly with small quantities of other metals. Practically,

infusible; but the point of a wire of extreme tenuity may be rounded in a well-sustained flame. Not attacked by the blowpipe fluxes.

(15) *Gold.*—Occurs principally in the metallic state, alloyed with variable proportions of silver. Also, but far less commonly, combined with mercury in some varieties of native amalgam, and with tellurium in some rare tellurides. In the metallic condition, or perhaps as an arsenide or sulphide, it is present likewise as an accidental component in many examples of arsenical pyrites, iron pyrites, copper pyrites, zinc blende, &c., in the proportions of a pennyweight or two, to several ounces, per ton. Fuses readily on charcoal before the blowpipe, and retains its bright surface in an oxidating flame. Not attacked by the blowpipe fluxes. Separated from silver by fusion with bisulphate of potash in a platinum spoon, the silver becoming dissolved, or (if the silver be not in too small a quantity) by dilute nitric acid moderately warmed. In the latter treatment, the gold separates as a dark mass or powder. This assumes a yellow colour and metallic lustre by compression with a glass rod or other hard body. An alloy of gold containing but little silver is merely blackened by the acid. In this case it may be folded in a small piece of pure sheet lead with a piece of silver of about twice or three times its size, and cupelled before the blowpipe (Operation 9, page 19). The alloy is then readily attacked by the acid, and the silver is dissolved out.

(16) *Silver.*—This metal occurs in nature under various conditions: principally in the simple state, as an amalgam with mercury, and as a sulphide, sulphantimonite, sulpharsenite, and chloride; less commonly as a selenide, telluride, antimonide, sulpho-bismuthite, bromide and iodide. It occurs also as an "accidental component" in many varieties of iron pyrites, &c., and in almost every example of galena.*
Metallic silver melts readily before the blowpipe, and the fused globule retains a bright surface after exposure to an oxidating flame. In a prolonged blast a slight brownish-red sublimate is deposited on the charcoal, the sublimate being more distinctly red in the presence of lead or antimony, but in the latter case it is scarcely observable until these metals become for the greater part volatilized. Silver oxide becomes rapidly reduced on charcoal. It is dissolved by borax and phosphor salt, forming glasses which are indistinctly yellowish whilst

---

* For its detection in this mineral, see the foot note on page 19.

hot, and opaline or opaque-white on cooling. Metallic silver is attacked with similar results by these fluxes, and also by bisulphate of potash. In all ordinary cases the presence of silver in minerals is best detected by reduction and cupellation with lead, as described under Operations 8 and 9, pages 16–20, above. Or a kind of scorification process may be employed, by mixing the unroasted ore (to avoid loss of silver) with a little borax, and fusing it in a small cylindrical case of pure lead-foil,' made by folding a piece of foil round the end of a common pencil, and flattening down the projecting edges. The mixture is inserted into this little case by a folded slip of glazed paper, or a small scoop of horn or thin brass. The upper edges of the foil being then pressed or flattened down, the case with its contents is sunk in a sufficiently deep charcoal-cavity, and exposed for a few minutes, first to a reducing, then to an oxidating, and then again to a reducing flame, until the rotating globule shew a clean, bright surface. If the metallic button, after separation on the anvil from accompanying slag, be too large to be cupelled in one operation, it may be flattened out and cut into several pieces. These can be concentrated on separate cupels, and then cupelled together as described at page 19.

III.—VOLATILIZABLE METALS.

The metals of this group are characterized (tin excepted) by the emission of more or less copious fumes when ignited before the blowpipe. Tin becomes rapidly coated with a crust of oxide, and is only slightly volatile. In arsenic and osmium the evolved fumes are accompanied by a marked odour. Tellurium, antimony, arsenic, bismuth, lead, thallium, cadmium and zinc, form characteristic sublimates on charcoal, and (cadmium and bismuth excepted) these metals impart a marked coloration to the flame-border. Tin forms only a slight sublimate. Lead, thallium and tin give malleable globules; tellurium, antimony and bismuth, brittle globules. The other metals of the group volatilize without fusion, or without yielding metallic globules on charcoal.

(17) *Tellurium.*—This metal is of rare occurrence. It is found occasionally in the simple state, and also combined with gold, silver, lead, and other bases in the small group of tellurides. The metal fuses easily, volatilizes, tinges the flame green, and forms a white

deposit of TeO^2 on charcoal. In the open tube, TeO^2 is also deposited as a white coating, but this, when the flame is directed upon it, melts into small colourless drops, a character by which it is distinguished from the sublimate formed by antimony and antimonial compounds. Tellurides produce the same general reactions. The presence of tellurium may also be recognized by fusing the test-matter with carb. soda on charcoal, cutting out the fused mass, and dissolving the resulting alkaline telluride in hot water. The solution has a distinct reddish-purple colour. A purple (or reddish) coloration is also obtained by warming the test-substance, in powder, with concentrated sulphuric acid.

(18) *Antimony.*—Occurs in nature (though rarely) in the simple state, and in one or two rare antimonides. Also much more abundantly in combination with sulphur; and as a sulphur-acid in combination with lead, copper, and other bases, in the somewhat extensive group of sulphantimonites. It also occurs in an oxidized condition, but in that state is comparatively rare. The presence of antimony is revealed in these minerals by the emission of copious white fumes, with deposition of a white coating on charcoal, and green coloration of the flame. The white coating if moistened with nitrate of cobalt, and gently ignited, assumes on cooling a greenish colour. By treatment in the open tube, a dense white, or greyish-white, uncrystalline sublimate is produced. This is soluble in tartaric acid. If a bead of sulphide of sodium (obtained by the fusion of some carb. soda with a little borax and some bisulphate of potash in a reducing flame on charcoal) be placed in the solution, an orange-red precipitate (Sb^2S^3) is produced. (*See* Appendix, No. 13.) Sulphantimonites are partially dissolved by a solution of caustic potash. Hydrochloric acid throws down from the solution the same orange-coloured precipitate of Sb^2S^3. Antimonial oxides dissolve readily in borax and phosphor-salt, forming beads which are slightly yellowish or colourless after exposure to an oxidating flame, and grey, from reduced particles of metal, when exposed to the R. F. Prolonged blowing, however, causes the metal to volatilize, and the glass becomes clear. The phosphor-salt bead treated with tin, becomes on cooling dark grey or black, and quite opaque. This reaction is characteristic of antimony and bismuth compounds.

(19) *Arsenic.*—Occurs, more especially, under the following conditions: In the simple metallic state (usually impure from the presence of small quantities of Sb, Fe, Co, &c.). In various arsenides, combined chiefly with cobalt, nickel, and iron. In combination with sulphur, alone, and combined with bases (Ag, Cu, &c.), forming a small series of sulpharsenites. In combination with oxygen, as arsenic acid, alone, and combined with CuO, NiO, and other bases forming the various arseniates. In these conditions its presence, as a rule, is easily recognized by the strong odour of garlic evolved during the ignition of the mineral on charcoal. In substances of a non-metallic aspect, the odour is more strongly developed, if the test-matter be mixed with carb. soda. Metallic arsenic sublimes, without melting, in copious fumes, which form a white or grayish deposit on the charcoal. A clear blue tint is communicated at the same time to the flame-border. Similar fumes are also emitted (though less copiously) by most arsenides and sulpharsenites, as well as by oxidized compounds, as the arsenic acid of the latter is readily reduced on charcoal. Non-oxidized arsenical bodies when ignited in the open tube (Operation 5, page 11), evolve arsenic, which becomes oxidized into arsenious acid As^2O^3, by the current of air passing up the tube; and this compound is in great part deposited in the form of minute crystals (octahedrons), a short distance above the test-matter. If the tube be of very narrow diameter, however, or if it be held too horizontally, a gray or black deposit of metallic arsenic, or a yellow or red deposit of sulphide of arsenic, may also be formed. The crystals, although very minute, can generally, from their glittering facets, be recognized by the unaided eye, but a strong magnifying glass or small microscope is required for their proper observation. All arsenical bodies, either *per se*, or when mixed with dry carb. soda, neutral oxalate of potash, or other reducing agents, and ignited in a narrow tube closed at one end, form a dark shining "mirror" on the inside of the neck of the tube. The reaction is assisted in the case of oxidized bodies which contain merely a small amount of arsenic, by placing a charred match or slip of charcoal in the tube, above the assay-mixture, and igniting first the charcoal and then the mixture, so as to drive the fumes over the charcoal. A dark metallic ring is formed by this method, even if the test-substance contain only traces of arsenic; and if the charcoal be shaken out of the tube, held against

the side of the flame until ignited, and then brought quickly under the nose, the presence of the slightest trace becomes revealed by the characteristic garlic-like odour which is then emitted.

Non-oxidized arsenical minerals possess a metallic aspect, or, in default of this, are readily inflammable. Arseniates, on the other hand, never present a metallic lustre, and none are inflammable. Many cupreous arseniates deflagrate strongly when ignited on charcoal. Arsenic acid, As^2O^5 (both alone, and in some arseniates), gives off oxygen on strong ignition, and becomes volatilized in the condition of As^2O^3.

(20) *Osmium.*—This metal is of quite exceptional occurrence. It is found in only one mineral, Osmium-Iridium, and is thus often classed as a so-called "platinum metal;" but its general characters and reactions give it a place near arsenic. Osmium-Iridium remains unchanged before the blowpipe, unless the osmium greatly preponderate (as in the variety known as sisserskite), in which case part of the osmium is volatilized. All varieties when fused with nitre in the closed tube or on charcoal, emit the penetrating disagreeable odour of osmic acid. Osmium, itself, volatilizes without fusing, emitting necessarily the same odour; and in a finely divided state it is inflammable. If volatilized in the pale flame of alcohol, or that of the Bunsen burner, it renders the flame highly luminous.

(21) *Mercury.*—Occurs sparingly in the simple state; in silver and gold amalgams; and in certain selenides. More abundantly as a sulphide—Cinnabar, the only ore of mercury.[*] Sparingly, also, in some varieties of grey copper ore (tetrahedrite); and in combination with chlorine, in native calomel. In these compounds, its presence may be readily ascertained by mixing the test-matter with some perfectly dry carb. soda, iron filings, neutral oxalate of potash, or other reducing substance, and igniting the mixture in a closed tube of narrow diameter. The metal volatilizes, and deposits itself on the neck of the tube in the form of a dark grey sublimate. If this be rubbed by an iron wire, it runs into fluid globules which can be

---

[*] Red ochre is frequently mistaken by explorers for cinnabar. Apart from the high sp. gr. of the latter, the two may be easily distinguished by an ignited lucifer match. Held (in the form of a small fragment) in the match flame, cinnabar takes fire and volatilizes; red ochre blackens and becomes magnetic.

poured out of the tube, and which are easily recognized as metallic mercury. Without the reducing agent, many of these mercurial compounds (cinnabar, calomel, &c.) sublime without or with only partial decomposition. When mercury is present in traces only, a piece of gold-leaf, twisted round an iron wire or glass rod, may be inserted into the mouth of the flask. The gold is whitened by a mere trace of the volatilized metal.

(22) *Bismuth.*—Occurs in nature chiefly in the simple metallic state. Found also, but more sparingly, in combination with tellurium, selenium, and sulphur, and with bases in sulpho-bismuthites. Occasionally, likewise, in an oxidized condition (Bi^2O^3) as bismuth ochre (commonly mixed with some carbonate of bismuth), and in a single rare silicate, arseniate, and vanadiate. Metallic bismuth fuses readily, and gradually volatilizes, depositing a dark yellow ring of oxide on the charcoal. The latter volatilizes in the inner flame without colouring the flame-border. Bismuth oxide is at once reduced and volatilized on charcoal. It dissolves in carb. soda in an oxidating flame, very readily, if a platinum wire or other non-reducing support be used. The glass is yellow or yellowish-brown whilst hot, pale yellow and opaque when cold. In borax and phosphor-salt, it dissolves also readily. The borax glass in the O. F. is yellowish, hot, and very pale yellow or white and opaline when cold. In the R. F. the glass becomes clear from separation of the reduced metal. The phosphor-salt glass in the O. F. may be rendered milk-white by flaming or saturation. In the R. F., with tin, it is transparent whilst hot, and very dark-grey or black on cooling. In this respect, the reaction resembles that produced by antimony. The presence of bismuth, in bodies generally, is detected by the dark-yellow coating or ring-deposit formed on charcoal by the fusion or ignition of the test-substance with carb. soda. This deposit is distinguished from that formed by lead, by its deeper colour, and by imparting no colour to the flame. Also, by the black bead formed by it (or by another portion of the test-substance) with phosphor-salt and tin in a reducing flame, as described above. The button of reduced bismuth, moreover, is brittle; that of lead, malleable. These metals may also be distinguished by the sublimates which they form when ignited on charcoal with iodide of potassium, according to the method of Merz;

or by fusion, first with sulphur, and then with iodide of potassium, according to the more delicate process of Von Kobell. With lead, the sublimate is lemon-yellow, or in thin layers, greenish-yellow; whilst with bismuth it presents a vivid scarlet colour, or a ring of this around the outer edge of a yellowish deposit. When a very small amount of bismuth oxide is associated with excess of lead oxide, Cornwall recommends a modification of the process, as follows: the substance, mixed with about an equal quantity of a mixture of five parts sulphur and one part iodide of potassium, is ignited in a test-tube by the spirit-flame or bunsen burner. The presence of bismuth is indicated by a scarlet or orange-coloured band, which forms above the yellow sublimate occasioned by the lead. (*See*, also, page 67, the characteristic reaction with hydriodic acid, lately discovered by Dr. Haanel.)

(23) *Lead.*—The occurrence of native lead is quite exceptional. The metal occurs most commonly as a sulphide (galena), and not uncommonly as a sulphantimonite (and to some extent as a sulpharsenite). Also, frequently in an oxidized condition, as a sulphate, carbonate, phosphate and arseniate. Among rarer (natural) compounds, it occurs as a selenide, telluride, chloride, oxide, chromate, vanadiate, tungstate, molybdate, antimoniate. The presence of lead in bodies generally is made known in blowpipe testing by the two following characters: the formation of a yellow ring-deposit on charcoal, and the ready formation of a malleable metallic globule— these reactions requiring, however, in some few cases, the assistance of carb. soda or other reducing flux for their proper manifestation.* Lead oxide is immediately reduced on charcoal, colouring the flame light-blue. It dissolves readily in the blowpipe fluxes if the fusion be performed on a non-reducing support. The glasses, produced by an oxidating flame, are colourless or yellowish, and become opaque by saturation or flaming. (See Appendix, No. 6.)

(24) *Thallium.*—This new metal is only known to occur (in very minute quantities) in certain examples of iron pyrites, copper pyrites, zinc blende, native sulphur, and some few other minerals. Its chief characteristic is its property of imparting a brilliant green coloration

---

* In the presence of sulphur, more especially, the reduction is facilitated by the addition of a small piece of iron wire. See note at foot of page 19.

to the Bunsen or blowpipe flame. In other respects its reactions much resemble those of lead, but the oxidized ring-deposit (best seen on a porcelain support or on the surface of a boneash cupel) is dark brown. (*See* Appendix, No. 14).

(25) *Cadmium.*—As an essential component, this metal occurs only in a rare sulphide, greenockite. It is present, however, in small quantity in many examples of zinc blende, and in certain varieties of the carbonate and silicate of zinc. Metallic cadmium, on charcoal before the blowpipe, shrinks somewhat together, blackens, takes fire slightly, and becomes volatilized in dense brown fumes. These deposit themselves in the form of a brownish-black and reddish-brown coating (CdO), with a tinge of brownish-yellow towards the outer edge. The deposit is at once reduced and dissipated by either flame, without communicating any colour to the flame border. In both the closed and open tube, if the latter be of narrow diameter, a metallic sublimate is formed near the assay-matter, and a dark-brown sublimate, with yellowish edge, higher up the tube. Fused with phosphor-salt on charcoal, metallic cadmium (like metallic zinc) gives rise as the bead cools to slight detonations and flashes of light. Cadmium oxide on a non-reducing support is infusible, and remains unvolatilized. With borax and phosphor-salt it forms colourless beads which become milk-white and opaque by saturation or flaming. On charcoal the oxide is rapidly reduced and volatilized, but yields no metallic globule. The dark red-brown sublimate, formed on charcoal or better on a porcelain support by the fusion of a cadmiferous substance with carb. soda, is the principal blowpipe-reaction of the metal. In the presence of much zinc, the blast must not be continued too long, otherwise the dark deposit of cadmium oxide, formed before the deposition of the zinc oxide, may be obscured by the latter. For the detection of cadmium in the presence of zinc generally, *see* Appendix, No. 17.

(26) *Zinc.*—Of doubtful occurrence in the native state. Found principally as a sulphide, oxy-sulphide, oxide, sulphate, carbonate, silicate and aluminate. Metallic zinc, when ignited on charcoal, burns vividly with transient flashes of green, blue and greenish-white flame, and throws off dense fumes which become oxidized and deposited as a coating on the charcoal. This coating (ZnO) is pale-

yellow and phosphorescent when hot, and white when cold. It is not driven off by the reducing flame, unless the blast be long continued. If moistened with a drop or two of nitrate of cobalt, and ignited by an oxidating flame, it becomes of a light-green colour on cooling. Zinc oxide forms with borax and phosphor-salt colourless beads, which become milk-white and opaque by saturation or when flamed. Metallic zinc fused with a bead of phosphor-salt on charcoal, detonates slightly and emits flashes of light after removal from the flame—a reaction first noticed by Wöhler, and considered to arise from the formation of a zinc phosphide.* It is manifested, however, not only by zinc, but also by cadmium, aluminium and magnesium, and to some extent by iron pyrites, arsenical pyrites and several other minerals; but it is not produced by tin, lead or thallium. The presence of zinc, in bodies, is best detected by fusing the substance, in powder, with two or three parts of carb. soda, and a little borax on a clean piece of charcoal. A characteristic ring-deposit (lemon-yellow and phosphorescent, hot; white, cold; and green, on cooling, after ignition with cobalt solution) is readily obtained as a rule by this treatment. In the case of silicates (and indeed in all cases) the deposition of this ring-coating is facilitated by first fusing the test-substance with phosphor-salt, and then crushing the saturated bead on the anvil, and re-melting it with carb. soda on charcoal.

(27) *Tin.*—Native tin is of doubtful occurrence. The metal of commerce is obtained entirely from the binoxide, known in its natural occurrence as cassiterite or tinstone. Tin occurs also, but rarely, as a sulphide in tin pyrites; and the binoxide is present in small quantities in tantalates generally, and in certain titaniates, silicates and other compounds. Metallic tin melts easily, without colouring the flame. Before the outer flame it rapidly oxidizes and gives off slight fumes, which form a coating on the fused globule and on the charcoal immediately around the latter. The coating is slightly-yellowish whilst hot, and white or greyish-white when cold, and it is not driven off by the flame, but in a long continued blast it may become reduced. When moistened with a drop of cobalt solution

---

* I have tried, but without success, to make this reaction available for the detection of phosphates by fusing these, in powder, with boracic acid, borax and other reagents, and then adding a piece of metallic zinc to the glass. The reaction, although sometimes produced by this treatment, is too uncertain to serve as a test.

and ignited, it becomes on cooling bluish-green. SnO and Sn^2O^3 (neither of any interest, mineralogically) burn on ignition, and become converted into binoxide. The latter SnO^2, is infusible by the blowpipe, but on charcoal, in a well-sustained blast, it is reduced to metal. The reduction is greatly facilitated by the addition of carb. soda, neutral oxalate of potash, or a mixture of carb. soda and cyanide of potassium, the latter acting most rapidly. In borax, the binoxide is very slowly attacked and dissolved; and phosphor-salt acts upon it still more slowly. With both reagents the glass remains clear when flamed. With soda in the outer flame, it forms, with effervescence, a greyish-white infusible mass. In a good reducing flame (especially if a little borax be added to promote fusibility) it yields reduced metal. As pointed out by Berzelius, a small portion of borax should always be added to the soda in the examination of tantalates and infusible bodies, generally, for the presence of tin. A malleable, easily oxidizable, metallic globule is then, as a rule, obtained without difficulty; but when a trace only, or very small percentage of tin is present, the regular reducing process (explained on page 17) must be resorted to. A button of metallic tin may be distinguished by its malleability, feeble sublimate and ready oxidation, from other metallic globules as obtained by the blowpipe. In nitric acid it becomes converted into a white insoluble powder (SnO^2), behaving in this respect like antimony; but the latter metal gives a brittle button, and also a copious sublimate or ring-deposit which volatilizes wholly or in chief part, and communicates to the flame a greenish coloration. From silver, the tin globule is distinguished by its ready oxidation, and its conversion into insoluble binoxide by nitric acid—silver, in that reagent, dissolving rapidly. From lead and bismuth, it is distinguished also by this acid reaction, and by the non-formation on charcoal of a yellow sublimate. When small pieces of tin and lead (or tin and thallium, or tin and bismuth), are melted together, a remarkable oxidation ensues—the fused mass becoming rapidly encrusted, and continuing, after withdrawal from the flame, to push out excrescences of white and yellow oxides. (*See* Appendix, No. 21).

IV.—FLUX-COLOURING METALS.

The oxides of the metals of this group possess, in common, the property of communicating distinct and more or less characteristic colours to borax and phosphor-salt glasses before the blowpipe. By some, also, a colour is imparted to the soda bead; but most of these oxides are insoluble in carb. soda. They fall into two leading sections, as in the following arrangement:

A.—Reducible from an oxidized or other condition by the blowpipe.
A^1.—Fusible, and therefore obtained by reduction in metallic globules:
Copper.
A^2.—Infusible (practically), and therefore obtained by reduction in the form of separate grains or scales:
† Magnetic:
Nickel. Cobalt. Iron.
†† Non-magnetic:
Tungstenum. Molybdenum.

B.—Not reducible from an oxidized or other condition by the blowpipe.
B^1.—The borax-glass not rendered opaque by flaming:
Manganese. Chromium. Vanadium.
B^2.—The borax-glass converted by flaming into a dark or light enamel:
Uranium. Cerium. Titanium.

(28) *Copper.*—This metal occurs frequently in the native state. Also as a base in numerous sulphides, and in certain arsenides, selenides, sulpharsenites and sulphantimonites. In combination likewise with chlorine. Also in an oxidized condition as Cu^2O and CuO; and in the latter form, as a base, very commonly in arseniates, phosphates and carbonates; and less commonly as a sulphate, chromate, vanadiate and silicate. Metallic copper, on charcoal, melts before the blowpipe into a malleable globule, the surface of which, if exposed to the outer flame, becomes quickly tarnished by a black coating of oxide. This oxide imparts to the flame-border a rich green colour. Cupreous sulphides, arsenides and related compounds become converted by careful roasting, with avoidance of fusion (see the Operation, page 11), into the same black oxide; and a roasting of this kind is always necessary as a preliminary to the reduction of the copper, and its detection by fusion with borax. Both the red and black oxides fuse readily and become reduced on charcoal. With borax and phosphor-salt, the glass after exposure to an oxidating flame, is green whilst hot, and clear-blue when quite cold—unless much iron or nickel be present, in which case it retains its green colour on cooling. In a reducing flame, especially on charcoal, the glass becomes almost colourless, and on cooling turns brick-red and opaque. This reaction (which serves for the detection of copper in the presence of most other flux-colouring bodies) is developed more

easily with borax than with phosphor-salt, but when very little copper oxide is present in the glass, it is not always obtained without long blowing. If, however, a small piece of tin or iron-wire be stuck through the soft glass, and the bead be then again submitted for a few moments to a reducing flame, the opaque red glass (due to the reduction of the CuO to Cu^2O) is readily produced. In place of iron-wire, a small fragment of any substance containing FeO (as iron-vitriol, magnetic iron ore, spathic iron, &c.) may be used to promote the reduction, the FeO becoming converted into Fe^2O^3 at the expense of some of the oxygen of the copper compound. The fusion may then be performed on platinum wire; but, in any case, the bead must not be kept too long in the flame, as the whole of the copper oxide might be reduced to metal, and the glass become colourless by prolonged fusion. By this reaction, the presence of copper in bodies generally (after the preliminary roasting of those which contain sulphur, antimony, &c.) is unmistakably revealed. Another characteristic reaction is the bright azure-flame produced by chloride of copper. The slightly-roasted substance may be moistened with a drop of hydrochloric acid—or fused with chloride of silver—and held just within the point of an oxidating flame. If copper be present, the flame around the test-substance will exhibit a brilliant azure coloration. The test may also be made by simply fusing the substance on platinum wire with phosphor-salt, and then adding some chloride of sodium to the bead. (*See*, also, Appendix, No. 12).

(29) *Nickel.*—Occurs in small and variable proportions in most examples of meteoric iron, and also in some meteoric stones as a phosphide and sulphide. In minerals proper, it is found more especially as an arsenide, antimonide, sulphide and sulpharsenite. It occurs also in an oxidized condition, at times as a simple oxide in coatings on nickel ores, but more commonly as an arseniate, carbonate, sulphate and silicate. In some (mostly magnesian) silicates, and in the apple-green variety of calcedony, known as chrysoprase, it is present in minute quantity as the colouring material of the substance. Metallic nickel is infusible in the blowpipe flame. As obtained by reduction of the oxide NiO by carb. soda or other reducing agent on charcoal, it forms numerous minute particles of a shining white colour. These are strongly magnetic. Sulphides,

arsenides and related compounds, become converted by roasting into this oxide. The latter is unaltered *per se* by the blowpipe flame. With borax, it forms in the O.F. a glass which is amethystine in colour whilst hot (if the NiO be in moderate quantity), and pure brown or yellowish-brown when cold. If not too deeply coloured, the glass on the addition of a carbonate or other salt of potash in excess, is rendered more or less distinctly blue or greyish-blue. The reaction, however, is not very strongly marked, and except under special conditions it can scarcely be regarded as characteristic. In the R.F., the borax glass becomes grey and opaque on cooling, from precipitation of reduced particles of metal. This is the characteristic blowpipe-reaction of nickel. It serves for the detection of that metal (when occurring in more than a very small percentage) in the presence of cobalt and iron oxides, but it is masked by the presence of copper. When copper and nickel occur together, however, the presence of the latter may be suspected by the borax glass, after exposure to an oxidating flame, remaining green when cold; whereas with copper oxide alone, it becomes clear blue on cooling. The reaction, nevertheless, is merely suggestive, as it is produced by other metals, Fe, Cr, &c., when associated with copper. With phosphor-salt, NiO produces much the same reactions as with borax, only the glass in the oxidating flame is less distinctly coloured. With carb. soda on charcoal, as stated above, it is reduced to minute shining particles of magnetic metal.

(30) *Cobalt.*—This metal, as an essential constituent, occurs only in a small number of minerals, and chiefly as an arsenide and sulphide, separately and combined. More rarely it is found as a selenide and oxide, and occasionally as an arseniate; but it is present in traces, as an accidental component, in many sulphides and arsenides, as in varieties of arsenical pyrites, cubical pyrites, &c. The metal itself is practically infusible. Sulphides, arsenides, &c., become converted by roasting into the oxide CoO. This, with carb. soda on charcoal, is readily reduced to shining, magnetic particles of metal. With both borax and phosphor-salt, and in both flames, the oxide forms glasses of a deep blue colour, even when present in traces only. This is the characteristic reaction. When much iron, nickel, or copper is present, the glass however is dark green; but copper and

nickel may be removed by reduction in the inner flame (especially if a small piece of tin be added to the glass on charcoal), and the tint derived from iron is generally overpowered in the outer flame by the much stronger reaction of the cobalt.

(31) *Iron.*—Occurs in the simple state in meteoric iron, though commonly alloyed with a small percentage of nickel. Occurs also, and in numerous localities, in various sulphides, arsenides and sulphur-salts; and in an oxidized condition as $FeO + Fe^2O^3$ in magnetic iron ore, as Fe^2O^3 in hæmatite, &c.; and as FeO or Fe^2O^3 in numerous silicates and other oxygen salts. Metallic iron is practically infusible in the blowpipe-flame, but the extremity of a very thin wire may be oxidized and then fused. Hard wires fuse in general the most easily, and the fusion is accompanied by a rapid scintillation or emission of sparks, whilst very frequently a thin green flame streams from the point of the wire. The latter reaction is due to the presence of phosphorus. (*See* Appendix, No. 11.) Sulphides, arsenides, &c., become converted into the sesquioxide Fe^2O^3 (often termed "red oxide") by roasting. This oxide, by fusion with carb. soda and a little borax on charcoal, is easily reduced to shining particles of metal, strongly attractable by the magnet. On platinum wire or other non-reducing support, it forms with soda a slaggy infusible mass. It dissolves readily, on the other hand, in borax and phosphor-salt, forming glasses which are reddish or yellowish whilst hot, and very pale-yellow or almost colourless when cold, after exposure to the OF; and more or less of a bottle-green colour after treatment in the R. F., especially if a small piece of tin be added to promote reduction, Fe^2O^3 becoming thus converted into FeO. All minerals which contain 5 or more per cent. of iron become magnetic after ignition or fusion. By this reaction, ferruginous substances may be easily recognized, as although cobaltic and nickeliferous bodies also become more or less magnetic on ignition, these latter bodies are of rare occurrence. They are readily distinguished, moreover, from ferruginous substances by the colours, &c., of the glasses which they form with borax. When the presence of iron has been recognized in a silicate or other body, it is often desirable to ascertain whether the iron is present as sesquioxide Fe^2O^3, or partly or wholly as protoxide, FeO. This may be determined by adding some of the test-substance,

in powder, to a bead of borax coloured blue by previous fusion with
a few particles of oxide of copper, and exposing the bead (in a loop
of platinum wire) to the point of the blue flame until the substance
begins to dissolve. If any FeO be in the substance, it will become
converted into Fe^2O^3 at the expense of some of the oxygen of the
copper oxide, and the latter will thus become reduced to suboxide,
Cu^2O, causing red streaks and spots to appear in the glass, as this
cools. If no FeO be present, the glass will, of course, become green
on cooling, but will remain transparent. (*See* Appendix, No. 5.)
A very minute trace of iron may be detected by the following
process: Fuse into a bead of phosphor-salt, on platinum wire, as
much of the substance, in powder, as the bead will take up. Then
saturate the bead with successive portions of bisulphate of potash (or
treat the crushed bead with that reagent in a platinum spoon), and
dissolve out the soluble matters in warm water. Finally, place in
the solution a very small particle of ferrocyanide of potassium
("yellow prussiate"). If iron be present, a deep-blue precipitate
will necessarily ensue.

(32) *Tungstenum or Wolframium.*—This comparatively rare metal
is known in nature only in an oxidized condition, as WO^3, a compound
which occurs occasionally alone, but more commonly in combination
with bases, thus forming the small group of tungstates. Tungstic
acid or anhydride WO^3, is scarcely affected by the blowpipe-flame;
but on charcoal, after long ignition in the R. F., it becomes blackened,
by conversion into W^2O^5. With carb. soda or neutral oxalate of
potash, it is reduced on charcoal to minute particles of metallic
tungstenum; but if much soda be used, the portion of test-matter
absorbed by the charcoal is generally obtained (by washing in the
agate mortar, page 17) in the form of minute yellow specks, of
metallic lustre, consisting of a compound of soda and tungstic oxide.
On platinum wire, with carb. soda, it dissolves more or less readily
into a yellowish glass, which becomes opaque and somewhat crystalline on cooling. Borax dissolves it readily. After exposure to the
O. F. the glass is yellowish and clear, but becomes enamelled by
flaming. In the R. F., with excess of test-matter, the glass is
yellowish-brown, and by flaming or on cooling it becomes opaque.
With phosphor-salt, in a reducing flame, a deeply coloured greenish-

blue glass is obtained. This is the characteristic blowpipe reaction of tungstenum compounds; but if much iron be present, the glass becomes deep-red. The presence of tungstenum may also be detected by fusing the powdered test-substance with 3 or 4 parts of carb. soda and a little nitre in a platinum spoon or loop of thick platinum-wire, dissolving out the soluble alkaline tungstate (as explained on page 20), decanting the clear solution, acidifying it with a few drops of hydrochloric acid, and placing in it a piece of zinc. A dark-blue coloration (from reduction of the WO^3 to W^2O^3) will rapidly result.

(33) *Molybdenum.*—This metal occurs in nature most commonly in combination with sulphur, in the sulphide molybdenite, a mineral which presents a curious resemblance to graphite in many of its properties (foliated or scaly-granular texture, softness and flexibility, soapy feel, detonation with nitre, infusibility, &c.). It occurs also, though rarely, in an oxidized condition as MoO^3, this latter compound being found at times alone, but more commonly combined with lead oxide in the molybdate wulfenite. Molybdic acid or anhydride, MoO^3, melts easily on charcoal, tinges the flame yellowish-green, and becomes gradually volatilized, forming a deposit which is slightly yellowish whilst hot, and white when cold. When touched by the reducing flame, this deposit assumes a dark-bluish tinge from partial conversion into Mo^2O^3. In addition to the white coating, an indistinct reddish deposit is also formed near the test-matter. With carb. soda, reduction to minute steel-grey particles is easily effected on charcoal. On platinum wire, solution takes place with effervescence. With borax, before the O. F., a yellowish glass, which becomes grey and opaque by flaming, is formed; and in the R. F., a brown or grey glass, with separation of dark flecks, the latter best seen by pressing the bead flat before it cools. With phosphor-salt, on cooling, and especially after exposure to a reducing flame, a fine green glass results. By this reaction (combined with the property of colouring the flame pale yellowish-green,* and yielding *per se* or with carb. soda a white sublimate and reduced particles of non-magnetic metal), molybdenum compounds are chiefly recognized in blowpipe practice. Molybdic acid and molybdates, as first made known by Von Kobell,

---

* Although molybdenum compounds colour the Bunsen flame very distinctly, they give no coloured bands in the spectroscope, but merely a continuous spectrum.

when warmed with sulphuric acid, produce a rich blue solution on the addition of alcohol. If the test-substance be fused with carb. soda and nitre, and the solution of the alkaline molybdate be treated with hydrochloric acid and metallic zinc, a bluish colour may appear at first, but this quickly changes to dark-brown. (*See* under Tungstenum, No. 32, above.)

(34) *Manganese.*—Does not occur, in nature, in the metallic state. Occurs occasionally as an arsenide and sulphide, but is chiefly found in an oxidized condition—mostly as MnO^2 and Mn^2O^3 (these compounds occurring alone, combined together, or as hydrates); and as MnO in various silicates, carbonates, phosphates, tungstates, &c. As an accidental or inessential component it is present in the latter state in very numerous minerals. In these, the MnO generally replaces small portions of MgO, CaO, or FeO. Manganese oxides are not reduced by carb. soda on charcoal. Very little of the oxide dissolves in the flux, but this communicates to the bead a green colour whilst hot, and a blue or greenish-blue colour when cold. The reaction is brought out more prominently by the addition of a little borax to the soda, as this promotes solution (*see* Appendix, No. 9); and it is also increased in intensity by melting a small portion of nitre into the bead, or by pressing the hot bead upon a small fragment of nitre. A greenish-blue bead of this kind is known technically as a "turquoise enamel." Manganese oxides dissolve readily in borax and in phosphor-salt, and the solution in the case of the higher oxides (MnO^2 especially) is accompanied by great effervescence or ebullition, due to the escape of oxygen from the test-matter. Oxygen is also evolved when these oxides are strongly ignited *per se*, as in a closed tube, &c. (*See* under "Oxygen," above.) The borax glass after exposure to an oxidating flame presents a beautiful amethystine colour. In a reducing flame it becomes colourless, but if allowed to cool slowly it absorbs oxygen, and the amethystine or violet colour is restored. This may be prevented by urging a stream of air from the blowpipe upon the bead, directly the latter is removed from the flame. When very little manganese is present in the test-matter, the formation of a violet-coloured glass is facilitated by the use of a small fragment of nitre. The phosphor-salt glasses resemble those produced with borax, only the amethystine

colour is paler, and when very little manganese is present it is scarcely developed without the aid of nitre. The great test for the presence of manganese in bodies, is the formation of a turquoise enamel by fusion on platinum wire or foil with carb. soda and a little borax. Less than one part in a thousand may be easily detected by this reaction; and by the addition of nitre, as described above, the reaction becomes still more delicate. Chromium compounds when fused with carb. soda in a reducing flame form a yellowish-green mass, which might in some cases be thought to arise from the presence of manganese. But if a greenish mass of this kind be fused with sufficient boracic acid or silica to form a clear glass, the latter in the case of manganese will present an amethystine colour, whilst in that of chromium it will be emerald-green. (*See* Appendix, No. 16.)

(35) *Chromium.*—Traces of this metal occur in some varieties of meteoric iron, but otherwise chromium is found in nature only in an oxidized condition, as Cr^2O^3 and as CrO^3. In the former state it occurs occasionally alone, as in chrome ochre; but more commonly in combination with iron in chromic iron ore, or, as a base, in certain silicates, and in varieties of spinel. In many silicates it is present as an inessential component, as in the emerald, proper. In the condition of CrO^3, it occurs in combination with lead oxide or copper oxide in the small group of chromates. The leading blowpipe reactions of chromic oxide are as follows: *Per se*, the oxide is practically unchanged. With carb. soda, it dissolves more or less readily, forming a yellowish, opaque bead in the outer flame, and a yellowish-green bead in a reducing flame. If a particle or two of nitre be fused into the bead, the latter becomes blood-red whilst hot, and light-yellow when cold—a soluble alkaline chromate resulting. With borax and phosphor-salt, clear, emerald-green glasses are produced, especially by treatment in a reducing flame, and after complete cooling. Whilst hot, the glass is yellowish or red, as in many other cases. The production of an emerald-green glass with borax generally serves for the detection of chromium compounds; but the character becomes necessarily masked to some extent by the presence of other flux-colouring bodies, as iron, copper, and cobalt oxides, for example. In the presence of bodies of this kind, chromium is best detected by fusing the test-matter (in powder) with three or four parts of carb. soda,

and a little nitre in a platinum spoon or loop of stout platinum-wire. A soluble alkaline chromate then results. The solution (*see* page 20), filtered or carefully decanted from the insoluble residuum, may be divided into two portions. One portion may be evaporated to dryness, and the resulting deposit tested by fusion with borax. The other portion may be carefully neutralized by a drop or two of dilute nitric acid, or acetic acid, and tested with a fragment of nitrate of silver: a red precipitate should be produced. Chromates, also, when treated with sulphuric acid and alcohol, form a rich green solution which remains green on dilution. *Chromic acid*, CrO^3, *per se*, blackens when ignited, gives off oxygen, and becomes converted into chromic oxide. Bichromates, and many chromates also (but not neutral alkaline salts), produce the same reaction.

(36) *Vanadium.*—Occurs, in nature, only in an oxidized condition, as V^2O^5, combined with lead-oxide, and more rarely with other bases, in the small group of vanadates. On charcoal, vanadic acid, fuses and becomes in part reduced to dark-grey or black shining scales of suboxide. If heated on a fragment of porcelain or other non-reducing support, it fuses without decomposition, and congeals with vivid emission of light, on removal from the flame, into a red or dark orange-coloured crystalline mass. With borax, it forms a clear yellowish-green glass, and with phosphor-salt a yellow glass, on cooling, after exposure to the outer flame; and emerald-green glasses with both fluxes, on cooling, after exposure to a reducing flame. With hydrochloric acid and alcohol, vanadates give a green solution which becomes light-blue on dilution (Von Kobell). In addition to this test, it may be observed that whilst chromium compounds give in the O. F. with phosphor-salt (on cooling) a green glass, the glass formed by vanadium remains yellow when cold— in the absence, at least, of copper or other flux-colouring bodies.

(37) *Uranium.*— Occurs only in an oxidized condition : chiefly as UO, U^2O^3 in the mineral pitchblende, and as U^2O^3 in uran ochre and a few comparatively rare phosphates, sulphates, carbonates, and silicates. The sesquioxide is infusible *per se*, but is blackened in the R. F. from partial reduction to UO. It is insoluble in soda, and is not reduced to metal by that reagent, but it is readily dissolved by borax and phosphor-salt. The borax glass is deep-yellow in the

O. F., and dingy brownish-green, when cold, after subjection to a reducing flame; and, if thoroughly saturated, it may be rendered black by flaming. The phosphor-salt glasses present a striking contrast, in being brightly coloured: yellowish-green in the O. F., and clear chrome-green in the R. F., especially when cold. This reaction serves to distinguish uranium compounds from those of chromium, &c.; but in the presence of other flux-colouring bodies uranium is not readily detected.

(38) *Cerium.*—Occurs in only a few comparatively rare minerals—chiefly as a fluoride, or in an oxidized condition in certain silicates, phosphates, &c. On ignition, CeO becomes converted into yellow or reddish Ce^2O^3. This remains unchanged. With carb. soda, on charcoal, it is reduced to grey CeO, but gives no metal. With borax in the O. F. a reddish or yellowish glass is obtained, and in the R. F. a colorless glass. Both glasses become opaque when flamed, if tolerably saturated. With phosphor-salt, the glasses on cooling are colorless, but they are not rendered opaque by flaming, even if strongly saturated. As a rule, the presence of cerium in minerals cannot be safely proved by the blowpipe alone.

(39) *Titanium.*—Occurs, in nature, in an oxidized condition only —as TiO^2 in three separate forms (Rutile, Octahedrite, Brookite), and combined with lime, yttria, zirconia, &c., in the small group of titaniates. In this condition it is present also in certain silicates; and as Ti^2O^3 it partly replaces Fe^2O^3 in titaniferous iron ores. TiO^2 becomes yellowish on ignition, but remains infusible, and reassumes its white colour on cooling. Moistened with nitrate of cobalt, and ignited, it becomes green when cold. With soda, on charcoal, it is not reduced to metal, but it fuses with effervescence, and on cooling the surface of the bead shoots into broad crystalline facets of a pearly-grey colour. With borax, it forms in the O. F. a yellowish glass which loses its colour on cooling, and when saturated becomes on cooling or by flaming milk-white and opaque. In the R. F., the glass, moderately saturated, assumes on cooling a brownish-amethystine colour, and with more of the test-matter it becomes blackish-blue and opaque on congealing. When flamed, a light greyish-blue film spreads over the surface of the bead. The dark-blue tint (Plattner calls it "brown") arises from Ti^2O^3; the light-blue

surface-film from the partial oxidation of this into TiO^2. With phosphor-salt, the glass in the O. F. is colorless or pale yellowish and in the R. F., on cooling, it assumes a fine amethystine colour. When titanium compounds contain iron, however, the glass is deep red-brown or blood-red. In the case of Menaccanite or Titaniferous Iron Ore, proper, this reaction is very marked; but it is not sufficiently definite to serve for the detection of small quantities of titanium in ordinary iron ores. In these, the presence of titanium is most readily detected as follows:—Reduce a portion of the ore to as fine a powder as possible; warm this with hydrochloric acid in a small covered beaker-glass for about half-an-hour on a sand-bath, keeping the acid just at the boiling-point; add a little water, and filter from the insoluble rock-matter, &c.; place a piece of metallic tin in the filtrate, and boil for ten or fifteen minutes. Thus treated, the deep-yellow solution will quickly become greenish and then colorless, and on the boiling being continued, a pink tinge will appear and gradually deepen into a distinct amethystine colour. In the absence of titanium, the solution will of course remain colorless, but the boiling must not be discontinued too soon. The presence of titanium in iron ores, &c., may also be detected by fusing the test-matter, in fine powder, with six or eight parts of bisulphate of potash (added in successive portions) in a platinum spoon; treating the fused mass with a very small quantity of warm water; decanting or filtering from insoluble matters; adding a few drops of nitric acid, and then five or six volumes of water; and, finally, boiling for ten or twelve minutes. Titanic acid, if present, is precipitated in the form of a white or pale-yellowish powder. This may be fused with phosphor-salt, in a reducing flame, for the production of a characteristic amethystine glass. As pointed out by Gustav Rose, a glass of this kind, rendered colorless or nearly so by the O. F., and then slightly flamed, becomes opalescent from the precipitation of numerous crystals of TiO^2. These are best examined, in the flattened bead, by a microscope with object glass of moderate but not too low power.

V.—EARTH METALS.

This group is to a great extent conventional. Tantalum is placed in the group, because in a scheme of this kind it can scarcely be placed elsewhere. The representatives of the group are separated from those of the preceding series by their property of forming uncoloured glasses with the blowpipe fluxes; and from those of the next series by their non-alkaline character. With reference purely to blowpipe characters, it would perhaps be a more satisfactory arrangement if magnesium were also referred to this group, the other metals of Group 6 and those of Group 7 being placed together in a single group under the name of *Flame-colourers*. Keeping, however, to the present distribution, it may be pointed out that aluminum compounds are distinguished from those of the associated metals by not forming an opaque glass with borax, and by the blue colour assumed after ignition with nitrate of cobalt. Compounds of the other metals belonging to the group are of comparatively rare occurrence.

(40) *Tantalum.*—Occurs only in an oxidized condition as tantalic acid (Ta^2O^5) commonly associated with columbic or niobic acid (Nb^2O^5) and combined with iron oxide and other bases, in a few minerals of exceptional occurrence. Tantalic acid becomes pale yellowish on ignition, but resumes its white colour on cooling, and remains infusible. After treatment with cobalt-solution it becomes pale flesh-red. With carb. soda it dissolves with effervescence, but is not reduced. With borax, it dissolves easily, the saturated glass becoming opaque on cooling or by flaming. With phosphor-salt it forms a permanently clear bead. Its presence in minerals cannot be safely detected by the blowpipe alone.

(41) *Aluminum.*—Occurs in nature as a fluoride (in cryolite, &c.), but essentially as an oxide, Al^2O^3. The latter compound occurs alone and in a hydrated condition (corundum, diaspore, gibbsite); and in combination with magnesia and other bases as the electro-negative principle of the small group of aluminates. It occurs also, and more frequently, as a base, in various silicates, phosphates, and sulphates. Exceptionally, also, as an arseniate; and in combination with an organic acid in the mineral mellite. Alumina presents the following blowpipe reactions: (1) *Per se*, it is infusible and unchanged. (2) Moistened with nitrate of cobalt, and ignited, it assumes, on cooling, a fine blue colour. The reaction is exhibited by all aluminous silicates, phosphates, &c., which are free from iron oxides or other strongly coloured bases (*See* page 12.) (3) Alumina is not attacked by carb. soda. (4) It is very slowly dissolved by borax and phosphor-salt, forming colorless, permanently clear beads. (5) It is

dissolved, in fine powder, by fusion in a platinum spoon with five or six parts of bisulphate of potash (page 20). The aqueous solution of the fused mass yields a white precipitate (soluble in caustic potash) with ammonia. Silicates resist this treatment, but in fine powder many are soluble in hydrochloric acid, and nearly all may be rendered soluble by previous fusion with a mixture of carb. soda and borax. The solution (with slight addition of nitric acid) must be evaporated slowly to dryness, the residuum moistened with a couple of drops of hydrochloric acid, water added, and the clear supernatant liquid decanted or filtered from the insoluble silica. If the precipitate formed in the filtrate by ammonia be brown in colour, it must be separated and boiled with caustic potash. This will take up any alumina that may be present, leaving Fe^2O^3 undissolved.

(42) *Glucinum*, or, *Beryllium*.—Occurs only in an oxidized condition, BeO, as a base in a small number of silicates (Phenakite, Beryl, Euclase, &c.), and in a single aluminate (Chrysoberyl). Glucina is infusible *per se*, and is not dissolved by carb. soda. With cobalt solution it becomes pale bluish-grey; with borax and phosphor-salt it dissolves more or less readily, the saturated glass becoming opaque on cooling or when flamed. When glucina is combined with other bodies, its blowpipe reactions are not sufficient for its detection.

(43) *Zirconium*.—Occurs only oxidized, as ZrO^2 in combination with silica and various bases in a small number of minerals. The zircon (ZrO^2, SiO^2), distinguished chiefly by its hardness, high sp. gr. (=4·2–4·8), Tetragonal crystallization, and infusibility, is the only representative species of tolerably common occurrence. Zirconia when ignited, glows with more than ordinary brightness, but remains unfused. After treatment with cobalt solution, it assumes a dull violet tinge. It is not dissolved by carb. soda, but dissolves freely in borax and phosphor-salt, forming a colourless glass which on saturation becomes opaque on cooling or by flaming. Zircon and other silicates in which zirconia is present become decomposed by fusion in fine powder with carbonate of soda, and they are then soluble or partially soluble in hydrochloric acid. The dilute solution, as first pointed out by BRUSH, imparts an orange-yellow or reddish-brown colour to turmeric paper, seen most distinctly as the paper dries.

(44) *Yttrium.*—This rare metal (almost always associated with Erbium) occurs in the mineral of Yttrocerite as a fluoride; but in general it is found in an oxidized condition (YO) as a base in certain silicates, titanates, tantalates, niobates and phosphates, all of more or less exceptional occurrence. The blowpipe reactions of yttria agree in all essential respects with those of glucina. It is thus infusible *per se*, and also with carb. soda; but soluble in borax and phosphor-salt, the saturated glass becoming opaque by flaming or on cooling. Practically, its presence in minerals escapes detection by the blowpipe.

VI.—ALKALINE EARTH-METALS.

This group includes magnesium, calcium, strontium, and barium. The two first by the insolubility of their oxides (before the blowpipe) in carb. soda, are allied to the metals of the preceding group, whilst the general solubility of strontium and barium compounds in that reagent, connects the latter metals with those of Group VII. The carbonates, sulphates, fluorides, &c., of all the representatives of the group, react alkaline after strong ignition, and thus restore the blue colour of reddened litmus-paper; but in other compounds (silicates, &c.), the reaction is less clearly marked or is not observable. All the oxides belonging to this group dissolve freely in borax and phosphor-salt, forming clear glasses which on saturation become opaque by flaming or when cold. Magnesium compounds impart no colour to the flame; compounds containing calcium and strontium colour the flame red or crimson, and barium compounds communicate to it an apple-green coloration.

(45) *Magnesium*—Occurs, though rarely, as a chloride, and still more rarely as a fluoride; very abundantly, on the other hand, as an oxide, magnesia, MgO. This compound, though occuring alone in Periclase, and as a hydrate in Brucite, is chiefly met with as a base in various aluminates, silicates, sulphates, carbonates, borates, phosphates and arseniates. Magnesia is infusible *per se*, and insoluble in carb. soda. After ignition with nitrate of cobalt it assumes on cooling a pale flesh-red colour. This reaction is manifested by magnesium carbonates, silicates, &c., in the absence of iron or other colouring oxides, but in many cases it is not very distinct. Magnesia does not colour the blowpipe-flame, and its compounds, when ignited in a Bunsen-burner, give no spectrum bands. With borax and phosphor-salt it dissolves

very readily, the saturated glass becoming opaque on cooling or when flamed. The non-coloration of the flame and the reaction with nitrate of cobalt generally serve to distinguish magnesian compounds, except in the case of certain silicates. In these, and in other doubtful cases, the test-substance, in fine powder, may be dissolved in a small quantity of hydrochloric acid in a porcelain capsule over the spirit lamp or Bunsen flame ; or, if insoluble in acids, it may be rendered soluble by previous fusion with a mixture of carb. soda and borax. The fusion is best performed in a paper cylinder (according to Plattner's method), the cylinder being made and filled as directed in the case of the lead cylinder on page 32. The solution is then to be diluted, a drop of nitric acid added, the whole evaporated to dryness (to separate silica), the residuum re-moistened with hydrochloric acid, distilled water added, and the solution filtered. In the filtrate, Al^2O^3 and Fe^2O^3, if present, are thrown down by ammonia in slight excess ; lime is next precipitated by oxalic acid or oxalate of ammonia ; and finally the magnesia is separated by some dissolved phosphor-salt. Care of course must be taken in each case to see that the precipitation be complete.

(46) *Calcium.*—Occurs frequently as a fluoride, and occasionally as a chloride ; but principally in an oxidized condition (CaO) as a base in silicates, carbonates, sulphates, phosphates and other oxygen compounds. Lime glows strongly on ignition, and imparts to the flame-border a distinct red colour, but this is less intense than the crimson coloration produced by strontium and lithium compounds. The characteristic bands in its spectrum are two in number—an orange-red band (a little farther from the sodium line than the orange strontium band), and a clear green band.* This flame reaction is

---

* In these examinations, a small, direct-vision spectroscope—such as Browning's pocket spectroscope with attached scale and extra prism—will be found most suitable. By a little practice, the student will readily recognize the positions of the red and orange lines, without the assistance of the scale, by their relative distance from the sodium line. A small fragment of lepidolite will give the sodium and lithium lines very distinctly. Strontianite, and also celestine, after a short exposure to the flame, give the orange, red, and blue lines characteristic of strontium ; heavy spar and witherite, the characteristic barium bands ; and fluor-spar, gypsum, calcite, &c., the red and green calcium lines. The effect is heightened by moistening the calcined test-matter with a drop of hydrochloric acid, but as regards the above (and various other) minerals, the distinctive lines come out very vividly by a sufficiently prolonged ignition of the substance *per se*. The small sharp-edged fragment is conveniently held in the platinum-tipped forceps, and these can be fixed at the proper height by thrusting their opposite ends across the stem of one of the ordinary wire supports used in spectroscope examinations.

given by carbonates and sulphates, as well as by fluor spar, after prolonged ignition in the Bunsen flame, but as a rule it is best obtained by moistening the test-substance with hydrochloric acid. *Per se*, lime is infusible. It is not dissolved by carb. soda, but dissolves readily by fusion with borax and phosphor-salt, the saturated glasses becoming opaque by flaming or on cooling. With nitrate of cobalt a dark-grey coloration is obtained. For the detection of lime in silicates, see under Magnesium, No. 45.

(47) *Strontium.*—Occurs only, among natural compounds, in an oxidized condition, as SrO, combined with sulphuric acid and with carbonic acid; more rarely with silica. Both the sulphate and carbonate become caustic on ignition, and then give the crimson flame-coloration and other reactions of pure strontia,—dissolving, like the latter, very readily and completely in carbonate of soda, a character by which strontium and barium compounds (with those of the alkali metals proper) are at once distinguished from other alkaline earths. With borax and phosphor-salt strontia dissolves freely, the colorless glass becoming opaque (if sufficiently saturated) on cooling or when flamed. After ignition with nitrate of cobalt, strontia becomes dark-grey or black. In the strontium spectrum the distinctive lines comprise (1) a broad orange-red line, quite close to the sodium line, (2) a group of several crimson lines, and (3) a single blue line. A small fragment of strontianite or celestine shews these lines very distinctly after a short exposure to the edge of the Bunsen flame. If a strontium compound be fused on platinum wire with chloride of barium, the crimson flame-coloration is destroyed. By this character --as well as by the spectrum—strontium compounds are readily distinguished from those of lithium. (*See* Appendix, No. 4.)

(48) *Barium.*—Occurs in nature in an oxidized condition only, and chiefly as a sulphate and carbonate, more rarely as a silicate. Present also in some of the naturally-occurring oxides of manganese. Baryta dissolves entirely in carb. soda, and resembles strontia in its other blowpipe reactions, except as regards the coloration of the flame and the reaction with nitrate of cobalt. It communicates to the flame-border an apple-green or yellowish-green colour, and becomes reddish-brown after treatment with the cobalt solution (page 12), but the latter reaction is of little moment. The spectrum of

barium compounds is essentially characterized by a group of green lines, four or five in number, of which two are especially vivid and distinct; with a line or two, often ill-defined, in the orange and yellow, and one or two more or less indistinct lines near the commencement of the blue, the whole at nearly equal distances apart. The group of green lines is the characteristic portion of the spectrum. In the calcium or lime spectrum there is only a single well-pronounced green or yellowish-green line, whilst the spectra of Sr, Na, Li, and K, show no green lines. See also Appendix, Nos. 1 and 2.

VII.—ALKALI METALS.

This group includes Lithium, Sodium, Potassium, and Ammonium. Compounds of these alkali metals much resemble strontium and barium compounds in their general blowpipe reactions. They impart a colour to the flame, and dissolve readily, by fusion, in carb. soda. The flame coloration of lithium compounds is crimson; of sodium compounds, yellow; of potassium compounds, clear violet; of ammonium compounds, pale or dull green.

(49) *Lithium.*—This metal as an essential mineral-component occurs only in an oxidized condition (Li^2O) in a few silicates and phosphates; but in minute quantities it appears to be widely distributed throughout nature. The presence of lithia in most compounds is readily detected by the crimson coloration imparted to the blowpipe flame or that of the Bunsen burner, especially on prolonged ignition. When lithia is merely present, however, as an accidental or inessential constituent, the flame-coloration is best brought out by moistening the test-matter in powder with a drop or two of hydrochloric acid. The mixtures of bisulphate of potash and fluor-spar, or gypsum and fluor-spar, recommended in books for this purpose, often bring out by themselves a vivid red coloration. By fusion with chloride of barium, the intensity of the lithium flame is increased, whereas by this treatment the red flame of strontium is destroyed (see Appendix, No. 4). The spectrum of lithium is also exceedingly characteristic. It consists practically of a single crimson line, much farther from the sodium line than the characteristic orange-red line of strontium, or the red calcium line. Most examples of lepidolite give the lithium and sodium lines together.

(50) *Sodium or Natrium.*—Widely distributed as a chloride, and occurring also as a fluoride. Present also abundantly in an oxidized condition (Na^2O) in various, silicates, sulphates, and carbonates, and in the nitrate soda-nitre. Distinguished very readily in most cases by the strong yellow coloration which its compounds impart to the Bunsen and blowpipe flame. Its spectrum consists of a single yellow line (as seen in ordinary spectroscopes) corresponding in position with the line (or double line) D of the solar spectrum. This yellow line is exceedingly characteristic; and its very constant presence in spectra, generally, serves as a convenient index to the position of other lines, as those of calcium, strontium, &c. The yellow flame-coloration is completely hidden if viewed through a deep-blue glass.

(51) *Potassium or Kalium.*—Occurs as a chloride; but more commonly in an oxidized condition (K^2O) as a sulphate and nitrate, and in various (chiefly aluminous) silicates. Potash (if perfectly free from soda) imparts to the outer flame a clear violet tint, but this coloration is masked or rendered more or less invisible by the least trace of soda or of any sodium compound, and also as a rule by other flame-colouring bodies. If the flame be viewed however, as first shewn by Cartmell, through a deep-blue glass or a solution of indigo, the yellow coloration due to sodium becomes entirely obliterated, and the potash-flame exhibits a bluish-red colour. The indigo-solution (1 part indigo, 8 concentrated sulphuric acid, 1500 water) is best contained in a prism-shaped or wedge-shaped bottle, so that different thicknesses may be conveniently brought between the eye and the flame. Cornwall has recommended a solution of permanganate of potash in place of the indigo solution. When the potash flame is obscured by lithium, it will be rendered visible, according to Merz, if viewed through a green glass, the lithium flame becoming then obliterated. A good deal depends, however, on the shade of colour of these glasses and solutions, and the results are not always entirely satisfactory. Whenever therefore recourse can be had to the spectroscope, the latter should always be employed. The potassium spectrum consists essentially of two lines, far apart—a red line, almost at the commencement of the normal spectrum (it coincides, practically, with the solar line A), and a violet line near the other extremity of the spectrum proper. The latter line, however, is not generally visible, and in small spectroscopes the two lines can rarely be seen together. The red

line is the characteristic one. It lies about (but not quite) as far from the red lithium-line as this lies from the sodium-line. Starting therefore from the latter, the characteristic orange and red spectrum lines of the common alkaline and earthy bodies succeed each other in the following order: (Na)—Sr—Ca—Sr (group of lines)—Li—K: one of the red Sr-lines coinciding with the solitary Li-line.* If the student be uncertain, at any time, as regards the red K-line, he should insert into the edge of the Bunsen flame a small scale of lepidolite (or other lithium-containing body), when the relative positions of the two will at once become apparent; or, if his spectroscope be fitted with an extra prism, he can, of course, examine the two spectra separately. The nitrate, and the natural sulphates and chlorides (as well as the ordinary potassic salts of the laboratory, phosphates, bromides, &c.), give the reaction very distinctly, but it is not always produced directly by natural silicates. To detect potash in the latter, a small portion of the silicate, in fine powder, must be fused on a loop of stout platinum wire with a mixture of carb. soda and borax, and the fused bead (crushed to powder) must be boiled with a few drops of hydrochloric acid. The solution, evaporated nearly to dryness, or a small portion of the pasty mass, may then be examined by the spectroscope. The presence of sodium does not interfere with the production of the red potassium-line, but the supporting wire should be kept, as a rule, just at the edge of the Bunsen-flame, and the observations should be made in a darkened room.

(52) *Ammonium.*—Occurs in Inorganic Nature chiefly as a chloride; more rarely in an oxidized condition as a sulphate and borate. Accidentally present also in many bog iron ores and other minerals which contain traces of intermixed organic matter. Its presence is recognized more or less readily by the odour evolved on moderate ignition, especially if the substance, mixed with dry carb. soda, be ignited in a test-tube. A slip of red litmus-paper, slightly moistened and placed at the top of the tube, will be rendered blue by the evolved vapours; and these will also manifest themselves in white fumes if a glass rod moistened with hydrochloric acid be brought

---

* The ash of a cigar or of ordinary tobacco, if moistened with hydrochloric acid, will show the green and red calcium lines and the red K-line very distinctly. The lithium-line is also shewn by some kinds of tobacco.

over the opening of the tube. Most ammonium compounds impart a feeble blueish-green or brownish-green colour to the flame, but none give a distinctive spectrum.

§ 6.
PLAN OF ANALYSIS.

In the examination of a mineral substance with a view to determine its general nature by the blowpipe—aided by such liquid reagents and processes as are available in blowpipe practice—it is advisable, in the first place, to determine the electro-negative element or compound in the substance (or, in other words, to ascertain the chemical group to which the substance belongs), and afterwards to determine the base or bases that may be present in it.

The methods of Blowpipe Analysis usually followed, although well adapted to convey a knowledge of the special reactions of bodies, have two essential defects : they draw no line of separation between electro-negative substances and bases, but mix up the two together in a loose and confusing manner ; and they exact the performance of a great number of experiments, by which many substances are detected over and over again, whilst others may easily escape detection altogether.

In the plan now proposed, these defects are in a great measure remedied, and a knowledge of the chemical nature of an unknown mineral—so far as this can be obtained by the Blowpipe—is arrived at without unnecessary trouble or delay. If the electro-negative principle in the substance be not detected by one or the other of the eight easily and rapidly performed experiments given under the first section of the scheme, the substance—unless it be a telluride, tantalate or other rare compound, properly omitted from consideration in an outline of the present character—will be either a simple basic-oxide or metal, and its true nature will be revealed in the examination for bases, as given under TABLE B. It will, of course, be understood, that, as a rule, the entire series of experiments for the detection of electro-negative bodies need not be carried out. Sulphates, for example, will be recognized by the first experiment, carbonates and silicates by the second, and so on as regards representatives of other groups. Except, therefore, in certain rare cases indicated in the text (as in the combination of a phosphate and fluoride, &c.), it will only be necessary to continue the experiments until the chemical group to which the substance essentially belongs has been ascertained. The base or bases, present in the substance, may then at once be sought for.

A.—DETECTION OF ELECTRO-NEGATIVE BODIES.

Experiments.	Results more Especially to be Looked For.	Substances Indicated.
1. Fuse the test-substance, in powder, with carb. soda (and a small addition of borax) in R. F. on charcoal. Moisten fused mass, and place on lead test-paper or silver-foil. N.B.—If the fusion be effected by a gas flame, the gas should be tested previously for presence of sulphur. See under "Sulphur" in § 5.	(1) Emission of arsenical odour. (2) Emission of copious fumes, and deposition of dense white coating on the charcoal. (3) Formation of "hepar," or alkaline sulphide. Other results (if any) such as reduction to metal yellow coating on charcoal, &c., may be noted down for after reference.	(1) As., Arsenides, Arseniates. (1 and 3) AsS., As^2S^3, Sulpharsenites. (2 and 3) Sb^2S^3; Sulphantimonites. (3) S. Sulphides, Sulphates, also the rare Selenides. See special reactions § 5, for distinctive and confirmatory characters.
2. Fuse solid particle of test-substance with (previously fused) bead of phosphor-salt on platinum wire.	(1) Very slow solution, with formation of silica-skeleton or opalescent bead. (2) Rapid solution, accompanied throughout by effervescence. Other results (as rapid solution without effervescence, &c.), may be noted down, but are not to be taken into account here.	(1) Silica, Silicates generally. (2) Carbonates (also bodies which evolve oxygen, as MnO^2, Bichromates, Chlorates, &c.). *Confirmatory tests.*—For SiO^2, fuse with carb. soda. Heat, in test-tube, with HC acid (for gelatinization, &c.). For Carbonates, warm, in test-tube, with dilute HC acid (for effervesence).
3. Fuse test-substance in powder with phosphor-salt and copper oxide on plat. wire, or with phosphor-salt alone on copper wire.	Rich azure blue flame. Note.—If a blue and green, or an intensely vivid green flame be produced, Br. and I may be suspected, but natural Bromides and Iodides are of very rare occurrence. Test with (dry) bisulphate of potash in closed tube over Bunsen flame (for yellow or violet fumes).	Chlorides. Also, chloro-phosphates (as pyromorphite, many apatites, &c.) Confirm by Experiment 4.

DETECTION OF ELECTRO-NEGATIVE BODIES—(Continued).

EXPERIMENTS.	RESULTS MORE ESPECIALLY TO BE LOOKED FOR.	SUBSTANCES INDICATED.
4. Boil the substance, in fine powder, with a few drops of nitric acid in a test-tube. Half-fill the tube with water, drop into the solution a fragment of amm. molybdate, and warm gently.	A canary-yellow precipitate.	Phosphates. NOTE.—Most phosphates, especially if moistened with sulphuric acid, impart a green tinge to the flame. Many natural phosphates are combined with chlorides or fluorides, or with both. Cl., if present, will have been detected by Expt. 3 ; Fl. must be sought for by Expt. 6.
5. Warm the test-substance, in powder, with a few drops of sulphuric acid, add a little alcohol, stir and inflame the mixture.	(1) A deep-green solution. (2) A rich blue solution. (3) A green coloration of the flame. NOTE.—A green flame is produced by most borates *per se*, in all, by moistening the test-substance with sulphuric acid, or with glycerine. Phosphates, however, produce the same reaction when thus treated, but do not give a green flame with alcohol.	(1) Chromates. (2) Molybdates. (3) Borates, also "Boro-Silicates." NOTE.—Small portions of B^2O^3 in silicates, &c., may escape detection by this Expt. but the object of the present scheme is not to detect minute or inessential components, but to determine the chemical group to which the test-substance may belong. See under Reactions, § 5.
6. Heat the substance, in powder, with a few drops of strong sulphuric acid in a narrow test-tube.	(1) Corrosion of inside of tube. (Wash out thoroughly, and dry before coming to conclusion.) (2) Evolution of ruddy (nitrous) fumes.	(1) Fluorides, also combinations of Fluorides and Phosphates (see under Expt. 4 above). (2) Nitrates. *Confirmatory test for nitrates.*—Ignite on charcoal (for deflagrescence).
7. Fuse test-substance, in fine powder, with about 3 parts of carb. soda and 2 nitre in a platinum spoon or loop of platinum wire. Dissolve resulting soluble matters in hot water; decant clear solution into a small porcelain capsule, add a few drops of hydrochloric acid, and place in the solution a piece of zinc.	A dark-blue coloration. NOTE.—Molybdenum compounds when thus treated may also produce a blue coloration at first, but this, on standing, becomes rapidly dark brown.	(1) Tungstates. If much MnO be present (as in Wolfram), the solution will at first be green, but this disappears rapidly on heating, and the solution becomes nearly colorless and then deep indigo-blue.

DETECTION OF ELECTRO-NEGATIVE BODIES—(Continued).

Experiments.	Results more Especially to be Looked For.	Substances Indicated.
8. Fuse test-substance, in fine powder, with 5 or 6 parts of bisulphate potash (added successively), in platinum spoon or wire loop. Dissolve out in slightly warm water, decant and boil.	A white or pale-yellowish precipitate, changing to a violet or amethystine colour if warmed with hydrochloric acid and a piece of zinc or tin-foil. (See page 51.)	Titanic Acid. Titaniates. *Confirmatory test.* — Fuse a portion of the principitate with phosphor-salt on plat. wire. See Reactions, § 5.

B.—DETECTION OF BASES.

In many minerals, the so-called base—lead, for example, in sulphide of lead, copper in red or black oxide of copper, baryta in carbonate of baryta, and so forth—may be easily recognized by the use of the blowpipe. This is especially the case, when the base consists of a single and easily reducible metal or metallic oxide, such as silver, lead, copper, tin, &c.; or where it imparts a colour to borax or other reagent, as in the case of copper, iron, cobalt, nickel, manganese, &c.; or where it forms a deposit on charcoal, communicates a colour to the flame, or exhibits other characteristic reactions. Even when several bodies of this kind are present, their recognition, as a general rule, is easily effected. Earthy and alkaline bases, when in the form of carbonates, sulphates, phosphates, fluorides, &c., can also be made out, in general, without difficulty, unless several happen to be present together, in which case it is not always possible, by the simple aid of the blowpipe, to distinguish them individually. When these bases are combined with silica, on the other hand, the blowpipe alone is rarely sufficient for their detection. This, however, so far as practical purposes are concerned, is of little consequence, as no economic value, in silicates of this character, is dependent on the base. In general cases, four experiments only will be required. These comprise : Testing for water by ignition in the bulb-tube; fusion or ignition of the substance *per se*; fusion with carb. soda ; and fusion with borax. It will thus be seen that, in many cases, the nature of the base will be sufficiently revealed by the reactions which ensue during the determination of the electro-negative character of the substance.

Experiments.	Results more especially to be looked for.	Substances Indicated.
1. Ignite in bulb-tube. Note.—This experiment may be omitted as a rule in the case of minerals of metallic aspect.	(1) Presence of moisture (2) Assumption of dark colour and magnetism. Other results (if any) may be disregarded.	1. Water. Test with blue and red litmus papers. 2. Iron, probably as FeO.
2. Ignite or fuse *per se* in platinum forceps, or, if metallic, on charcoal.	(1) Coloration of flame: 1^a Red flame; 1^b Yellow flame; 1^c Green flame; 1^d Blue flame; 1^e Violet flame.	$(1)^a$ Lithia, strontia, lime. $(1)^b$ Soda $(1)^c$ Copper, antimony, zinc, molybdenum, baryta, ammonia. 1^d Lead. (Also CuCl, &c.) 1^e Potash. See Reactions, § 5, and Addendum to Table B, below. The student must remember that certain electro-negative bodies, S, P^2O^5, B^2O^3, &c., also give coloured flames.
	(2) Ring-deposit on charcoal: 2^a White dep.; 2^b Red-brown dep.; 2^c Yellow dep.	2^a Antimony (yellowish, hot); arsenic; zinc (yellow and phosphorescent, hot); molybdenum (yellowish, hot); tin (very slight). 2^b Cadmium. 2^c Bismuth; lead; zinc (whilst hot). See Addendum, below.
	(3) Assumption of magnetism. (4) Assumption of causticity. (Page 9.) Other results (if any) may be disregarded.	3. Iron. 4. Alkaline earths (CaO, &c.) in carbonates, sulphates, fluorides, &c.
3. Fuse (after thorough roasting, if necessary) with carb. soda and a little borax on charcoal; or, if the substance present a non-metallic aspect, on platinum wire.	(1) White or yellow ring-deposit on charcoal. (2) Reduced metal: 2^a Fusible, non-oxidizable globule; 2^b Infus., non-ox. particles; 2^c Infusible, oxidizable, magnetic particles; 2^d Fusible, oxid., non-volatile globules; 2^e Fusible, volatilizable globules. (3) A green-blue turquoise enamel. (4) Complete solution (with absorption, if on charcoal).	(1) See under Expt. 2; also the Addendum below. $(2)^a$ Gold; Silver. $(2)^b$ Platinum. $(2)^c$ Iron, Nickel, Cobalt 2^d Copper; Tin (practically). 2^e Bismuth; Lead; Antimony. (3) Manganese. (4) Baryta; Strontia; Alkalies. See Addendum, below.

PLAN OF ANALYSIS.

DETECTION OF BASES—(Continued).

Experiments.	Results more especially to be looked for.	Substances Indicated.
4. Fuse with borax on platinum wire (after thorough roasting, if necessary).	(1) A coloured bead which becomes turbid or opaque (from reduction or partial reduction) in the RF.	(1) Copper; Nickel; Cerium; Uranium (the glass becomes black in RF). Also Molydenum (to some extent), Tungstenum and Titanium; but these metals occur mostly in minerals as oxidized electro-negatives, and thus come under detection in Table A.
	(2) A coloured bead, not becoming opaque in RF.	(2) Manganese; Chromium (see Table A); Iron; Cobalt.
	(3) A colourless bead, not affected by flaming.	(3) Alumina; Tin oxide (to some extent). Both very slowly attacked.
	(4) A colourless bead which becomes opaque on saturation or by flaming.	(4) Zirconia; Glucina; Yttria; Zinc oxide; Alkaline earths (MgO, CaO, etc.); Alkalies.
5. Additional experiments—(as ignition with cobalt solution; testing for Hg. with reducing agents in closed tube; cupellation, &c.,) if thought necessary by physical characters of the test-substance, or by indications resulting from the above blowpipe trials.		

ADDENDUM TO TABLE B.

A Classification, according to their Blowpipe Characters, of the more commonly occurring Mineral Bases.

SECTION 1.—GIVING *per se*, OR WITH CARB. SODA, ON CHARCOAL, METALLIC GLOBULES OR METALLIC GRAINS.

Group 1.—*Yielding malleable metallic globules, without deposit on the charcoal.*

Gold. Silver. Copper.

Gold is insoluble in the fluxes. *Silver* is not oxidized *per se*, but retains a bright surface after exposure to an oxidating flame. *Copper* becomes encrusted on cooling with a black coating. It imparts a green colour to the flame-border; and forms strongly coloured glasses with borax and phosphor-salt: (green (hot), blue (cold), in O F; red-brown, opaque, in R F: see above). Gold and silver may be separated from copper, &c., by fusion with lead, and subsequent cupellation. If gold and silver be present together, the bead is generally more or less white. By fusing it in a small platinum-spoon with bisulphate of potash, the silver dissolves, and the surface of the globule becomes yellow. If the globule be flattened out into a disc on the anvil, before treatment with bisulphate of potash, the silver is more rapidly extracted. The sulphate of silver must be removed by treating the spoon, in a porcelain or platinum capsule, with a small quantity of water, over the spirit-lamp. By evaporation, and fusion of the residuum with carb. soda on charcoal, metallic silver can be again obtained.

Group 2.—*Yielding infusible metallic grains, without deposit on the charcoal.*

Platinum. Iron. Nickel. Cobalt. Molybdenum. Tungstenum.

Platinum is not attacked by the blowpipe fluxes. *Iron, Nickel,* and *Cobalt,* or their oxides, are readily dissolved by fusion with borax or phosphor-salt, producing a coloured glass. (*See* under "Borax," pages 13, 14, above.) These metals are also magnetic. As a general rule if a substance become attractable by the magnet after exposure to the blowpipe, the presence of iron may be inferred, cobalt and nickel compounds being comparatively rare. The presence of cobalt is readily detected by the rich blue colour of the borax and phosphor-salt glasses, in both an oxidating and reducing flame; but if much iron be present also, the glass is bluish-green. With borax in the R F, nickel compounds give reduced metal, and the glass becomes gray and troubled. *Molybdenum* and *Tungstenum* give non-magnetic grains of reduced metal. They are commonly present in minerals as the electro-negative principle, and their presence is best detected by the method given under Experiment 7, Table A, above.

Group 3.—*Yielding metallic globules, with white or yellow deposit on the charcoal.*
Tin. Lead. Bismuth. Antimony.

Tin and *Lead* give malleable globules.* The sublimate formed by tin is white, small in quantity, and deposited on, and immediately around, the globule. The lead sublimate is yellow, and more or less copious. *Bismuth* and *Antimony* give brittle globules. The bismuth sublimate is dark yellow; the antimony sublimate, white, and very abundant. Lead imparts a clear blue colour to the flame-border; antimony, a greenish tint. As a general rule, a yellow deposit on the charcoal may be regarded as indicative of the presence of lead; whilst, the emission of copious fumes, and deposition of a white coating on the charcoal, may be safely considered to indicate antimony. The coating or sublimate formed by zinc (see below), although white when cold, is lemon-yellow whilst hot. The rare metal, tellurium, closely resembles antimony in its reactions, but if warmed with concentrated sulphuric acid, it forms a reddish-purple solution.

NOTE.—An excellent method of distinguishing the blowpipe-sublimates of lead, bismuth, antimony, and also cadmium, has been recently discovered by Dr. Eugene Haanel, of Victoria College, Cobourg (Ontario). Moistened with a drop of hydriodic acid, and ignited, the lead sublimate becomes bright canary-yellow; the bismuth sublimate, chocolate-brown; the antimony sublimate, bright red; and the cadmium sublimate, white. The hydriodic acid is obtained by steeping iodine in water, and passing through the liquid a current of sulphuretted hydrogen until it becomes clear. The reactions produced by this method are remarkably distinct.

SECTION 2.—REDUCIBLE, BUT YIELDING NO METAL ON CHARCOAL. (This arises from the rapid volatilization of the reduced metal.)

Group 1.—*Volatilizing without odour, and without formation of a deposit on the charcoal.*
Mercury.

For the proper detection of this metal, a small portion of the test-substance in powder must be mixed with some previously dried carb.

---

* *See* in the Appendix, No. 21, the striking reaction manifested by alloys of these metals.

soda, and the mixture strongly ignited at the bottom of a small tube or narrow flask. If mercury be present, a gray sublimate will be formed. This may be collected by friction with a wire, &c., into small metallic globules, and poured out of the tube. If some iron filings be mixed with the carb. soda, the mercurial sublimate is more readily obtained.

Group 2.—*Volatilizing without odour, but forming a deposit on the charcoal.*

Cadmium. Zinc.

The deposit produced by cadmium is dark yellowish-brown or reddish-brown. That produced by zinc is lemon-yellow and phosphorescent whilst hot, and white when cold. If moistened with a drop of nitrate of cobalt and ignited, it becomes bright green.[*]

Group 3.—*Volatilizing with strong odour of garlic.*

Arsenic (more commonly present in minerals as an electro-negative body. See Table A, above).

The alliaceous or garlic like odour is most readily developed when the test-matter is mixed with some carb. soda, or other reducing flux, and exposed on charcoal to the action of a reducing flame.

The presence of arsenic may also be proved as follows: (1) By roasting a fragment of the substance in an open glass tube, when minute octahedrons of arsenious acid (easily recognized by their triangular faces if examined by a common lens) will be deposited at the upper end of the tube; and (2), by igniting the test-substance, mixed with some dry oxalate of potash or cyanide of potassium, at the bottom of a small flask or closed tube, when a dark, shining sublimate of metallic arsenic will be produced. Without the reducing flux, a yellow or yellowish-red sublimate of arsenical sulphide might be formed in certain cases.

SECTION 3.—NOT REDUCIBLE BEFORE THE BLOWPIPE.

Group 1.—*Imparting a colour to borax.*

Manganese. Chromium. Titanium. (The two latter are commonly present in minerals as electro-negative bodies.)

---

[*] In testing a substance supposed to contain cadmium, a little chalk-powder or bone ash may be rubbed over the surface of the charcoal. If cadmium be present, its reddish-brown sublimate (CdO) is then more readily seen.

Manganese compounds impart, before an oxidating flame, a violet colour to borax; *Chromium compounds*, a clear green colour. (*See* also under "Carbonate of Soda," page 15, above.) *Titanium compounds* form, with borax in the R F, a brownish-amethystine glass, which becomes light blue and opaque by flaming. The presence of titanium in minerals is most readily detected by fusing the substance in very fine powder with 3 or 4 parts of carb. soda in a platinum spoon, dissolving the fused mass in hydrochloric acid, diluting slightly, and then boiling with a slip of tin or zinc. The solution, if titanium be present, will gradually assume an amethystine tint. Or, the substance, in fine powder, may be fused with bisulphate of potash in successive portions. The titanic acid by this treatment becomes soluble in water, from which it may be precipitated as a white or slightly yellowish powder by boiling. The precipitate can then be fused before the blowpipe in a reducing flame with some phosphor-salt, when a violet-coloured or amethystine bead will result. If iron be present in the substance, a drop or two of hydrochloric acid should be added to the solution before the precipitation of the titanic acid.

The rare metals, cerium, uranium, &c., belong also to this group. Reference should also be made to iron, nickel, cobalt and copper, as the oxides of these latter metals, if in small quantity, might escape detection by the reducing process. (*See* under Operation 5, pp. 13, 14, the colours imparted by these oxides to borax.)

Group 2.—*Imparting no colour to the fluxes. Slowly dissolved by borax, the glass remaining permanently clear.*

Alumina.

Moistened with nitrate of cobalt and then ignited, this base assumes on cooling a fine blue colour.

Group 3.—*Imparting no colour to the fluxes. Rapidly dissolved by borax, the glass becoming opaque on cooling or when flamed. Insoluble in carb. soda.*

Magnesia. Lime.

Moistened with nitrate of cobalt, and ignited, *Magnesia* becomes pale-red in colour; *Lime*, dark gray.

Group 4.—*Entirely dissolved by fusion with carb. soda.*
Baryta. Strontia. Lithia. Soda. Potash.

Baryta compounds impart a distinct green colour to the point and border of the flame. *Strontia* and *Lithia* colour the flame deep carmine-red. The crimson coloration is destroyed in the case of strontia if the substance be fused with chloride of barium. *Soda* colours the flame strongly yellow. *Potash* communicates to it a violet tint; but this colour is completely masked by the presence of soda, unless the flame be examined through a deep blue glass. See also the spectroscope reactions of these bodies given under their respective heads in § 5.

APPENDIX.

ORIGINAL CONTRIBUTIONS TO BLOWPIPE ANALYSIS.

BY E. J. CHAPMAN.

1.—REACTION OF MANGANESE SALTS ON BARYTA.

When moistened with a solution of any manganese salt, and ignited in an oxidating flame, baryta and baryta compounds, generally, assume on cooling a blue or greenish-blue colour. This arises from the formation of a manganate of baryta. Strontia and other bodies (apart from the alkalies) when treated in this manner, become brown or dark-gray. A mixture of baryta and strontia also assumes an indefinite grayish-brown colour. If some oxide of manganese be fused with carbonate of soda so as to produce a greenish-blue bead or "turquoise enamel," and some baryta or a baryta salt be melted into this, the colour of the bead will remain unchanged; but if strontia be used in place of baryta, a brown or grayish-brown enamel is produced.

NOTE.—Some examples of witherite, barytine, and baryto-calcite, contain traces of oxide of manganese. These, after strong ignition, often assume *per se* a pale greenish-blue colour. 1846.

2.—DETECTION OF BARYTA IN THE PRESENCE OF STRONTIA.

This test is chiefly applicable to the detection of baryta in the natural sulphate of strontia; but it answers equally for the examination of chemical precipitates, &c., in which baryta and strontia may be present together. The test-matter, in fine powder, is to be melted in a platinum spoon with 3 or 4 volumes of chloride of calcium, and the fused mass treated with boiling water. For this purpose, the spoon may be dropped into a test-tube, or placed (bottom upwards) in a small porcelain capsule. The clear solution, decanted from any residue that may remain, is then to be diluted with 8 or 10 times its volume of water, and tested with a few drops of chromate (or bichromate) of potash. A precipitate, or turbidity, indicates the presence of baryta. 1846.

3.—DETECTION OF ALKALIES IN THE PRESENCE OF MAGNESIA.

In the analysis of inorganic bodies, magnesia and the alkalies (if present) become separated from other constituents towards the close of the operation. In continuation of the analysis, it then becomes desirable to ascertain, at once, whether magnesia be alone present, or whether the saline mass, produced by the evaporation of a portion of the solution, consist of magnesia and one or more of the alkalies, or of the latter only. By fusing a small quantity of the test-matter with carbonate of soda, the presence of magnesia is readily detected, as this substance remains undissolved; but the presence or absence of alkalies is not so easily determined, the coloration of the flame being frequently of too indefinite a character to afford any certain evidence on this point. The question may be solved, however, by the following simple process. Some boracic acid is to be mixed with the test-matter and with a few particles of oxide of copper, and the mixture is to be exposed for a few seconds, on a loop of platinum wire, to the action of an oxidating flame. In the absence of alkalies, the oxide of copper will remain undissolved; but if alkalies be present, an alkaline borate is produced, forming a readily fusible glass, in which the copper oxide is at once dissolved, the glass becoming green whilst hot, and blue when cold. If magnesia also be present, white specks remain for a time undissolved in the centre or on the surface of the bead. Any metallic oxide which imparts by fusion a colour to alkaline borates, may, of course, be employed in place of oxide of copper; but the latter has long been used in other operations, and is therefore always carried amongst the reagents of the blowpipe-case. 1847.

4.—METHOD OF DISTINGUISHING THE RED FLAME OF LITHIUM FROM THAT OF STRONTIUM.

It has been long known that the crimson coloration imparted to the blowpipe-flame by strontia, is destroyed by the presence of baryta. This reaction, confirmed by Plattner (see, more especially, the third edition of his "Probirkunst," page 107), was observed as early as 1829 by Butzengeiger ("Annales des Mines," t. v., p. 36). The latter substance, however, as first indicated by the writer, does not affect the crimson flame-coloration produced by lithia. Hence, to distinguish the two flames, the test-substance may be fused with 2 or 3 volumes of chloride of barium on a loop of platinum wire, the fused mass being kept just within the point or edge of the blue cone. If the original flame-coloration proceeded from strontia (or lime), an impure brownish-yellow tinge will be imparted to the flame-border; but if the original red colour were caused by lithia, it will not only remain undestroyed, but its intensity will be much increased.

This test may be applied, amongst other bodies, to the natural silicates, lepidolite, spodumene, &c. It is equally available, also, in the examination of phosphates. The mineral triphylline, for example, when treated *per se*, imparts a green tint to the point of the flame, owing to the presence of phosphoric acid; but if this mineral be fused (in powder) with chloride of barium, a beautiful crimson coloration in the surrounding flame-border is at once produced. 1848.

5.—METHOD OF DISTINGUISHING THE MONOXIDE OF IRON (FeO) FROM THE SESQUIOXIDE (Fe^2O^3) IN SILICATES AND OTHER COMPOUNDS.

If iron be recognized in an oxidized body, its presence or absence as ferrous oxide (FeO) is readily indicated by this test: assuming, of course, that no other reducing body be present, a point easily ascertained by the blowpipe. The test is performed as follows: A small quantity of black oxide of copper (CuO) is dissolved in a bead of borax on platinum wire, so as to form a glass which exhibits, on cooling, a decided blue colour, but which remains transparent. To this, the test-substance in the form of powder is added, and the whole is exposed for a few seconds, or until the test-matter begin to dissolve, to the point of the blue flame. If the substance contain Fe^2O^3 only, the glass on cooling will remain transparent, and will exhibit a bluish-green colour. On the other hand, if the test-substance contain FeO, this will become at once converted into Fe^2O^3 at the expense of some of the oxygen of the copper compound; and opaque red streaks and spots of Cu^2O will appear in the glass as the latter cools. 1848.

NOTE.—Although this test is quoted by Plattner—perhaps the best criterion of its accuracy—it is passed over, without mention, in many works on chemical analysis. The writer may therefore be allowed to call to mind, in proof of its efficacy, that by its use in 1848 he pointed out the presence of FeO in the mineral staurolite ("Chem. Gaz.," July 15, 1848; *see* also Erdmann's "Journal für pract. Chem.," XLVI., p. 119), nearly thirteen years before this fact—now universally admitted—was discovered and announced by Rammelsberg, "Berichte d. Kongl. preuss. Akad. d. Wiss. zu Berlin," Marz, 1861.

6.—DETECTION OF LEAD IN THE PRESENCE OF BISMUTH.

When lead and bismuth are present together, the latter metal may be readily detected by its known reaction with phosphor-salt in a reducing flame—antimony, if present, being first eliminated; but the presence of lead is less easily ascertained. If the latter metal be present in large quantity, it is true, the metallic globule will be more or less malleable, and the flame-border will assume a clear blue colour when made to play upon its surface, or on the sublimate of lead-oxide as produced on charcoal; but in other cases this reaction becomes exceedingly indefinite. The presence of lead may be detected, however, by the following plan, based on the known reduction and precipitation of salts of bismuth by metallic lead, a method which succeeds perfectly with brittle alloys containing from 85 to 90 per cent. of bismuth. A small crystal or fragment of nitrate of bismuth is placed in a porcelain capsule, and moistened with a few drops of water, the greater part of which is afterwards poured off; and the metallic globule of the mixed metals, as obtained by the blowpipe, having been slightly flattened on the anvil until it begins to crack at the sides, is then placed in the midst of the sub-salt of bismuth formed by the action of the water. In the course of a minute or even less, according to the amount of lead that may be present, an arborescent crystallization of metallic bismuth will be formed around the globule. The reaction is not affected by copper; but a precipitation of bismuth would ensue, in the absence of lead, if either zinc or iron were present. These metals, however, may be eliminated from the test-globule by exposing this on charcoal for some minutes, with a mixture of carb. soda and borax to a reducing flame. The zinc becomes volatilized, and the iron is gradually taken up by the borax. If a single operation do not effect this, the globule must be removed from the saturated dark green glass, and treated with further portions of the mixture, until the resulting glass be no longer coloured. 1848.

7.—DETECTION OF LITHIA IN THE PRESENCE OF SODA.

This test may be applied to mixtures of these alkalies in the simple state, or to their carbonates, sulphates, nitrates, or other compounds capable of being decomposed by fusion with chloride of barium. The test-substance, in powder, is to be mixed with about twice its volume of chloride of barium, and a small portion of the mixture is to be exposed on a loop of platinum wire to the point of a well-sustained oxidating flame. A deep yellow coloration of the flame-border, produced by the volatilization of chloride of sodium, at first ensues. This gradually diminishes in intensity, and after a short time a thin green streak, occasioned by chloride of barium, is seen to stream from the point of the wire, as the test-matter shrinks further down into the loop. On the fused mass being then brought somewhat deeper into the flame, the point and edge of the latter will at once assume the rich crimson tinge characteristic of the presence of lithium compounds; and the colour will endure sufficiently long to prevent the slightest chance of misconception or uncertainty. The presence of strontium compounds does not affect this reaction, as these compounds, when fused with chloride of barium, cease to impart a red colour to the flame. (*See* No. 4.) In order, however, to ensure success in the application of this test, it is necessary, in some cases, to keep up a clear and sharply-defined flame for about a couple of minutes. If the red coloration do not appear by that time, the absence of lithia—unless the latter substance be present in minute traces only—may be safely concluded. 1850.

8.—ACTION OF BARYTA ON TITANIC ACID.

Fused with borax in a reducing flame, titanic acid, it is well known, forms a dark amethystine-blue glass, which becomes light blue and opaque when subjected to the flaming process. The amethystine colour arises from the presence of Ti^2O^3; the light blue enamelled surface, from the precipitation of a certain portion of TiO^2. The presence of baryta, even in comparatively small quantity, quite destroys the latter reaction. When exposed to an intermittent flame, the glass (on the addition of baryta) remains dark blue, no precipitation of titanic acid taking place. Strontia acts in the same manner, but a much larger quantity is required to produce the reaction. 1852.

9.—DETECTION OF OXIDE OF MANGANESE WHEN PRESENT IN MINUTE QUANTITY IN MINERAL BODIES.

It is usually stated in works on the blowpipe, that the smallest traces of manganese may be readily detected by fusion with carbonate of soda, or with a mixture of carbonate of soda and nitrate of potash: but this statement is to some extent erroneous. In the presence of much lime, magnesia, alumina, sesquioxide of iron, or other bodies, insoluble, or of difficult solubility, in carbonate of soda, traces of oxide of manganese may easily escape detection. By adding, however, a small portion of borax or phosphor-salt to the carbonate of soda, these bodies become dissolved, and the formation of a "turquoise enamel" (manganate of soda) is readily effected. The process may be varied by dissolving the test-substance first in borax or phosphor-salt, and then treating the fused bead with carbonate of soda: the latter being, of course, added in excess. By this treatment, without the addition of nitrate of potash, the faintest traces of oxide of manganese in limestone and other rocks, are at once made known. 1852.

NOTE.—This method of examining bodies for the presence of manganese, was recommended by Dr. Leop. H. Fischer in 1861 ("Leonh. Jahrbuch" [1861], p. 653), but the writer had forestalled him by nine years, having already described it in 1852.

10.—THE COAL ASSAY.

In the practical examination of coals, the following operations are essentially necessary:* (1) The estimation of the water or hygro-

---

* To these might be added, the determination of the heating powers or "absolute warmth" of the coal, but this may always be estimated with sufficient exactness for practical purposes by the amount of coke, ash, and moisture, as compared with other coals. Properly considered, the litharge test, resorted to for the determination of the calorific power of coals, is of very little actual value. The respective results furnished by good wood charcoal and ordinary coke, for example, are closely alike, if not in favour of the charcoal; and yet experience abundantly proves the stronger heating powers of the coke. In practice, moreover, the actual value of a coal does not always depend upon the "absolute warmth" of the latter, as certain coals, such as brown coals rich in bitumen, may possess heating powers of considerable amount (as estimated by the reduction of litharge) though only of brief duration. Thus, the lignites of the department of the Basses Alpes in south-eastern France, and those of Cuba, yield with litharge from 25 to 26 parts of reduced lead; whilst many caking coals, practically of much higher heating power, yield scarcely a larger amount. When pyrites also is present in the coal—a condition of very common occurrence—the litharge test becomes again unsatisfactory, the pyrites exerting a reducing action on the lead compound.

As described, however, by Bruno Kerl, in quoting the writer's coal assay ("Löthrohr-Untersuchungen:" Zweite Aufl. 1862, p. 146) the so-called absolute warmth or heating power of a coal sample may be determined, if desired, in blowpipe practice, by the following modification of Berthier's method: 20 milligrammes of the coal, in fine powder, are to be mixed intimately with 500 milligrammes of oxy-chloride of lead (consisting of three parts of litharge + 1 part of chloride of lead, fused together and finely pulverized). The mixture is to be placed in a blowpipe crucible, and covered with about an equal amount of the lead compound, a second cover of 8 blowpipe-spoonfuls of powdered glass + 1 spoonful of borax being spread over this. The crucible, covered with a clay capsule, is then to be fitted into a charcoal block in the ordinary blowpipe furnace, over which a charcoal lid is placed, and the flame directed against its under side, so as to keep it at a red heat for from 5 to 8 minutes. The weight of the reduced lead divided by 20 gives the amount of the lead mixture reduced by one part of the coal. One part of pure carbon reduces 34 parts of this mixture; one part of charcoal, 30 to 33 parts; one part of bituminous coal, 19 to 33; one part of brown coal, 11 to 26; one part of peat, 8 to 27; and one part of wood, 12 to 15 parts.

metric moisture present in the coal; (2) the determination of the weight and character of the coke; (3) the estimation and examination of the ash or inorganic matters; and (4) the estimation of the sulphur, chiefly present in the coal as FeS^2.

Estimation of Moisture.—This operation is one of extreme simplicity. Some slight care, however, is required to prevent other volatile matters from being driven off during the expulsion of the moisture. Seven or eight small particles, averaging together from 100 to 150 milligrammes, are to be detached from the assay specimen by means of the cutting pliers, and carefully weighed. They are then to be transferred to a porcelain capsule with thick bottom, and strongly heated for four or five minutes on the support attached to the blowpipe-lamp, the unaided flame of the lamp being alone employed for this purpose. It is advisable to place in the capsule, at the same time, a small strip of filtering or white blotting-paper, the charring of which will give indications of the temperature becoming too high. The coal, whilst still warm, is then to be transferred to the little brass capsule in which the weighings are performed, and its weight ascertained. In transferring the coal from one vessel to the other, the larger pieces should be removed by a pair of fine brass forceps, and the little particles or dust afterwards swept into the weighing capsule by means of the camel's-hair pencil or small colour-brush belonging to the balance case. The weighing capsule should also be placed in the centre of a half sheet of glazed writing-paper, to prevent the risk of any accidental loss during the transference. After the weighing, the operation must always be repeated, to ensure that no further loss of weight occur. In place of the blowpipe-lamp, the spirit-lamp may be employed for this operation; but, with the former, there is less danger of the heat becoming too high. By holding a slip of glass for an instant, every now and then, over the capsule, it will soon be seen when the moisture ceases to be given off. It should be remarked that some anthracites decrepitate slightly when thus treated, in which case the porcelain capsule must be covered at first with a small watch-glass.

In good samples of coal, the moisture ought not to exceed 3 or 4 per cent., but in coals that have been long exposed to damp it is often as high as 6 or 7, and even reaches 15 or 20 per cent. in certain lignites. Where large quantities of coal are consumed, therefore, a serious loss is entailed on the purchaser unless the moisture be properly determined and allowed for.

Estimation, &c., of Coke.—In this operation, a small crucible of platinum is most conveniently employed. The crucible may consist of a couple of rather deep spoons—the larger one without a handle, so as to admit of being placed over the smaller spoon, thus serving as a lid. The long handle of the crucible-spoon must be bent as shewn in the annexed figure, in order that the spoon may retain an upright position when placed on the pan of the balance. About 150 milligrammes of coal are detached as before, in several small fragments, from the assay-specimen. These may be weighed directly in the crucible, the latter being placed in the little weighing capsule of horn or brass, with its handle-support projecting over the side of this. The crucible, with its cover on, is then taken up by a pair of spring forceps, and is brought gradually before the blowpipe to a red heat. The escaping gases will take fire and burn for a few seconds around the vessel, and a small amount of carbonaceous matter may be deposited upon the cover. This rapidly burns off, however, on the heat being continued. As soon as it disappears, the crucible is to be withdrawn from the flame, and placed on the blowpipe-anvil to cool quickly. Its weight is then ascertained, always without removing the cover. The loss, minus the weight of moisture as found by the first process, gives the amount of volatile or gaseous matter. The residue is the coke and its contained ash. The coke in some anthracites exceeds 89 or 90 per cent. In anthracitic or dry coals it usually varies from 70 to 80 per cent., and the fragments are sometimes slightly agglutinated. In ordinary bituminous or caking coals, it amounts in general to about 65 or 70 per cent., and presents a fused and mammillated surface. In cannel or gas coals, the percentage of coke may be assumed to equal 50 or 60, but it is sometimes as low as 30. The coke fragments are often partially agglutinated, but they never present a fused, globular aspect. Finally, in lignites or brown coals, the coke may vary from 25 to 50 per cent. It forms sharp-edged fragments of a dull charcoal-like appearance, without any sign of fusion.

Estimation of Ash or Inorganic Matters.—A platinum capsule is employed for this operation. One of about half an inch in diameter, with a short ear or handle, is sufficiently large. A somewhat smaller capsule, with its handle cut off, may be fitted into this (in reversed position) to serve as a lid. The coal must be reduced to a coarse powder, and about 150 milligrammes weighed out for the experiment. The platinum capsule is then to be fixed in a slightly-inclined position above the spirit-lamp, and heated as strongly as possible. If the wick of the spirit-lamp be raised sufficiently, and the capsule be light and thin, the temperature will be sufficient to burn off the carbon, at least in the majority of cases. The lid of the capsule must be placed above the coal powder until combustion cease, and the more gaseous products are driven off, as otherwise a portion of the powder might very easily be lost. During the after combustion the powder must be gently stirred, and if agglutination take place the particles must be carefully broken up by a light steel spatula, or by a piece of stout platinum wire flattened at one end. If the carbonaceous matter be not burnt off by this treatment, the blowpipe may be used to accelerate the process; but the operator must blow cautiously, and direct the flame only against the under side of the capsule, in order to avoid the risk of loss. Finally, on the ash ceasing to exhibit in any of its particles a black colour, the lid of the capsule is to be carefully replaced, and the whole cooled and weighed.*

In good coals, the amount of ash is often under 2 per cent., and it rarely exceeds 4 or 5 per cent. In coals of inferior quality, however, it may vary from 8 or 10 to even 30 per cent. As regards its composition, the ash may be—(1) argillaceous, consisting essentially of a silicate of alumina; (2) argillo-ferruginous; (3) calcareous; and (4) calcareo-ferruginous. If free from iron, it will be white or pale gray ; but if more or less ferruginous, it will present a red, brown, or yellowish colour. Phosphor-salt, so useful in general cases for the detection of siliceous compounds, cannot be safely used to distinguish the nature of the ash obtained in blowpipe assays. Owing to their fine state of division and to the small quantity at command, argillaceous ashes dissolve in this reagent with as much facility as those of a calcareous nature, and without producing a characteristic silica skeleton, or causing the opalization of the glass. With calcareous ashes also, the

---

* If the ash be very ferruginous—in which case it will present a red or tawny colour—the results, as thus obtained, will require correction, the original iron pyrites of the coal being weighed as sesquioxide of iron. In ordinary assays, however—as distinguished from analyses—this may be fairly neglected. When also the ash happens to be calcareous and to occur in large quantity, it should be moistened with a drop or two of a solution of carbonate of ammonia, and gently heated, previous to being weighed.

amount obtained is rarely sufficient to saturate even an exceedingly minute bead of phosphor-salt or borax, and hence no opacity is produced by the flaming process. The one kind of ash may be distinguished, nevertheless, from the other, by moistening it, and placing the moistened mass on reddened litmus paper. Calcareous ashes always contain a certain amount of caustic lime, and thus restore the blue colour of the paper. The calcareous ashes, also, though principally composed of carbonate of lime, sometimes contain small portions of phosphate and sulphate of lime. The presence of the latter may be readily detected by the well-known production of an alkaline sulphide by fusion with carbonate of soda in a reducing flame—the fused mass exhibiting a reddish colour, and imparting when moistened a dark stain to a plate of silver or piece of lead test-paper. The latter may be replaced by a glazed visiting-card. In examining earthy sulphates by this method, a little borax ought always to be added to the carbonate of soda, in order to promote the solution of the test-matter. If oxide of manganese be present in the ash, the well-known manganate of soda, or "turquoise enamel," will also be obtained by this treatment.

Estimation of Sulphur.—The following plan is perhaps the most simple that can be employed for the determination of sulphur in coal samples. It is merely an adaptation to blowpipe practice of the process very generally employed for that purpose :

As large an amount of coal as practicable, several pounds at least, taken from different parts of the same heap or bed, must be broken into powder and well stirred together. About 150 milligrammes are to be weighed out for the assay. This amount is to be intimately mixed with about 450 milligrammes of nitrate of potash and an equal quantity of carbonate of potash, and the mixture, with a good covering of salt, is to be fused in a small platinum crucible of about a quarter of an ounce capacity. The crucible may be fixed in an ordinary blowpipe-furnace, in the centre of an already used charcoal-block, as the cavity of the latter will require to be larger than usual ; or it may be ignited by the flame of a Bunsen burner, without the aid of the blowpipe. The heat at first must be very moderate, as the mixture swells up greatly ; but after a couple of minutes, or thereabouts, a tolerably strong blast may be kept up for from two to three minutes in addition, when the operation will be finished. The alkaline sulphate, thus produced, is dissolved out by boiling water, and the filtered solution, acidified by a few drops of hydrochloric acid, is then treated with chloride of barium. The weight of the precipitate divided by 7.28 gives the amount of sulphur. An ordinary blowpipe-crucible of clay may be employed for this operation; but it is always strongly attacked by the mixture during fusion, and is otherwise less convenient for the purpose than one of platinum.

When the iron pyrites in the coal is not in a state of semi-decomposition, its amount, and consequently the amount of sulphur, may be arrived at, far more nearly than might at first thought be supposed, by the simple process of washing in the agate mortar. Each single part of pyrites corresponds to 0.533 of sulphur. Some large pieces of the assay-coal should be selected, and broken up into powder; and on this, several trials must be made. About 500 milligrammes may be taken for each trial, and washed in three or four portions. In the hands of one accustomed to the use of the mortar in reducing experiments, the results, owing to the lightness of the coal particles, and the consequent ease with which they are floated off, come out surprisingly near to the truth. In travelling, we may dispense with the washing bottle, by employing, in its place, a piece of straight tubing drawn out abruptly to a point. This is to be filled by suction, and the water expelled with the necessary force by blowing down the tube. A tube 6 inches long and the fourth of an inch in diameter will hold more than a sufficient quantity of water to be used between the separate grindings. The mortar should be but slightly inclined, and the stream of water must not be too strong: otherwise, especially if the coal be ground up very fine, portions of the pyrites may be lost. The proper manipulation, however, is easily acquired by a little practice. 1858.

11.—PHOSPHORUS IN IRON WIRE.

Many years ago, it was stated by GRIFFIN that thin iron wire exhibits, in burning, a green light. This statement is repeated by Prof. GALLOWAY in various editions of his useful little work on chemical analysis: iron wire being placed in one of the tables, given in that manual, among the substances which impart a green coloration to the blowpipe-flame. On the other hand, neither BERZELIUS, PLATTNER, RICHTER, VON KOBELL, DR. HARALD LENZ *(Die*

Löthrohrschule, 1848), SCHEERER, BRUNO KERL, nor any other of the numerous workers with the blowpipe on the continent of Europe have ever alluded to the reaction. LENZ gives a minute description of the action of the blowpipe-flame on iron wire, and points out that the fusion is always accompanied by oxidation; but he makes no allusion to any coloration of the flame. Struck by this apparent omission, I have examined a number of samples of iron wire by the blowpipe. All the light-coloured and comparatively hard wires exhibited the reaction very distinctly. A bright green flame streamed from the point of the wire during the oxidation and fusion of the latter, and a rapid scintillation or emission of sparks accompanied the phenomenon. On the other hand, the soft and dark wires fused much less readily, and did not occasion the slightest coloration of the flame. The green flame-coloration, occasioned by the harder wires, arises, I find, from the presence of a minute amount of phosphorus, this being converted into phosphoric acid during the combustion of the wire. As iron-wire is often employed in blowpipe practice as a reagent for phosphoric acid in phosphates, and as it is also occasionally used in preparing a solution of iron oxide (Fe^2O^3) for the estimation of phosphoric acid in bodies generally, the publication of the present note may not be altogether superfluous. 1864.

12.—DETECTION OF MINUTE TRACES OF COPPER IN IRON PYRITES AND OTHER BODIES.

Although an exceedingly small percentage of copper may be detected in blowpipe experiments by the reducing process, as well as by the azure-blue coloration of the flame when the test-matter is moistened with hydrochloric acid, these methods fail in certain extreme cases to give satisfactory results. It often happens that veins of iron pyrites lead at greater depths to copper pyrites. In this case, according to the experience of the writer, the iron pyrites will almost invariably hold minute traces of copper. Hence the desirability, in exploring expeditions more especially, of some ready test, by which, without the necessity of employing acids or other bulky and difficultly portable reagents, these traces of copper may be detected. The following simple method will be found to answer the purpose: The test substance, in powder, must first be roasted on charcoal, or, better, on a fragment of porcelain,* in order to drive off the sulphur. A small portion of the roasted ore is then to be fused on platinum wire with phosphor-salt; and some bisulphate of potash is to be added to the glass (without this being removed from the wire) in two or three successive portions, or until the glass becomes more or less saturated. This effected, the bead is to be shaken off the platinum loop into a small capsule, and treated with boiling water, by which either the whole or the greater part will be dissolved; and the solution is finally to be tested with a small fragment of ferrocyanid of potassium ("yellow prussiate.") If copper be present in more than traces, this reagent, it is well known, will produce a deep red precipitate. If the copper be present in smaller quantity, that is, in exceedingly minute traces, the precipitate will be brown or brownish-black; and if copper be entirely absent, the precipitate will be blue or green—assuming, of course, that iron pyrites or some other ferruginous substance is operated upon. In this experiment, the preliminary fusion with phosphor-salt greatly facilitates the after solution of the substance in bisulphate of potash. In some instances, indeed, no solution takes place if this preliminary treatment with phosphor-salt be omitted. 1865.

13.—DETECTION OF ANTIMONY IN TUBE-SUBLIMATES.

In the examination of mineral bodies for antimony, the test-substance is often roasted in an open tube for the production of a white sublimate. The presence of antimony in this sublimate may be detected by the following process—a method more especially available when the operator has only a portable blowpipe-case at his command. The portion of the tube to which the chief portion of the sublimate is attached is to be cut off by a triangular file, and dropped into a test-tube containing some tartaric acid dissolved in water. This being warmed or gently boiled, a part at least of the sublimate will be dissolved. Some bisulphate of potash—either alone, or mixed with some carb. soda and a little borax, the latter to prevent absorption—is then

---

* In the roasting of metallic sulphides, &c., the writer has employed, for some years, small fragments of Berlin or Meissen porcelain, such as result from the breakage of crucibles and other vessels of that material. The test-substance is crushed to powder, moistened slightly, and spread over the surface of the porcelain; and when the operation is finished, the powder is easily scraped off by the point of a knife-blade or small steel-spatula. In roasting operations, rarely more than a dull red heat is required; but these porcelain fragments may be rendered white-hot, if such be necessary, without risk of fracture. They are held, most conveniently, by a pair of spring-forceps.—"Canadian Journal," September, 1860.

to be fused on charcoal in a reducing flame; and the alkaline sulphide, thus produced, is to be removed by the point of a knife-blade, and placed in a small porcelain capsule. The hepatic mass is most easily separated from the charcoal by removing it before it has time to solidify. Some of the tartaric acid solution is then to be dropped upon it, when the well-known orange-coloured precipitate of Sb^2S^3 will at once result.

In performing this test, it is as well to employ a somewhat large fragment of the test-substance, so as to obtain a thick deposit in the tube. It is advisable also to hold the tube in not too inclined a position in order to let but a moderate current of air pass through it; and care must be taken not to expose the sublimate to the action of the flame—otherwise it might be converted almost wholly into a compound of Sb^2O^3 and Sb^2O^5, the greater part of which would remain undissolved in the tartaric acid solution. A sublimate of arsenious acid, treated in this manner, would, of course, yield a yellow precipitate, easily distinguished by its colour, however, from the deep orange antimonial sulphide. The crystalline character, etc., of the sublimate, would also effectually prevent any chance of misconception.

14.—ON THE REACTIONS OF METALLIC THALLIUM BEFORE THE BLOWPIPE.

The following reactions are given from direct experiments by the writer:*

In the closed tube, thallium melts easily, and a brownish-red vitreous slag, which becomes pale yellow on cooling, forms around the fused globule.

In the open tube, fusion also takes place on the first application of the flame, whilst the glass becomes strongly attacked by the formation of a vitreous slag, as in the closed tube. Only a small amount of

---

* The reactions given by Crookes are as follows: "The metal melts instantly on charcoal, and evolves copious brown fumes. If the bead is heated to redness, it glows for some time after the source of heat is removed, continually evolving vapours which appear to be a mixture of metal and oxide. A reddish amorphous sublimate of proto-peroxide surrounds the fused globule. When thallium is heated in an open glass tube, it melts and becomes rapidly converted into the more fusible protoxide, which strongly attacks the glass. This oxide is of a dark red colour when hot, solidifying to a brown crystalline mass. The fused oxide attacks glass and porcelain, removing the silica. Anhydrous peroxide of thallium is a brown powder, fusing with difficulty and evolving oxygen at a red heat, becoming reduced to the protoxide. The phosphate and sulphate will stand a red heat without change."

sublimate is produced. This is of a grayish-white colour, but under the magnifying-glass it shews in places a faint iridescence.

On charcoal, *per se*, thallium melts very easily, and volatilizes in dense fumes of a white colour, streaked with brown, whilst it imparts at the same time a vivid emerald-green coloration to the point and edge of the flame. If the heat be discontinued, the fused globule continues to give off copious fumes, but this action ceases at once if the globule be removed from the charcoal. A deposit, partly white and partly dark brown, of oxide and teroxide is formed on the support; but, compared with the copious fumes evolved from the metal, this deposit is by no means abundant, as it volatilizes at once where it comes in contact with the glowing charcoal. If touched by either flame, it is dissipated immediately, in imparting a brilliant green colour to the flame-border. The brown deposit is not readily seen on charcoal; but if the metal be fused on a cupel, or on a piece of thin porcelain or other non-reducing body, the evolved fumes are almost wholly of a brownish colour, and the deposit is in great part brownish-black. It would appear, therefore, to consist of TlO^3, rather than of a mixture of metal and oxide. On the cupel, thallium is readily oxidized and absorbed. It might be employed, consequently, as suggested by Crookes, in place of lead in cupellation; but, to effect the absorption of copper or nickel, a comparatively large quantity is required. When fused on porcelain, the surface of the support is strongly attacked by the formation of a silicate, which is deep red whilst hot, and pale yellow on cooling.

The teroxide, as stated by Crookes, evolves oxygen when heated, and becomes converted into TlO. The latter compound is at once reduced on charcoal, and the reduced metal is rapidly volatilized with brilliant green coloration of the flame. The chloride produces the same reaction, by which the green flame of thallium may easily be distinguished from the green copper-flame; the latter, in the case of cupreous chlorides, becoming changed to azure-blue. With borax and phosphor-salt, thallium oxides form colourless glasses, which become gray and opaque when exposed for a short time to a reducing flame. With carb. soda, they dissolve to some extent, but on charcoal a malleable metallic globule is obtained. The presence of soda, unless in great excess, does not destroy the green coloration of the flame.

Thallium alloys more or less readily with most other metals before the blowpipe. With platinum, gold, bismuth, and antimony, respectively, it forms a dark-gray brittle globule. With silver, copper, or lead, the button is malleable. With tin, thallium unites readily, but the fused mass immediately begins to oxidize, throwing out excrescences of a dark colour, and continuing in a state of ignition until the oxidation is complete. In this, as in other reactions, therefore, the metal much resembles lead. 1876.

15.—ON THE OPALESCENCE PRODUCED BY SILICATES IN PHOSPHOR-SALT.

It is well known that most silicates when fused with phosphor-salt are only partially attacked; the bases, as a rule, gradually dissolving in the flux, whilst the silica remains in the form of a flocculent mass technically known as a "silica skeleton." Very commonly, almost invariably indeed, if the blast be long continued, the bead becomes more or less milky or opalescent on cooling. This latter reaction was apparently regarded by Plattner as essentially due to the presence of alkaline or earthy bases, such as exhibit the reaction *per se*. He states, "Probirkunst," Dritte Auflage, p. 468: "Da man nun von mehreren Silikaten ein Glas bekommt, welches, so lange es heiss ist, zwar klar erscheint, aber unter der Abkühlung mehr oder weniger opalisirt, so muss man sich von der ausgeschiedenen Kieselsaure überzeugen, so lange das Glas noch heiss ist, und dabei die Loupe zu Hülfe nehmen. Die so eben erwähnte Erscheinung tritt gewöhnlich bei solchen Silikaten ein, deren Basen, Kalkerde, Talkerde, Beryllerde oder Yttererde sind, die für sich mit Phosphorsalz, bei gewisser Sättigung des Glases, unter der Abkühlung oder durch Flattern milchweiss oder opalartig werden." Dr. Theodor Richter, the editor of the 4th edition of Plattner's work, leaves out the "gewöhnlich" of the above quotation, and so makes the implication still stronger. In this vierte Auflage, the statement runs: "Bei solchen Silikaten deren Basen für sich mit Phosphorsalz, bei gewisser Sättigung des Glases, unter der Abkühlung oder durch Flattern milchweiss oder opalartig werden (Kalkerde, Talkerde, Beryllerde, oder Yttererde) wird die Perle unter der Abkühlung mehr oder weniger trübe." It is true enough that silicates in which these bases are present exhibit the reaction; but as other silicates, practically all, indeed, exhibit the reaction also, the inference implied in the above statement is not admissible. The opalescence of the glass arises entirely from precipitated silica. If the blast be sufficiently kept up, a certain amount of silica is almost always dissolved, but this becomes precipitated as the glass cools. A simple experiment will shew that this is the true cause of the opalescence. If some pure silica (or a silicate of any kind), in a powdered condition, be dissolved before the blowpipe-flame in borax until the glass be nearly saturated, and some phosphor-salt be then added, and the blowing be continued for an instant, a precipitation of silica will immediately take place, the bead becoming milky—or, in the case of many silicates, opaque-white—on cooling. This test may be resorted to for the detection of silica in the case of silicates which dissolve with difficulty in phosphor-salt alone, or which do not give a well-pronounced "skeleton" with that reagent.* 1876.

16.—ON THE REACTIONS OF CHROMIUM AND MANGANESE WITH CARBONATE OF SODA.

When a mineral substance is suspected to contain manganese, it is commonly tested by fusion with carbonate of soda. But chromium compounds form with that reagent a green or greenish-yellow enamel, much resembling that formed by some compounds of manganese.

The chromate-of-soda enamel, however, is yellowish-green after exposure to an oxidating flame, and the green colour never exhibits any tinge of blue.

The manganate-of-soda enamel, on the other hand is generally greenish-blue when quite cold.

To avoid, however, any risk of error in the determination, the bead may be saturated with vitrified boracic acid, until all the carbonic acid is expelled, and a clear glass is obtained. The chrome glass will retain its green colour, whilst the manganese glass will become amethystine or violet. In place of boracic acid, silica may

---

* By whom was the formation of a "silica skeleton" first made known? There is no reference to it in the early treatise of Von Engestrom attached to his translation of Cronstedt's "Mineralogie," 1st edition, 1770; 2nd edition, by John Hyacinth de Magellan, 1788), although phosphor-salt is mentioned as a reagent under the term of *sal fusibile microcosmicum*, and was indeed used by Cronstedt before 1758, the year in which his "Mineralogie" was anonymously published. Bergmann, who followed as a blowpipe worker, states that "siliceous earth" is very slowly attacked by microcosmic salt, but he does not seem to have remarked the skeleton formation in the case of any silicate. The reaction appears to have been first definitely pointed out by Berzelius in his standard work on the blowpipe, published in 1820. It was therefore most probably discovered by him, or perhaps—as he lays no claim to its discovery, whilst claiming to be the originator of other tests—it may have been communicated to him by Gahn?

be used if more convenient. In this case the reaction is assisted by the addition of a very small amount of borax. 1871-76.

17.—ON THE DETECTION OF CADMIUM IN THE PRESENCE OF ZINC IN BLOWPIPE EXPERIMENTS.

When cadmiferous zinc ores, or furnace-products derived from these, are treated in powder with carb. soda on charcoal, the characteristic red-brown deposit of cadmium oxide is generally formed at the commencement of the experiment. If the blowing be continued too long, however, this deposit may be altogether obscured by a thick coating of zinc oxide. When, therefore, the presence of cadmium is suspected in the assay-substance, it is advisable to employ the following process for its detection. The substance, if in the metallic state, must first be gently roasted on a support of porcelain or other non-reducing body. Some of the resulting powder is then fused with borax or phosphor-salt on a loop of platinum wire, and bisulphate of potash in several successive portions is added to the fused bead. The latter is then shaken off the wire into a small porcelain capsule, and treated with boiling water. A bead of alkaline sulphide is next prepared by fusing some bisulphate of potash on charcoal in a reducing flame, and removing the fused mass before it hardens. A portion of the solution in the capsule being tested with this, a yellow precipitate will be produced if cadmium be present. The precipitate can be collected by decantation or filtration, and tested with some carb. soda on charcoal. This latter operation is necessary, because if either antimony or arsenic were present, an orange or yellow precipitate would also be produced by the alkaline sulphide. By treatment with carb. soda on charcoal, however, the true nature of the precipitate would be at once made known. 1876.

18.—ON THE SOLUBILITY OF BISMUTH OXIDE IN CARBONATE OF SODA BEFORE THE BLOWPIPE.

Neither in the treatise of Berzelius, nor in the more modern and advanced work of Plattner, is any reference made to the behaviour of oxide of bismuth with carb. soda in an oxidating flame. In Plattner's "Tabellarishe Uebersicht des Verhaltens der Alkalien, Erden, und Metalloxyde für sich und mit Reagentien im Löthrohrfeuer," whilst oxide of lead is stated, correctly, to be soluble in carb.

soda in an oxidating flame, the reference to oxide of bismuth is, simply, that with carb. soda on charcoal it becomes immediately reduced to metailic bismuth; and none of his translators seem to have thought it necessary to supply the omission. In Hartmann's tabular "Untersuchungen mit dem Löthrohr," in the handy little work of Bruno Kerb ("Leitfaden bei qualitativen und quantitativen Löthrohr-Untersuchungen"), in the "Löthrohr-Tabellen" of Hirschwald, and all other blowpipe books that I have met with, the same singular omission occurs. This seems to bear out very forcibly the somewhat cynical adage that "books are made from books." To supply the omission, it may be observed that bismuth oxide dissolves in carb. soda very readily in an oxidating flame, if the supporting agent be platinum wire or other non-reducing body. The glass is clear yellow whilst hot, but on cooling it assumes an orange or yellowish-brown colour, and becomes pale yellow and opaque when cold. As regards their solubility by fusion in carb. soda, metallic oxides fall into three groups: (1) *Easily soluble*, e.g.. PbO, Bi^2O^3, BaO, &c.; (2) *Slightly or partially soluble.* e.g., Mn^2O^3, CoO, &c.; and (3), *Insoluble*, e.g., Fe^2O^3, Ce^2O^3, NiO, CaO, MgO, &c. 1876.

19.—ON THE DETECTION OF CARBONATES IN BLOWPIPE PRACTICE.

A mineral substance of non-metallic aspect, in nine cases out of ten, will be either a silicate, sulphate, phosphate, borate, carbonate, fluoride, or chloride: more especially if the streak be uncoloured or merely exhibit some shade of green or blue, or if the substance evolve no fumes when heated on charcoal.

Simple fusion with phosphor-salt on a loop of platinum wire serves at once to distinguish a silicate from any of the other bodies enumerated above, as, whilst the silicate is but slowly attacked, these other bodies are readily and rapidly dissolved. Among the latter, again, the carbonates are distinguished very readily by the marked effervescence which they produce in the bead by the evolution of carbonic acid during fusion—the phosphates, sulphates, &c., dissolving quietly. The reaction is quite as distinctive as that produced by the application of an ordinary acid; but, of course, it may arise in both cases not only from a carbonate proper, but from the presence of intermixed calcite or other carbonate in the substance under exami nation; and it is also occasioned by bodies which evolve oxygen on ignition; but these latter, manganese oxides excepted, are of rare occurrence among minerals proper. By this reaction, upwards of twenty years ago, the writer detected the presence of carbonate of lime in certain specimens of Wernerite (the "Wilsonite" variety, portions of which had previously been analyzed without the impurity having been discovered. It need scarcely be stated that the test-substance must be added to the phosphor-salt, on the platinum loop, only after the quiet fusion of the flux into a transparent glass. The reaction is, of course, manifested equally well with borax. 1871–76.

20.—ON THE DETECTION OF BROMINE IN BLOWPIPE EXPERIMENTS.

When fused with phosphor-salt and copper oxide, the bromides, it is well known, impart an azure-blue coloration to the flame, much like that produced by chlorides under similar treatment, although streaked more or less with green, especially at the commencement of the operation. To distinguish these bodies more closely, Berzelius recommended the fusion of the test-substance with 6 or 7 volumes of bisulphate of potash in a closed tube. Bromides by this treatment become decomposed as a rule, and give off strongly-smelling brownish or yellowish-red vapours of bromine. But this process does not always give satisfactory results, as in some instances the bromide is very slightly attacked. In this case, the following method, based on a peculiar reaction of bromide of silver, first pointed out by Plattner, may be resorted to: If insoluble, the bromide is fused with 2 or 3 volumes of carb. soda. A soluble bromide of sodium is thus formed, with separation of the base. To the filtered or decanted solution of the fused mass, a small fragment of nitrate of silver is added, in order to precipitate bromide of silver. This, collected by decantation, is fused with a small quantity of bisulphate of potash in a little flask or test-tube. The bromide of silver will quickly separate from the flux in the form of a blood-red globule, which becomes pale-yellow when cold. The little globule, washed out of the tube by dissolving the fused bisulphate in some warm water, is carefully dried by being rubbed in a piece of blotting or filtering paper, and is then placed in the sunlight. After a short time it will turn green. Chloride of silver, as obtained in a similar manner, melts into an orange-red globule, which changes to clear-yellow on cooling, and finally becomes white, or nearly so. Placed in sunlight, it rapidly assumes a dark-gray colour. Iodide of silver, under similar treatment, forms whilst hot an almost black globule, which becomes amethyst-red during cooling, and dingy-yellow when cold. In the sunlight it retains the latter colour. A mixture of chloride and iodide of silver assumes a greenish tint somewhat resembling the colour acquired by the bromide globule. This, however, can scarcely give rise to any error, as the presence of iodine is revealed—even if no violet-coloured fumes be emitted—by the dark amethystine colour of the bead whilst hot. 1876.

21.—BLOWPIPE REACTIONS OF METALLIC ALLOYS.

In examining these reactions, about equal portions of the metals (forming the alloy) may be placed together, on charcoal, and subjected to the action of a reducing flame.

1. *Platinum and Tin* unite with violent deflagration and emission of light, forming a hard, brittle, and infusible globule.

2. *Platinum, Zinc and Tin* unite with violent action, the zinc throwing off long flakes of oxide.

3. *Platinum and Zinc, per se,* do not combine, the zinc burning into oxide.

4. *Platinum and Lead* unite quietly, forming a brittle globule.

5. *Platinum and Thallium* unite quietly; the resulting globule is dark externally, gray internally, and quite brittle.

6. *Platinum and Bismuth* unite quietly, or with merely slight spitting, into a dark, brittle globule.

7. *Platinum and Copper* combine quietly, though not very readily, into a hard, light-coloured, malleable globule.

8. *Platinum and Silver* unite quietly, but not very readily, unless the silver be greatly in excess, into a white malleable globule.

9. *Platinum and Gold* unite quietly, forming (if the gold be somewhat in excess) a yellow malleable globule.

10. *Gold and Tin* unite quietly into a very brittle globule.

11. *Gold and Zinc* do not combine *per se;* the zinc burns into oxide.

12. *Gold and Lead* combine quietly, forming a gray brittle bead.

13. *Gold and Thallium* unite quietly, but separate again to some extent during cooling. The globule may thus frequently be flattened out, but not without cracking at the sides. If the metals remain united, the button is dark blackish-gray, and quite brittle.

14. *Gold and Bismuth* unite quietly and readily, forming a very brittle globule.

15. *Gold and Copper*, and 16, *Gold and Silver*, unite, and form a malleable globule.

17. *Silver and Tin* unite quietly into a malleable globule.

18. *Silver and Lead* unite readily into a malleable globule.

19. *Silver and Thallium* combine readily : the globule is malleable.

20. *Silver and Bismuth* unite readily and quietly : the globule is brittle, but admits of being slightly flattened out.

21. *Silver and Copper*, and 22, *Silver and Gold*, form malleable globules. The gold alloy, even with gold largely in excess, is quite white. If it be flattened out and heated in a platinum spoon with some bisulphate of potash, it will become yellow from the silver on the surface being dissolved. On re-melting the flattened disc, a silver-white globule is again obtained.

23. *Copper and Tin* unite into a gray and partially malleable bead, the surface of which, in the O F, becomes more or less thickly encrusted with cauliflower-like excrescences of oxide.

24. *Copper and Zinc* do not unite, *per se*, into a globule, the zinc burning into oxide. Under carb. soda, or carb. soda and borax, brass is readily formed.

25. *Copper and Lead* form a dark gray globule, which is sufficiently malleable to admit of being extended on the anvil.

26. *Copper and Thallium* melt into a dark gray malleable globule.

27. *Lead and Tin* unite readily, but the globule commences immediately to oxidize, throwing out excrescences of white and yellow oxide. On removal from the flame it still continues in ignition, and pushes out further excrescences. The unoxidized internal portion (if any remain) is malleable.

28. *Lead and Bismuth* unite readily : the molten globule acquires a thin dark coating of oxide on the surface only, and admits of being flattened out, more or less, upon the anvil.

29. *Lead and Thallium* form a malleable globule.

30. *Bismuth and Tin* unite readily, but the fused mass immediately throws out excrescences, and becomes covered with a dense crust of oxides. The reaction, however, is not so striking as with lead and tin.

31. *Thallium and Tin* exhibit the same reaction as lead and tin, but the cauliflower-like excrescences are brownish-black. 1876.

PART II.

ORIGINAL TABLES

(BASED ESSENTIALLY ON BLOWPIPE CHARACTERS)

FOR THE

DETERMINATION OF ALL KNOWN MINERALS.

PART II.

INTRODUCTION.

In these Tables for the Determination of Minerals, an attempt has been made to place in the same Table, or under its secondary sub-divisions, those minerals only which are related to each other: related, that is, not by a single determinative character, but by their composition and characters generally. It is not, of course, possible to effect this with complete success in all cases; but the present Tables, it is thought, will be found for the greater part to be at least free from the startlingly incongruous, and hence objectionable, groupings seen in Determinative Mineral Tables hitherto published. At the same time, as regards ready application and efficacy in a purely determinative point of view, the present Tables will compare favourably, it is hoped, with other efforts in this direction. In using the Tables, the student is assumed to be familiar with the more common blowpipe-operations and reactions, as given in Part I of this Essay. It has not been thought necessary, therefore, in prefixing to subordinate sections the headings "Cu reaction," "Pb reaction," "Na reaction," &c., to give these reactions in full.

The present work is not, of course, intended to serve as a substitute for an ordinary text-book, but simply as an adjunct to the latter. To add, however, to its usefulness, the leading characters of each species, including Composition, System of Crystallization (with an occasional angle), Hardness, Specific Gravity, Colour, &c., are briefly given. The composition is stated in percentage values in most cases;

but in others merely the components, as separated by analysis—*e. g.*, CaO, FeO, Al^2O^3, Fe^2O^3, CO^2, SiO^3, &c.—are stated. The student will thus be able, after determining a mineral by the Tables, to verify its composition as a confirmatory test.

The names of the Crystal Systems are printed chiefly in abbreviated form, as follows:—*Reg.* (= Regular, Tesseral, Isometric, Monometric, &c.); *Tet.* (= Tetragonal, Quadratic, Dimetric, &c.); *Hex.* (= Hexagonal), or *Hemi-Hex.* (= Rhombohedral and other Hemi-Hexagonal forms); *Rh.* (= Rhombic, Ortho-Rhombic, Trimetric, &c.); *Clino-Rh.* (= Clino-Rhombic, Monoclinic, Oblique Rhombic, &c.); *Anorth.* (= Anorthic, Triclinic, Clino-rhomboidal, &c.) In Rhombic and Clino-Rhombic crystals, the prism angle ($= \infty P : \infty P$. Naumann) is sometimes given under the symbol of $V : V$, and other interfacial angles are occasionally stated.*

Hardness (= H) refers, of course, to the universally adopted Scale of Mohs. This scale is given below, together with a roughly corresponding scale (published by the author in 1843) to serve as a substi-

---

* In the system of crystallographic notation long followed by the author- one that possesses the advantage of allowing the symbols to be readily translated into words—all forms (apart from those of the Regular System, and certain special forms of the Hexagonal System, in the case of which it is more convenient to employ arbitrary symbols) are referred to one of three sets, namely: *Vertical forms* (parallel with the vertical axis); the *Basal form* (parallel with the basal or middle axes); and *Polar* or *Pyramidal forms* (inclined towards the vertical axis or principal poles of the crystal). Vertical forms, generally, are denoted by the common symbol V; the basal form, by B; and polar or pyramidal forms, by P. When a form lies parallel to any axis, the sign of the axis (where this is necessary to indicate the position of the form) is placed above the symbol. Thus V denotes a vertical form consisting of planes parallel with the vertical axis only; whilst $\acute{V}$ (in verbal language, a "Front Vertical," = a Macro-Vertical or Ortho-Vertical, according to the System) denotes a form parallel with the right-and-left transverse axis (= the macrodiagonal or orthodiagonal, as the case may be); and $\acute{V}$ or $\grave{V}$ denotes a "Side-Vertical," "Brachy-Vertical" or "Clino-Vertical," parallel with vertical and brachy-axis, or vertical and clino-axis, according to the System. B, the symbol of the basal form, needs no axial signs, as it cannot vary. The polar forms comprise: Polars or Pyramids proper, Front Polars, and Side Polars (or macro-polars, brachy-polars, &c.), and are indicated, respectively, by the symbols P, $\bar{P}$, and $\acute{P}$ or $\grave{P}$ (with secondary signs where necessary, as in the Clino-Rhombic and Anorthic Systems). Values placed before a symbol, as 2P, $\frac{1}{4}\bar{P}$, &c., refer to the vertical axis; those placed after a symbol, as V2 or V$\check{2}$, refer to one of the middle axes, either understood conventionally, or indicated by its sign above the figure. It is of course evident that no other forms than Vertical, Basal, or Polar forms can possibly be present in any crystal. Hence, by the employment of the symbols V, B, and P, with modifications as described above, the position of a given form becomes taken up by the eye at a glance, and without risk of misconception.

tute where the minerals of which the scale of Möhs consists may not be at hand.

SCALE OF MÖHS.	CHAPMAN'S CONVENIENT SCALE, TO CORRESPOND WITH THAT OF MÖHS.
1 ... Talc. 2 ... Rock Salt. 3 ... Calcite. 4 ... Fluor Spar. 5 ... Apatite. 6 ... Orthoclase. 7 ... Rock Crystal (Quartz). 8 ... Topaz. 9 ... Corundum. 10 ... Diamond.	1 .. Yields to the finger-nail. 2 .. Does not yield to the nail, but is scratched by a copper coin. 3 .. Scratches a copper coin (i. e., a copper coin proper, not a modern bronze coin), but is also scratched by one. 4 .. Not scratched by a copper coin, but easily scratched by a penknife. Does not scratch ordinary window-glass. 5 .. Scratches glass very feebly, leaving its powder on it. 6 .. Scratches glass strongly. Not scratched by a penknife, but yields to a hard file. Readily scratched by a piece of quartz. 7 .. Scarcely touched by a file. 8 – 9 – 10 .. Harder than quartz. Convenient objects for the comparison of minerals possessing a higher degree of hardness than No. 7, cannot readily be found ; but these minerals are few in number, and, as a rule, they are easily distinguished by other characters.

The sign G indicates specific gravity. This character is ascertained very expeditiously by the spring balance contrived by Professor Jolly of Munich; but where an instrument of this kind is not at hand, a small pair of ordinary scales may be conveniently used. The centre of one pan is perforated for the passage of a horse-hair with running noose (to hold the mineral), or is provided on its under-side with a small hook to which the hair is attached, and the strings of this pan should be somewhat shortened. The mineral—a small crystal or fragment of about a gramme or couple of grammes in weight—is weighed first in the ordinary way, and the weight is then taken whilst the mineral is suspended in distilled water. If a equal the the weight in air, and w the weight in water, $G = \dfrac{a}{a - w}$. Bodies which are soluble in water may be weighed in alcohol or other suitable liquid of known sp. gr. Calling this latter, G', and the weight of the mineral in the liquid, W', the true sp. gr. becomes $\dfrac{a}{a - w}, G'$.

In other words, the sp. gr. of the substance as found by the liquid, must be multiplied by the sp. gr. of the latter.

In testing the solubility, &c., of minerals in acids, a small fragment of the substance should be reduced to powder; and some of the latter (inserted into a test-tube by a narrow strip of glazed paper folded gutter-wise) may be covered to the depth of about half-an-inch with the acid to be employed. The tube may then be warmed, so as to bring the acid gently to the boiling-point, over the flame of a small spirit lamp or Bunsen burner. Or, in place of the test-tube, a small porcelain capsule, provided with a short handle, may be used.

In the examination of minerals for the presence of earths and alkalies, a small direct-vision spectroscope will be found very serviceable. The small pocket spectroscopes, $3\frac{1}{2}$ inches long, with attached scale, made by Browning of London, cannot be too highly recommended. Many minerals (Calcite, Gypsum, Polyhallite, Strontianite, Celestine, Barytine, Lepidolite, &c., &c.) give characteristic spectra by sufficiently prolonged ignition in the outer border of a Bunsen flame, but the reaction becomes in most cases greatly intensified by moistening the ignited substance with hydrochloric acid, as described at page 55 and in many of the following Tables. In the Tables proper, all, or practically all, known species are inserted; but each Table is followed by an Explanatory Note, in which the commonly occurring or important species of the Table are alone referred to. In these notes, crystallographic and other distinctive characters are given in somewhat greater detail.

INDEX TO THE TABLES.

A.—THE MINERAL PRESENTS A METALLIC LUSTRE.

A^1.—*A small fragment ignited, BB, on charcoal volatilizes wholly or partly.*

(1) It gives As fumes, but no sulphur-reaction with carb. soda.... TABLE I.
(2) It gives As fumes and sulphur-reaction .... TABLE II.
(3) It gives reaction of Sulphur or Selenium, but no fumes of Sb or Te. TABLE III.
(4) It gives sulphur-reaction, and fumes of Sb or Te .... TABLE IV.
(5) It gives fumes of Sb or Te, but no sulphur-reaction........... TABLE V.
(6) It gives no reaction of S, Se, Te, Sb, or As.................. TABLE VI.

A^2.—*A small fragment ignited on charcoal does not perceptibly volatilize.*

(1) It fuses, BB, on charcoal into a globule.................... TABLE VII.
(2) It is infusible, or fuses only on the thin edges .... TABLE VIII.

B.—THE MINERAL PRESENTS A SUB-METALLIC ASPECT.

(1) It is easily fusible or reducible *per se* .... TABLE IX.
(2) It is infusible, or fusible only on thin edges .... TABLE X.

C.—THE MINERAL PRESENTS A VITREOUS, PEARLY, EARTHY, OR OTHER NON-METALLIC ASPECT.

C^1.—*A small fragment takes fire when held against a candle or Bunsen-flame.*

(1) It burns with blue flame and sulphurous or alliaceous odour.. TABLE XI.
(2) It burns with bituminous or aromatic odour................ TABLE XII.

C^2.—*The mineral is not inflammable. It is readily dissolved or attacked by fusion with borax or phosphor-salt.*

(1) It is attacked with effervescence by dilute hydrochloric acid, TABLE XIII.
(2) It emits As fumes by fusion with carb. soda on charcoal.... TABLE XIV.
(3) It emits Sb fumes by fusion with carb. soda on charcoal...... TABLE XV.
(4) It gives sulphur-reaction with carb. soda .... TABLE XVI.

(5) Its solution in nitric acid* gives a canary-yellow pre. with molybdate of ammonia .... TABLE XVII.

(6) Its powder, moistened with sulphuric acid and alcohol, communicates a green colour to the flame of the latter........ TABLE XVIII.

(7) It gives chlorine (I or Br) reaction (azure or green flame) by fusion with phosphor-salt and copper oxide .... TABLE XIX.

(8) It evolves orange-red fumes when warmed with a few drops of sulphuric acid in a test-tube........................ TABLE XX.

(9) It corrodes the glass when warmed in powder with sulphuric acid in a test-tube .... TABLE XXI.

(10) It forms by fusion with carb. soda and nitre an alkaline mass partly soluble in water, the solution assuming a blue, brown, or green colour when boiled with addition of hydrochloric acid and a piece of tin or zinc .... TABLE XXII.

(11) It does not produce any of the above reactions.......... TABLE XXIII.

C^3.—*The mineral is very slowly dissolved, or is only partially attacked, BB, by borax or phosphor-salt.*

† It is infusible, or fusible only on the thinnest edges:

(1) It is hard enough to scratch ordinary glass distinctly, TABLE XXIV.

(2) It is not hard enough to scratch glass distinctly.... TABLE XXV.

†† It is more or less readily fusible:

(1) It yields no water (or merely traces) by ignition in bulb-tube .... TABLE XXVI.

(2) It gives off a distinct amount of water by ignition in bulb-tube... TABLE XXVII.

NOTE.—In order to appreciate the distinctive character of the respective sections C^2 and C^3, the student is recommended to add a small fragment of calcite, gypsum, fluor spar, barytine, or apatite, on the one hand,—and a small particle of orthoclase, pyroxene, amphibole, garnet, talc, quartz, or corundum, on the other—to a previously fused bead of phosphor-salt ; and to observe the rapidity with which the first-named minerals are dissolved under the action of the blowpipe, whilst the minerals of the latter group remain practically unaffected, or are very slowly or incompletely attacked.

---

* Crush a small fragment of the substance to powder. Place this, by a bent slip of paper, in a test-tube. Drop a little nitric acid upon it, and warm or boil. Then add some distilled water and a grain or two of the molybdate, and warm again.

TABLES
FOR THE DETERMINATION OF MINERALS.

TABLE I.

[Metallic aspect. Wholly or partly vol. with As fumes, but yielding no S reaction.]

A.—Entirely vol. (or leaving merely a feeble residuum).

NATIVE ARSENIC: Hemi-Hex.; H 3·5; G 6·0; tin-white with dark tarnish.

Allemontite: differs merely by having part of the As replaced by Sb.

NATIVE BISMUTH—ARSENIC-HOLDING VARIETIES. G 9·7, BB, a yellow deposit on charcoal. *See* TABLE VI.

B.—Partially vol., leaving distinct residuum.

B¹.—RESIDUUM MAGNETIC.

SMALTINE: (CoNiFe) 28, As 72. Reg.; H 5·5-6; G 6·5; greyish tin-white. Chloanthite (Chathamite) is a highly nickeliferous smaltine. Skutterudite is probably a mixture of smaltine and arsenic ($= CoAs^2 + As$).

LÖLLINGITE: Fe 27·2, As 72·8. Rh.; H 5-5·5; G 7-7·4; greyish silver-white. Leucopyrite (Fe 32·2, As 66·8 (?)) is closely related. In both, a little S is often present. (*See* below).

B².—RESIDUUM NOT MAGNETIC.

(*Ni reaction*).

RAMMELSBERGITE: Ni (CoFe) 28, As 72. Rh.; H 5·5; G 7·1; greyish silver-white.

NICKELINE: Ni (Fe, &c.) 43·6, As 56·4. Hex.; H 5·5; G 7·5-7·7; pale copper-red.

(*Cu reaction*).

DOMEYKITE: Cu 71·7, As 28·3; H 3-3·5; G 7-7·5; silver-white or tin-white, tarnished. ALGODONITE (Cu 83·5, As 16·5) and WHITNEYITE (Cu 88·4, As 11·6) are closely related, but with higher sp. gr. (8-8·3).

(*Ag reaction*).

RITTINGERITE ; Normally, AgAs (?) with 57·7 Ag, but commonly contains sulphur. Iron-black, red by transmitted light; streak orange-yellow, lustre, mostly, sub-metallic. Clino-Rh. ; H 2·5-3 ; G 5·63. (*See* TABLE IX).

NOTE ON TABLE I.

The only minerals of general occurrence belonging to this Table are Native Arsenic, Smaltine, and Nickeline. N. Arsenic is commonly in botryoidal masses with dark surface-tarnish, and is readily distinguished BB by volatilizing rapidly without fusing. Smaltine occurs most frequently in small tin-white octahedrons of sufficient hardness to scratch glass, but is also found in reticulated groups of minute indistinct crystals, and massive. After roasting, the smallest particle imparts BB a rich blue colour to borax. Nickeline is rarely found otherwise than massive. Its light copper-red or yellowish-red colour and high sp. gr. are its more salient characters. BB, it melts easily into a hard brittle non-magnetic globule with crystalline surface. The globule remains non-magnetic after long exposure to the flame.

TABLE II.
[Metallic aspect. As and S reactions.]

A.—Residuum magnetic.
(Co reaction).

COBALTINE: Co (Fe, &c.) 35·5, As 45·2, S 19·3. Reg. H 5·5; G 6·3; silver-white, greyish.

GLAUCODOT: (CoFe) 35, As 45·5, S 19·5. Rh.; H 5·5, G 6·2; silver-white, greyish. Strictly, a cobaltic Mispickel.

(Ni reaction).

GERSDORFFITE: Ni 35, As 45·5, S 19·5. Reg.; H 5·5; G 6-6·3; greyish tin-white.

ULLMANNITE: essentially antimonial: *See* TABLE IV. Corynite, with more As than Sb, is closely related. Also Wolfachite, but the latter is Rhombic in crystallization.

(Fe reaction).

MISPICKEL or ARSENICAL PYRITES: Fe 34·4, As 46, S 19·6. Rh.; H 5·5-6; G 6·0-6·3; silver-white, greyish. GLAUCODOT and DANAITE are cobaltiferous varieties. ALLOCLASE is a related steel-grey species, containing Co, Ni, Bi, &c. GEIERITE is also a related compound, but with higher percentage of arsenic (= Fe 33·6, As 60, S 6·4). PLINIAN is apparently a clino-rhombic mispickel.

LÖLLINGITE:—LEUCOPYRITE: Normally, iron arsenides free from sulphur, but frequently mixed with a little FeS^2. *See* TABLE I.

(Cu reaction).

TENNANTITE (Arsenical Tetrahedrite): Cu, Fe, As, S, with Cu averaging 50 p. c. Reg.; H 3·5-4·0; G 4·4-4·5; dark lead-grey, iron-black. Some examples of Tetrahedrite, proper, contain traces of As. *See* TABLE IV.

B.—Residuum non-magnetic.
(Cu reaction).

ENARGITE: Cu 48·5, As 19, S 32·5. Rh.; H 3; G 4·44; dark-grey, iron-black. Epigenite is closely related, but contains some iron. Also Clarite and Luzonite.

COPPER BINNITE (= Dufrenoysite of Damour, Kengott, &c.) Cu 39, As 31, S 30. Reg.; H 2·5; G 4·6; dark steel-grey, brownish-black; streak, red-brown.

(*Ag reaction*).

POLYBASITE (arsenical variety, *see* TABLE IV.) gives large silver-globule by cupellation. Iron-black; red in thin pieces by transmitted light.

RITTINGERITE: normally, Ag As, but sulphur commonly present. Iron-black, red by transmitted light; streak orange-yellow. *See* TABLES I., IX., XIV.

(*Pb reaction*).

DUFRENOYSITE (v. Rath): Pb 57, As 21, S 22. Rh.; H 3; G 5·5-5·6; dark lead-grey, streak red-brown. Jordanite is nearly related. Pb 51, As 25, S 24.

LEAD-BINNITE (= Binnite of Heusser, Scleroclase of V. Waltershausen, Sartorite of Dana): Pb 42·7, As 31, S 26·3. H 3; G 5·4; dark lead-grey, streak red-brown.

GEOCRONITE (occasional varieties, but the species is essentially antimonial. *See* TABLE IV.)

NOTE ON TABLE II.

Cobaltine, Mispickel, and Tennantite, are the only minerals of ordinary occurrence belonging to this Table. Cobaltine is commonly in small crystals of a silver-white colour with slightly reddish tinge. These crystals are most commonly combinations of the cube and pentagonal dodecahedron ∞^2, the latter predominating; or combinations of this pentag. dodecahedron with the octahedron. The crystals scratch glass easily. More rarely, cobaltine occurs massive. The smallest particle, after roasting, imparts BB a deep-blue colour to borax.

Mispickel or Arsenical Pyrites occurs commonly both in masses and in small prismatic crystals of the Rhombic System. Its colour is silver-white, but the surface soon assumes a greyish or other tarnish. The crystals, which, as a rule, scratch glass distinctly, are mostly rhombic prisms (with $V:V = 111° 12'$) terminated by two nearly flat and transversely striated planes (the brachydome or side-polar $\frac{1}{4}\bar{1}'$, with summit angle $= 146° 28'$). It fuses easily, with emission of copious arsenical fumes, and the fused globule (after sufficient exposure to the flame) attracts the magnet strongly. Many varieties contain cobalt, and in some, nickel is present. Nearly all varieties, moreover, hold a certain amount of gold or silver, varying from a few dwts. to several ounces per ton.

Tennantite is readily distinguished from the above by its dark colour and low degree of hardness, as well as by its strong copper-reaction. It occurs only in small crystals of the Regular System: mostly tetrahedral combinations, or these associated with the rhombic dodecahedron or cube.

TABLE III.

[S or Se reaction. No fumes of As, Sb, or Te.]

A—Fusible: fusion-product magnetic.

(Co and Ni reactions).

LINNÆITE (ZIEGENITE): (Co, NiFe) 58, S 42. Reg.; H 5·5; G 4·9; light steel-grey with reddish tarnish.

(Ni reaction).

MILLERITE: Ni 35·5, S 64·5. Hemi-Hex., acicular; H 3; G 5·3 (4·6 Kengott). Brass or bronze-yellow.

POLYDYMITE: Ni 59·5, S 40·5. Reg.; H 4·5; G 4·81; lead grey. SAYNITE is this species mixed with copper pyrites, galena, &c. BEYRICHITE is closely related.

(Fe reaction).

IRON PYRITES (MUNDIC): Fe 46·7, S 53·3. Reg.; H 6-6·5; G 4·8-5·2; pale brass-yellow. (*See* Note, below).

MARCASITE: Rhombic in crystallization; otherwise like ordinary Pyrites.

PYRRHOTINE (MAGNETIC PYRITES): Fe 60·5, S 39·5. Hex.; H 3·5-4·5; G 4·4-4·7; bronze-yellow; magnetic. Horbachite is a nickeliferous var.; Troilite, a meteoric pyrrhotine.

(Cu reaction).

COPPER PYRITES (Chalcopyrite): Cu 34·6, Fe 30·5, S 34·9. Tetr.; H 3·5-4; G 4·1-4·3; rich brass-yellow, often with variegated tarnish; streak greenish-black, or dark-green. HOMICHLINE is apparently a mixture of this sp. and the next. BARNHARDTITE is also closely related.

BORNITE (Purple Cop. Pyrites, Buntkupfererz): consists of Cu, Fe, S in somewhat variable proportions. The Cu averages 50-60 p. c. Many analyses shew: Cu 55·6, Fe 16·4, S 28. Reg.; H 3; G 4·5-5·2; brownish copper-red, rapidly tarnishing blue, green, &c.; streak black.

CUBANITE: Cu 20, Fe 41, S 39 (?). Reg.; H 4; G 4·1; brass-yellow, streak black.

STANNINE (Tin Pyrites). Fusion-globule in some cases magnetic. *See* B².

B —Fusible: fusion product non-magnetic.

B¹.—EVOLVING, BB, STRONG ODOUR OF SELENIUM.

(Cu reaction).

BERZELINE : Cu 61·6, Se 38·4. In thin coatings; very soft; silver-white, tarnishing black.

CROOKESITE : Cu 45·76, Th 17·25, Ag 3·71, Se 33·28. Compact; H 2·5-3 ; G 6·9 ; lead-grey. Colours flame intensely green.

EUKAIRITE : Ag 43·1, Cu 25·3, Se 31·6. Soft, lead-grey.

ZORGITE : Cu, Pb, Se, in variable proportions. Lead-grey, soft. Comprises, probably, several distinct species.

(Pb or Bi reaction).

CLAUSTHALLITE : Pb 72·4, Se 27·6. Reg.; H 2·5-3; G 8-8·8; lead-grey.

NAUMANNITE : Ag (Pb) 73, Se 27. Reg.?; H 2·5; G 8·0; iron-black. BB, yields large bead of silver.

LEHRBACHITE : contains Pb and Hg, with Se. H 2·5 ; G 7·9 ; lead-grey.

GUANAJUATITE : Bi 69·7, Se (S) 30·3. Apparently Rhombic, but very imperfectly known. Silaonite is a related compound. TETRADYMITE : Essentially a bismuth telluride, but sometimes contains Se and S. Refer to TABLES IV., V.

(Hg reaction).

TIEMANNITE : Hg 75, Se 25. H 2·5 ; G 7-7·4 ; dark lead-grey. BB, rapidly volatilized. Onofrite is an allied compound of Hg, Se, and S. Guadalcazarite, a sulphide of Hg and Zn, has part of its S replaced by Se, and should therefore be referred to here. It is iron-black in colour, with H 2, and G 7·15.

B².—NO SELENIUM ODOUR EVOLVED ON IGNITION.

(Cu reaction).

CHALKOSINE (Copper Glance): Cu 79·8, S 20·2. Rh.; H 2·5-3·0; G 5·5-5·8. Dark metallic-grey, usually with green or blue-green tarnish.

STANNINE (Tin Pyrites): Cu, Sn, Fe, &c., S. Reg.; H 3·5-4; G 4·4 ; yellowish steel-grey. Decomposed by nitric acid, leaving residuum of SnO^2.

STROMEYERINE: Ag 53, Cu 31·3, S 15·7. Rh.; H 2·5-3·0; G 6·25. Blackish lead-grey. BB, by cupellation, gives large silver-button.

ZALPAITE: Ag 71·8, Cu 14, S 14·2. Reg.; H 2-2·5; G 6·9. Blackish lead-grey; ductile.

(Cu and Pb or Bi reaction).

AIKINITE (Needle Ore): Cu 11, Pb 36, Bi 36, S 17. Rh.; H 2·5; G 6·7. Dark-lead or steel grey, with yellowish or other tarnish. Mostly acicular in quartz.

WITTICHENITE: Cu 38·5, Bi 42, S 19·5 (?). Rh.; H 2·5; G 4·3-4·6. Dark metallic-grey.

EMPLECTITE: Cu 19, Bi 62, S 19. Rh.; H 2-2·5; G 5·2; tin-white, yellowish. Acicular in quartz.

CUPRO-PLUMBITE: Cu 20, Pb 65, S 15 (= $Cu^2S + 2 PbS$). Massive, with cubical cleavage; dark lead grey. H 2·5; G 6·4. ALISONITE is a related compound, but with more copper (= $3 Cu^2S + PbS$).

(Pb or Bi reaction).

GALENA (Lead Glance): Pb 86·6, S 13·4. Reg.; cleavage cubical; H 2·5; G 7·3-7·6. Lead-grey.

BISMUTHINE (Bismuth Glance): Bi 81·25, S 18·75. Rh.; H 2-2·5; G 6·4-6·7. Light metallic-grey, often iridescent.

COSALITE: Pb (Ag) 41·7, Bi 42·2, S 16·1; Lead-grey; H abt. 2·0. Retzbanyite, a related compound.

(Ag reaction).

ARGENTITE (Silver Glance): Ag 87, S 13. Reg.; H 2; G 7·2-7·4. Blackish lead-grey, iron-black; malleable. ACANTHITE has the same composition, but is Rhombic in crystallization.

*** See, also, the Cu-Ag sulphides, Stromeyerine and Zalpaite, above.

(Hg reaction).

METACINNABARITE: Hg 86·2, S 13·8. Black, streak black. G 7·7. H 1·5-2. Guadalcazarite is identical or closely related.

C.—Infusible, or Fusible on edges only.

(Mo reaction. Flame tinged pale-green).

MOLYBDENITE: Mo 59, S 41. Hex. !; H 1-1·5; G 4·4-4·8. Light lead-grey. Mostly in flexible plates and scaly masses, which mark on paper and otherwise much resemble graphite, but easily distinguished by communicating a distinct yellowish-green colour to the outer flame, as well as by sulphur reaction, and higher sp. gr.

(*Zn reaction*).

SPHALERITE or ZINC BLENDE: Some varieties, only, are metallic or sub-metallic in lustre. Streak pale-brown. *See* TABLES X. and XVI.

(*Mn reaction*).

ALABANDINE: Mn 63·2, S 36·8. Black, brownish, dark steel-grey. Streak greenish. Lustre sub-metallic. *See* TABLE X.

HAUERITE: Mn 46·2, S 53·8. Dark red-brown, blackish-brown. Streak brownish. Lustre sub-metallic, only. *See* TABLE X.

NOTE ON TABLE III.

The minerals of comparatively general occurrence belonging to this Table, although more numerous than those of Table II., do not exceed ten or eleven in number. They may be arranged, as regards determination, under two leading groups, according to colour. In the first group, the colour is some shade of metallic yellow or red; and in the second, metallic grey or black. The first group includes Iron Pyrites, Marcasite, Pyrrhotine, Copper Pyrites, and Bornite. The second group includes Argentite, Molybdenite, Galena, Bismuthine, and Chalkosine, with, exceptionally, certain dark varieties of Zinc Blende, in which the lustre inclines to metallic.

(*Colour pale brass-yellow:* $H = 6·0$ *or more*).

Iron Pyrites and Marcasite belong to this section: they are sufficiently hard to scratch glass distinctly. Iron Pyrites occurs both massive and in crystals. The latter are commonly cubes (with faces marked by alternate striæ), or combinations of cube and octahedron, or combinations of the cube and the pentagonal dodecahedron $\frac{2\infty}{y}$, or this pentag. dodecahedron alone. Marcasite presents the same composition (FeS^2), but differs by its Rhombic crystallization, and its greater tendency to fall into decomposition. The crystals are commonly flat prismatic combinations, with largely developed basal plane, and V : V = 106°5′; and they are frequently in twinned forms, or grouped in crested rows; whence the name "spear pyrites," "cockscomb pyrites," etc., applied to the species.

(*Colour brass-yellow, bronze-yellow, or reddish:* H *under* 5·0).

Pyrrhotine or Magnetic Pyrites, Copper Pyrites, and Bornite, belong to this section; none scratch glass. Pyrrhotine is bronze-yellow, almost always massive, and more or less magnetic, sometimes showing polarity. Copper Pyrites is rich brass-yellow, often with variegated tarnish (= "Peacock Ore," etc.), and its streak is blackish-green. It is commonly massive; but occurs also in Tetragonal crystals, mostly small tetrahedrons or sphenoids, much resembling regular tetrahedrons. Bornite or Purple Copper Pyrites has properly a peculiar reddish colour (whence "horse-flesh ore"), but this becomes rapidly obscured by a blue or green tarnish. It is nearly always massive, and its streak is black without any shade of green in the colour.

(*Colour metallic grey or black: flexible in thin pieces, or malleable*).

This section includes Argentite and Molybdenite. Argentite is at once distinguished by its dark colour and its malleability; as well as by its high sp. gr. (over 7·0), and by yielding, BB, a large silver-globule. When crystallized, it is mostly in combinations of cube, octahedron, and rhombic dodecahedron, but the crystals are commonly distorted. It occurs also frequently in leafy and filiform examples. Molybdenite is light lead-grey, mostly in scaly or leafy masses, very soft and flexible, but not malleable. It is readily distinguished by the yellowish-green colour which it communicates to the outer edge of the Bunsen or blowpipe flame, and by its infusibility. It forms, BB, on charcoal a white deposit of MoO_3.

(*Colour metallic grey or black: BB, on charcoal a yellow deposit*).

This section includes Galena and Bismuthine,—the first of very common occurrence, the latter comparatively rare. Galena is distinguished by its rectangular or cubical cleavage, and its high sp. gr. (= 7·3-7·7).˙ When crystallized, it is commonly in cubes or in combinations of cube and octahedron. The fusion-globule is malleable, and it generally yields a little silver on cupellation. Bismuthine is mostly in fibrous masses or acicular crystals. It melts, if held (in the form of a thin splinter) against the outer edge of the flame, without the application of the blowpipe. Its nitric-acid solution yields a white precipitate on the addition of water.

(*Colour blackish metallic-grey. Surface usually encrusted here and there with a greenish, earthy efflorescence.*)

This section (as regards minerals of common occurrence) contains Chalkosine, Cu_2S, only. Easily distinguished by its marked copper-reactions. Forms BB no coating on charcoal, but boils, spirts, and yields a copper-globule. Commonly massive. When crystallized, mostly in small, Rhombic combinations of pseudo-hexagonal aspect.

(*Colour black or brownish-black. Lustre properly sub-metallic. Streak pale-brown. Infusible, or practically so*).

Certain dark varieties of Zinc Blende (Black Jack) may be referred to here, as these are sometimes mistaken for galena.* Their infusibility, brownish streak, and comparatively low sp. gr. (= about 4·0), constitute their more distinctive characters. Mixed, in powder, with carb. soda and a little borax, they yield, BB, on charcoal a ZnO sublimate.

---

* A practical illustration of this came under the author's notice in Colorado a few years ago. He was asked to look at a somewhat roughly constructed reverberatory that had been recently put up for the smelting of lead ore, but which had turned out a failure. The ore, it appeared, got into a pasty mass holding a little reduced lead, and would not work. After examining the furnace, and seeing nothing particularly amiss in it, the writer asked to look at the ore. This was regarded at the furnace as a tolerably clean galena, but was found to consist of nearly two-thirds "Black Jack" mixed with galena in a calcareous gangue. The "pasty stuff" which had given the furnace a bad name was thus easily accounted for. The old name Blende (and the newer Sphalerite) is based on this deceptive aspect. In general, however, the lustre is non-metallic, or at most, sub-metallic.

TABLE IV.

[S reaction. Sb or Te fumes.]

A.—On charcoal, BB, a white deposit.

A¹.—ENTIRELY AND RAPIDLY VOL.

STIBNITE (Antimony Glance, Grey Antimony Ore): S 28·24, Sb 71·76. Rh ; H 2 ; G 4·5-4·7. Lead-grey, often with iridescent or dark tarnish. Melts *per se* in outer edge of the flame without the aid of the blowpipe. *See* also the note below.

CINNABAR (HgS). Some dark or lead-grey varieties. Streak red; G 7·7–9. Inflammable. Lustre, as a rule, non-metallic. (*See* TABLE XI).

A².—PARTIALLY VOL., A LARGE SILVER-GLOBULE REMAINING.

(The minerals of this section present as a rule a sub-metallic aspect. The three first are slightly translucent in thin pieces, and have a red streak).

MIARGYRITE: S 21·8, Sb 41·5, Ag 36·7. Clino-Rh. ; H 2-2·5 ; G 5·18-5·26. Iron-black, streak dull-red.

PYRARGYRITE (Dark Red Silver Ore): S 17·7, Sb 22·5, Ag 59·8. Hemi Hex. ; H 2-2·5 ; G 5·75-5·85. Iron-black, reddish, streak red.

POLYBASITE: S, Sb (As), Ag 64-74 p. c. ; Cu sometimes present. Rh. ; H 2·5 ; G 6·0-6·2. Iron-black ; streak, black, red. Polyargyrite is closely related, but is Regular in crystallization.

STEPHANITE (Melanglanz, Brittle Silver Ore): S, Sb (As), Ag (Cu) 68 p. c. H Rh. ; H 2·5 ; G 6·3. Iron-black, dark lead-grey, often iridescent.

A³.—PARTIALLY VOL , THE RESIDUUM MAGNETIC.

BERTHIERITE: Average comp. S 30, Sb 57, Fe 13. Rh. (?); H 2·5-3 ; G 4-4·3. Dark steel-grey, often with variegated tarnish.

ULLMANNITE (Antimonial Nickel Glance): S 15, Sb 57·5, Ni 27·5. Reg. ; H 5·0-5·25 ; G 6·2-6·5 ; lead-grey or steel-grey, with dark or variegated tarnish. Some examples are arsenical.

A⁴.—PARTIALLY VOL., THE RESIDUUM GIVING STRONG COPPER-REACTION.

TETRAHEDRITE (Grey Copper Ore ; Fahlerz): S, Sb (As), Cu 33-44 p. c., Ag, Fe, &c. Reg. (tetrahedral); H 3-4 ; G 4·8-5·4. Steel-grey, iron-black.

CHALKOSTIBITE (Wolfsbergite): S 25·7, Sb 25·4, Cu 48·9. Rh.; H 3·5; G 4·7-5; dark lead-grey, iron-black, often with variegated tarnish.

B.—On charcoal, BB, a yellow (or white and yellow) deposit.

B¹.—PARTIALLY VOL, A GOLD OR SILVER GLOBULE FINALLY REMAINING.

(If the blowing be stopped too soon, a rich gold-lead or silver-lead globule will of course result. This may be freed from lead on the cupel).

FREIESLEBENITE (Donacargyrite). S 18·8, Sb 26·9, Ag 23·8, Pb 30·5. Clino-Rh. H 2·2·5; G 6·2-6·5; metallic grey. Diaphorite (v. Zepharovich) from Przibram is closely related, but is Rhombic in crystallization. G 5·9.

BRONGNIARDITE: S 19·5, Sb 29·5, Ag 26, Pb 25. Reg.; H 2·5; G 5·9-6·0; dark metallic-grey.

NAGYAGITE (Leafy Tellurium Ore, Blättererz); S, Te, Pb, Au, Ag, &c. Au, commonly, 6-9 p. c. Tet., but mostly in thin flexible laminæ. H 1-1·5; G 6·8-7·2. Blackish lead-grey. Melts *per se* in edge of candle-flame.

B².—PARTIALLY VOL., THE RESIDUUM GIVING STRONG COPPER-REACTION.

BOURNONITE: S 19·66, Sb 24·98, Pb 42·38, Cu 12·98. Rh.; H 2·5-3; G 5·7-5·9. Dark steel-grey, iron-black.

See also Tetrahedrite, some examples of which contain Pb or Bi; and Zinkenite and Jamesonite, which sometimes contain a small percentage of copper.

B³.—PARTIALLY VOL., BUT GIVING NO MARKED REACTION OF Ag, Au, or Cu.

(*Sp. gr. under* 6·0).

ZINKENITE: S 22, Sb 42, Pb 36. Rh., acicular; H 2·5-3·5; G 5·3-5·4. Steel-grey, lead-grey, often with variegated tarnish.

PLAGIONITE: S 21, Sb 37, Pb 42. Clino-Rh.; H 2·5; G 5·4. Dark lead-grey.

JAMESONITE: S 19·6; Sb 29·8, Pb 50·6. Rh. H 2·5; G 5·5-5·62. Metallic-grey. Cleavage basal, strongly marked.

BOULANGERITE: S 18, Sb 23, Pb 59. Crystn.?; H 2·5-3; G 5·7-5·95. Dark lead-grey.

(*Sp. gr. over* 6·0).

MENEGHINITE: S 17·3, Sb 18·8, Pb 63·9. Clino-Rh., acicular. H 2·5-3; G 6·34-6·4. Lead-grey. Some examples appear to be Rhombic in crystallization.

GEOKRONITE: S, Sb, Pb 65 p. c. Some examples contain also a little Cu. Rh. ; H 2·3 ; G 6·44-6·54 ; lead-grey, tarnishing darker. Kilbrickenite is identical or closely related.

KOBELLITE: S 16·8, Sb 10·7, Pb 54·3, Bi 18·2. Crystn. ?; H 2·5; G 6·15-6·30. Dark lead-grey.

TETRADYMITE : Normally, a compound of Bi and Te, but frequently containing small amounts of S or Se. Light steel-grey. H 1-3 ; G 7·4-7·9 ; flexible in thin pieces. Wehrlite is a var. containing S. Some vars. also contain a small percentage of Ag. *See* TABLE V.

NOTE ON TABLE IV.

The only minerals of common or general occurrence belonging to this Table, comprise: Stibnite, Tetrahedrite, Pyrargyrite, Bournonite, Zinkenite, and Jamesonite.

Stibnite or Antimony Glance (also known as Grey Antimony Ore) is distinguished (if pure: *id est*, if unmixed with lead sulphide, &c.) by its rapid volatilization before the blowpipe ; and by its powder becoming orange-yellow in a hot solution of caustic potash. It is generally in masses of a more or less fibrous structure and light lead-grey colour, or in small Rhombic prisms (with $V : V = 90°54'$) terminated by the planes of a rhombic octahedron. The prism-planes are longitudinally striated, but the crystals are usually acicular or more or less indistinct. The only species which somewhat resemble it are the sulphantimonites Zinkenite, Jamesonite, Bournonite, &c., but these give a lead sublimate on charcoal, and Bournonite gives also a strong copper-reaction. They are attacked but not rendered yellow by caustic potash, but an orange precipitate is thrown down if the potash solution be neutralized by hydrochloric acid. Jamesonite is chiefly distinguished by its ready cleavage in one direction ; and Bournonite by its copper-reaction. The latter mineral is often found in small, flat, Rhombic crystals with largely developed basal plane, and $V : V = 93°40'$. These crystals are frequently in cruciform or other twins.

Tetrahedrite is dark-grey or iron-black in colour, and when crystallized is in small tetrahedrons or tetrahedral combinations. It gives strong copper-reactions, and some examples (Rionite) contain zinc ; others, silver, mercury, &c.

Pyrargyrite or Dark Red Silver ore is iron-black or reddish lead-grey in colour, except in thin pieces by transmitted light, when the colour appears blood-red. The streak is red ; and the crystals are mostly combinations of the hexagonal prism with the planes of one or two rhombohedrons ($R : R = 108°42'$; $\frac{1}{2} R : \frac{1}{2} R = 137° 58'$), but the mineral is most commonly massive or in indistinct crystal aggregations. It melts *per se* in the outer edge of the flame without the aid of the blowpipe. On charcoal, BB, a silver globule is easily obtained. Like other sulphantimonites, it is attacked by hot caustic potash, and hydrochloric acid precipitates orange-red Sb^2S^3 from the solution.

TABLE V.

[Metallic Aspect. Sb or Te fumes, but no S reaction.]

A.—Entirely volatilizable, or leaving merely a minute globule of metal.

A¹.—ON CHARCOAL, BB, A WHITE DEPOSIT.

NATIVE ANTIMONY: Hemi-Hex., cleavable; H 3-3·5; G 6·7; tin-white. Converted by nitric acid into yellowish-white powder (Sb^2O^3 + Sb^2O^5).

NATIVE TELLURIUM: Hemi-Hex., cleavable; H 2-2·5; G 6·1-6·3: tin-white. Soluble in nitric acid. Warmed with strong sulphuric acid (the acid being used in excess) forms a purplish-red solution, which becomes colourless on addition of water—metallic Te falling as a dark-grey precipitate. Forms also a red solution when boiled in powder with caustic potash.

A².—ON CHARCOAL, BB, A YELLOW (OR WHITE AND YELLOW) DEPOSIT.

TETRADYMITE: Bi 52, Te 48, but S and Se often present in small proportions. Hemi-Hex.; H 1·5-2; G 7·4-7·9; pale metallic-grey; flexible in thin pieces.

ALTAITE: Pb 61·2, Te 38·8. Reg.; H 2·5-3·5; G 8 1·8·2; tin-white, yellowish.

B —Partially volatilizable.

B¹.—YIELDING, BB, ON CHARCOAL A LARGE GLOBULE OF Ag OR Au.

DYSCRASITE: Ag and Sb in several proportions: Ag 64-84, Sb 15·8-36. Rh.; H 3·5; G 9·4-10. Silver-white or tin-white, with dark or yellowish tarnish.

HESSITE: Ag 62·8, Te 37·2. Rh.; H 2-3·0; G 8·1-8·5. Dark metallic-grey. Petzite is a closely related mineral, but with a large part of the Ag replaced by Au (G 8·7-9·4).

SYLVANITE (Graphic Tellurium): Ag, Au, Te, in variable proportions. Sb and Pb also present in some examples, Au 25-45, Ag 1-15, Te 45-56. Clino-Rh. (or Rh.?); H 1·5-2; G 8-8·4. Light steel-grey inclining to silver-white or pale yellowish. Calaverite is a yellow var., with Au 44·5, Te 55·5. Müllerine is also an auriferous var., containing Pb and Sb in addition to the normal components.

B².—YIELDING, BB, A MAGNETIC (NICKELIFEROUS) GLOBULE.

BREITHAUPTITE (Antimonial Nickel Ore): Ni 32·2, Sb 67·8. Hex. ; H 5 ; G 7·5-7·6 ; pale copper-red, mostly with bluish tarnish. Commonly massive, or in small tabular crystals with striated base. Isomorphous with Nickeline, TABLE I. Part of the Ni usually replaced by Fe.

MELONITE: Ni 23·5, Te 76·5. Hex. ? Pale reddish-white.

NOTE ON TABLE V.

All the minerals of this Table are of exceptional or merely local occurrence. Those which contain gold or silver are easily recognized by the metallic globule which they yield, BB, on charcoal. The presence of antimony is revealed by the copious fumes emitted, BB; and by the formation of a yellowish-white powder (Sb^2O^3 or Sb^2O^5, or a mixture of the two) in nitric acid. The presence of tellurium, revealed by its blowpipe reactions, is readily confirmed by warming a small portion of the substance in a test-tube about half filled with strong sulphuric acid, when a reddish solution will result. On addition of water, a dark precipitate of metallic tellurium is thrown down.

TABLE VI.

[Aspect metallic. No S reaction. No fumes of As, Sb, or Te.]

A —On charcoal, BB, no sublimate. (In closed tube, Hg Reaction).

A¹.—ENTIRELY VOL.

NATIVE MERCURY. In small fluid globules of a tin-white colour. G 13·6.

A².—PARTIALLY VOL., A SILVER-GLOBULE REMAINING.

AMALGAM : Properly an isomorphous union of Ag and Hg : hence these components are present in variable proportions. Reg.; H 2-3·5 ; G 10·8-14·10 ; brittle. Arquerite is a variety containing $86\frac{1}{2}$ p. c. silver. Kongsbergite, a var. containing 95 p. c. silver. Some amalgams contain gold : in these the sp. gr. is usually 15 or more.

B.—On charcoal, BB, a yellow-sublimate.

B¹.—MALLEABLE.

NATIVE LEAD : Reg.; H 1·5 ; G 11·3-11·4 ; lead-grey ; ductile.

B².—CLEAVABLE (OR NOT MALLEABLE).

NATIVE BISMUTH : Hemi-Hex. H 2·5 ; G 9·6-9·8. Reddish silver-white, mostly with yellowish or variegated tarnish.

NOTE ON TABLE VI.

Native Bismuth and Native Amalgam are the only minerals of ordinary occurrence belonging to this Table. N. Bismuth is readily distinguished by its (practically) complete volatilization before the blowpipe, with formation of a yellow deposit of oxide on charcoal. It dissolves rapidly in nitric acid, the solution yielding a white precipitate on the addition of water. Some varieties contain traces of As, S, Te, &c. It occurs commonly in small cleavable masses, but occasionally in dendritic and other examples. When crystallized, it is mostly in small rhombohedrons with basal plane, the principal cleavage being parallel with the latter. Amalgam is often in small crystals of the Regular System, commonly in dodecahedrons or combinations of cube and octahedron. In ordinary varieties the sp. gr. exceeds 13·5. This latter character, together with the large bead of silver which it yields, BB, and its mercurial reaction, serve sufficiently to distinguish it.

TABLE VII.

[Lustre metallic. Not perceptibly vol. Fusible on charcoal into a globule.]

A.—Malleable.

NATIVE GOLD: Reg.; H 2-3; G 15·5-19·4. Gold-yellow. Not attacked by nitric acid, nor by blowpipe fluxes. Always contains a small amount of Ag.

NATIVE SILVER: Reg.; H 2-3; G 10·5 (or 10-11). Silver-white, often with black surface-tarnish. Easily dissolved by dilute nitric acid on heating: a white curdy pre. (turning dark-grey on exposure) is formed by hydrochloric acid, or any soluble chloride, in the solution.

NATIVE COPPER: Reg.; H 2·5-3; G 8·5-8·9. Copper-red, often with dull-brown tarnish. Easily sol. in nitric acid, forming a green solution, which becomes deep-blue on addition of ammonia. The fused bead blackens in the OF, its surface becoming encrusted with CuO. This tinges the flame green.

B —Not malleable.
(*Cu reaction*).

CUPRITE (Red Copper Ore). Colour and streak red. Lustre occasionally sub-metallic. (*See* TABLE IX).

TENORITE (Black Oxide of Copper): Cu 79·85, O 20·15. Rh.? in small, tabular crystals, massive, &c.; H 2-3; G 6·9-6·5. Steel-grey to black. Melaconite is an earthy or scaly var., sometimes pseudomorphous.

(*Mn and Fe reactions.*)

WOLFRAM. Brown, black. Lustre sub-metallic only. H 5-5·5; G over 7. An iron-manganese tungstate. (*See* TABLE IX).

NOTE ON TABLE VII.

All the minerals of this Table, Tenorite excepted, are of tolerably common occurrence. They are readily distinguished by the characters given above, or by the following condensed scheme :

  Malleable:
    Colour yellow—N. Gold.
    " white —N. Silver.
    " red —N. Copper.
  Brittle :
    Streak red—Cuprite.
    " black or brown :
      BB, Cu reaction —Tenorite.
     " Mn reaction—Wolfram.

These three latter minerals belong, properly, to other Tables. Exceptional varieties, only, come under notice here.

TABLE VIII.

[Lustre metallic. Not perceptibly vol. Infusible; or fusible at the extreme point or edges, only.]

A.—Not dissolved, BB, by borax or phosphor-salt.

A^1.—VERY SOFT, BLACK, MARKING OR SOILING.

GRAPHITE (Plumbago). Normally, pure carbon: usually slightly ferruginous, &c. Hex.; H 1-1·5; G 1·9-2·3. Black, lustrous, greasy-feeling.

A^2.—MORE OR LESS MALLEABLE. SP. GR. OVER 11 (IN MOST CASES, 17 OR HIGHER).

NATIVE PLATINUM: Reg.; H 4-5; G 17-18. Silver-white, pale steel-grey. Sol. in hot nitro-hydrochloric acid. Many examples contain a small percentage of Fe, and thus act slightly or strongly on the magnet. *See* under B^1.

NATIVE IRIDIUM or PLATINUM-IRIDIUM: Ir, Pt, Rh, &c. Reg. H 5·5-7; G 18-23, usually about 22. Greyish silver-white; scarcely malleable. Insol. in nitro-hydrochloric acid.

OSMIUM-IRIDIUM or NEWJANSKITE. IrOs, mostly with the Ir in excess. Hex.; H 6·5-7; G 19·5; tin-white. Emits disagreeable odour of osmic acid when fused with nitre in closed tube.

IRIDOSMIUM or SYSSERSKITE: IrOs, with Os predominating. G 21-21·2. Emits odour of osmic acid by ignition *per se* on charcoal. Otherwise like Osmium-Iridium.

NATIVE PALLADIUM. Reg. H 4·5-5; G 11·8-12·2. Light steel-grey or greyish tin-white. Malleable. Sol. in hot nitric acid, forming a reddish solution.

B.—Dissolved or readily attacked by fusion with borax or phosphor-salt.

B^1.—MAGNETIC BEFORE OR AFTER IGNITION.

(*Malleable*).

NATIVE IRON (Meteoric Iron): Fe combined in nearly all cases with a certain percentage of Ni. Reg.; H 4·5-5; G 7-7·8; steel-grey, iron-black.

PLATINUM-IRON: Pt with 10-20 p. c. Fe. Reg.; H 6; G 13-15. Dark steel-grey. Properly, a ferruginous var. of Native Platinum.

Some examples (unless in fine filings) are not readily attacked by borax. Some examples, also, are said to be non-magnetic.

(*Brittle: i.e. not malleable*).

MAGNETITE (Magnetic Iron Ore): Fe 72·4, O 27·6 (= FeO, Fe^{2}O^3). Reg.; H 5·5-6·5; G 4·9-5·2; iron-black, streak black; often exhibits magnetic polarity. Magno-ferrite (better named Ferro-magnesite) is a volcanic variety in which the FeO is essentially replaced by MgO. G 4·6-4·7. Jacobsite is another variety, containing both MgO and MnO. G 4·75. Many examples of Magnetite are also titaniferous. These might fairly rank as a distinct species, having the same relation to Magnetite proper that Ilmenite bears to Hæmatite.

FRANKLINITE : ZnO, FeO, MnO, Fe^{2}O^3, in variable proportions, but yielding the general formula RO, R^{2}O^3. Reg.; H 6-6·5; G 5-5·1; iron-black, streak dark reddish-brown. Usually, more or less magnetic. BB, in powder with carb. soda and borax, gives coating of ZnO on charcoal.

CHROMITE; normally, FeO, 32, Cr^{2}O^3 68. Reg.; H 5·5; G 4·3-4·6. Iron-black, streak dark-brownish. Lustre, in most examples, sub-metallic only.

HÆMATITE (Specular Iron Ore): Fe 70, O 30 (= Fe^{2}O^3). Hemi-Hex.; H 5·5-6·5;* G 5·0-5·3. Steel-grey, often with variegated tarnish; streak cherry-red. Sometimes feebly magnetic. Martite is a var. in small octahedrons altered from Magnetite.

ILMENITE or MENACANNITE (Titaniferous Iron Ore). Fe^{2}O^3, Ti^{2}O^3 in variable proportions. Hemi-Hex.; H 5·5-6; G 4·5-5·3. Iron-black, dark steel-grey; streak black to brownish-red. Dissolved or attacked in fine powder by hot hydrochloric acid, the diluted solution by boiling with tin becoming first colourless and then assuming an amethystine tint.

ARKANSITE (variety of BROOKITE). Black, sub-metallic lustre. See TABLE X.

(*Yield water in bulb-tube*).

TURGITE. Red, blackish-red; lustre sub-metallic. See TABLES X., XXIII.

GŒTHITE. Red, brown; lustre sub-metallic in some examples. See TABLES X., XXIII.

---

* As regards ordinary examples; but the scaly variety, although shewing metallic lustre, soils the hands.

LIMONITE (Brown Iron Ore). Brown, streak yellowish. Lustre occasionally sub-metallic. See TABLES X., XXIII.

B².—NON-MAGNETIC AFTER IGNITION.

(*Strong Mn reaction.* Anhydrous*).

PYROLUSITE (Black Manganese Ore): Mn 63·2, O 36·8. Rh.; H 1-1·5; G 4·7-4·9; iron-black, dark steel-grey, streak black; soils and marks. Ignited, and moistened with HC acid, shews Ba-lines in spectroscope.

POLIANITE: Identical with Pyrolusite as regards composition and general crystallization, but with H = 6-7.

BRAUNITE—HAUSMANNITE. Aspect commonly sub-metallic. See TABLE X.

CREDNERITE. Gives copper reactions. Aspect commonly sub-metallic. See TABLE X.

(*Strong Mn reaction, and yielding aq in bulb-tube*).

MANGANITE: MnO^2 90·9, H^2O 9·10. Rh.; H 3·5-4; G 4·3-4·5; dark steel-grey; streak, brown, black.

PSILOMELANE: MnO, BaO, etc., with about 4 or 5 p. c. H^2O. Amorphous (reniform, &c.); H 5-6; G 3·7-4·7. Iron-black, dark steel-grey; streak, brownish. Some examples show distinct K-line in spectroscope.†

(*No marked Mn reaction. No ebullition by fusion with borax*).

PITCHBLENDE — TANTALITE — COLUMBITE — YTTROTANTALITE — SAMARSKITE—EUXENITE. Lustre sub-metallic, only. See TABLE X.

MUSCOVITE—PHLOGOPITE—and some other MICAS. Lustre pearly-metallic (pseudo-metallic); foliated or scaly; streak white or greyish. See TABLES XXV., XXVI.

NOTE ON TABLE VIII.

Minerals of ordinary occurrence belonging to this Table comprise—in addition to Graphite—the iron ores, Magnetite, Hæmatite, and Ilmenite; and the

---

* Dissolved also, BB, by borax with strong ebullition, caused by liberation of oxygen.

† This is best seen by igniting the test-substance, and then moistening it with hydrochloric acid. Green Ba-lines first appear for a moment, after which the red K-line comes out very distinctly and is tolerably permanent. If a piece of deep-blue glass be held between the spectroscope and the Bunsen-flame, the yellow Na-line, always present with its accompanying glare, becomes entirely obliterated, and the red K-line alone remains visible. By ignition and treatment with HC acid, nearly all manganese oxides of natural occurrence give a momentary Ba-spectrum.

manganese ores, Pyrolusite and Manganite. The other minerals, mentioned in the Table, are either rarely met with, or otherwise they present merely a sub-metallic lustre, and therefore come properly under examination in a succeeding Table.

Graphite occurs chiefly in foliated or sub-granular masses, more rarely in hexagonal tables. Its dark colour, flexibility, greasy feel, and property of marking and soiling, are among its more salient characters. The only mineral which might be mistaken for it, is the sulphide Molybdenite. The latter is much lighter in colour, and is at once distinguished by the pale green or yellowish-green coloration which it imparts to the outer edge of a Bunsen or other flame.

Magnetite is sufficiently distinguished by its magnetism, and by its black colour and streak. When crystallized, it is commonly in octahedrons, more rarely in rhombic dodecahedrons. Franklinite and Chromite are closely related to it, but possess, as a rule, merely a sub-metallic lustre, and their streak is more or less brown in colour. Chromite, moreover, gives BB with borax a chrome-green glass; Franklinite, with carb. soda, a strong manganese-reaction.

Hæmatite presents many varieties, but that which properly belongs to this Table is the variety known as Specular Iron Ore. This is commonly in dark steel-grey, laminar, crystalline, or scaly masses, cherry-red in the streak. The crystals are rhombohedral combinations, often with largely developed basal plane. $R:R = 86°10'$; $B:R = 122°30'$. The scaly variety crumbles under the fingers; the massive and crystalline varieties scratch glass.

Ilmenite is closely related to Hæmatite, and closely resembles the latter in crystallization and general characters, but is usually darker in colour, with blackish, or indistinctly red, streak. It is best distinguished by the amethystine colour produced in its hydrochloric-acid solution by boiling with tin. The student must remember, however, that many examples of magnetite and hæmatite are titaniferous to some extent, and with these the reaction would also be obtained.

Pyrolusite occurs commonly in iron-black or dark steel-grey fibrous masses, sufficiently soft to soil the hands. It produces chlorine fumes when warmed with hydrochloric-acid, and the smallest fragment gives with carb. soda, BB, a strong reaction of manganese in the form of a turquoise-enamel.

Manganite is also of a dark steel-grey or iron-black colour. It occurs commonly in groups of prismatic crystals or in coarsely-fibrous masses. The crystals belong to the Rhombic System, and are frequently twinned. $V:V = 99°40'$. Its acid and blowpipe reactions, generally, are the same as in Pyrolusite, but it differs from the latter species by yielding water (9-10 per cent.) in the bulb-tube.

TABLE IX.

[Lustre sub-metallic. Readily fusible or reducible *per se*.]

A.—Wholly or partly volatilizable by ignition on charcoal.

A^1.—ENTIRELY VOL.

(Hg reaction).

CINNABAR (HgS). Some dark or lead-grey varieties. Streak red. G 8-9. Inflammable. *See* TABLE XI.

(Sb fumes and coating).

KERMESITE (Red Antimony Ore): $Sb^2S^3 70$, $Sb^2O^3 30$. Dark blueish-red, with cherry-red streak. Rh. (chiefly acicular and fibrous); H 1·5; G 4·5. Melts in candle-flame. *See*, also, TABLES XI., XV.

A^2.—PARTLY VOL., A LARGE SILVER-GLOBULE REMAINING.

(Sb fumes and coating).

MIARGYRITE: Ag 36·7, Sb 41·5, S 21·8. Clino-Rh. H 2-2·5; G 5·18-5·26. Iron-black with cherry-red streak.

PYRARGYRITE (Dark Red Silver Ore): Ag 59·8, Sb 22·5, S 17·7. Hemi-Hex.; H 2-2·5; G 5·75-5·85. Dark lead-grey, reddish-black; streak cherry-red. *See* Note, below.

(As fumes).

PROUSTITE (Light Red Silver Ore): Ag 65·46, As 15·15, S 19·39. Red, blueish-red; streak bright-red. Lustre, properly, non-metallic. *See* TABLE XIV.

RITTINGERITE: Ag (57·7 p. c.) with As, or with Sb, S or Se (?) Clino-Rh.; H 2·5-3·0; G 5-6·3; iron-black with variegated tarnish; reddish or yellow by transmitted light; streak, orange-yellow.

POLYBASITE: Ag, Cu, As, Sb, S. Iron-black; red in thin pieces by transmitted light. Streak, red, black. *See* TABLES III., IV.

A^3.—PARTLY VOL., A CUPREOUS GLOBULE REMAINING.

COVELLINE (Indigo Copper Ore): Cu 66·46, S 33·54. Hex. (but commonly massive, nodular, &c.); H 1·5-2; G 4-4·6. Dark coppery-blue, blackish-blue, with black streak. Inflammable.

CHALKOSINE (Copper Glance): Lustre sub-metallic in occasional examples, only. Dark iron-grey, usually with greenish coating in patches. G 5·6. *See* TABLE III.

A⁴.—PARTLY OR WHOLLY VOL., WITH PRODUCTION OF LEAD GLOBULE AND LEAD COATING ON CHARCOAL.

PLATTNERITE: Pb 86·6, O 13·4. Iron-black; Hex. (pseudomorphous after Pyromorphite ?); H 3-4?; G 9·4.

B.—Non-volatile on ignition.

B¹.—REDUCIBLE, BB, TO METALLIC COPPER.

CUPRITE (Red Copper Ore): Cu 88·8, O 11·2. Reg.; H 3·5-4; G 5·7-6. Dark red, sometimes with blueish or lead-grey tinge. Streak, red. Surface often altered to green carbonate. Tile-ore is an impure var. mixed with Fe^2O^3, &c.

TENORITE (Black Copper-oxide). Cu 79·85, O 20·15. Mostly massive. H 2-3; G 5·9-6·5. Blackish steel-grey, iron-black.

B².—FUSIBLE INTO A MAGNETIC BEAD.

(G 7·7·5. Readily dissolved, BB, by Phosphor-salt. With carb. soda, strong Mn reaction).

WOLFRAM: FeO, MnO, WO^3, in somewhat variable proportions: the WO^3, 76-76·5 p. c. Clino-Rh.; H 5·5·5; G 7·1-7·55. Dark brown, brownish black, with brownish streak. *See* Note, below.

Samarskite—Scarcely fusible. Black. *See* TABLE X.

(SiO^2 reaction with Phosphor-salt. Gelatinizing in hot hydrochloric acid).

ALLANITE—ILVAITE or LIEVRITE—FAYALITE: Black, brownish or greenish-black. Lustre, properly, non-metallic. *See* TABLE XXVI.

NOTE ON TABLE IX.

Omitting the silicates, Allanite, Ilvaite, &c., the lustre of which is properly non-metallic, the commonly occurring minerals of this Table comprise : Cinnabar, Kermesite, Pyrargyrite, and Proustite, all of which give a marked sulphur-reaction with carb. soda on charcoal ; the red, copper-suboxide Cuprite ; and the tungstate, Wolfram.

Cinnabar presents a sub-metallic lustre in occasional examples only. Most commonly it has a red colour and non-metallic aspect. Its ready inflammability and high sp. gr. (8 – 9) serve at once to distinguish it from the other red minerals of the Table. It forms no deposit on charcoal, but yields readily a grey sublimate of metallic mercury if strongly ignited in a closed tube with dry carb. soda, iron-filings, or other reducing agents. *See* also, the Note to TABLE XI.

Kermesite resembles Cinnabar as regards rapid volatilization, but it forms on charcoal a dense white coating of Sb^4O^3 or Sb^2O^5, and its sp. gr. does not exceed 4·6. It occurs commonly in tufted groups of acicular crystals, or in radiated fibrous examples. In a hot solution of caustic potash it is rapidly converted into an orange-red powder.

Pyrargyrite and Proustite are closely akin by crystallization and chemical formulæ; but Pyrargyrite is very dark in colour, and it emits, BB, dense antimonial fumes (commonly accompanied by arsenical odour); whilst Proustite is distinctly red, with commonly an adamantine or non-metallic lustre and certain degree of translucency, and it is essentially a sulpharsenite. Both occur commonly massive, or in small (usually indistinct) crystals of the Hexagonal System, the more frequent forms comprising a combination of hexagonal prism and rhombohedron, and scalenohedral combinations. Twins and hemimorphous examples are common. Both species fuse *per se* when held against the edge of a candle-flame. The powder becomes immediately black in a hot solution of caustic potash. Hydrochloric acid precipitates orange-brown Sb^2S^3, or yellow As^2S^3, from the solution. *See*, also, Notes to TABLES IV. and XIV.

Cuprite is separated from the preceding minerals by yielding no sulphur-reaction before the blowpipe. It occurs frequently in octahedrons and rhombic dodecahedrons, with green coating of malachite covering the entire surface of the planes; more rarely in acicular shapes arising from elongated cubes. It is also frequently in massive examples. It dissolves in nitric acid with strong effervescence and production of orange-red nitrous fumes, the Cu^2O being converted into CuO at the expense of some of the oxygen of the acid. The solution is of course green or blue in colour, and becomes intensely blue on sufficient addition of ammonia.

Wolfram is readily distinguished by its dark-brown or black colour, and high sp. gr. (over 7). It occurs massive, and very frequently in somewhat large crystals of the Clino-Rhombic System; mostly, flattened six-planed prisms (composed of the forms V and V́) terminated by a sharply sloping base and several polar planes. $V:V = 100°37'$; $V:V = 140°18'$; $B:V = 118°6'$. It fuses into a magnetic globule with crystalline surface. Melted, in powder, with carb. soda and nitre in a platinum spoon, it forms an alkaline tungstate soluble in hot water, the bases remaining for the greater part undissolved. The solution (which at first is green from some dissolved manganate of soda) when boiled with hydrochloric acid and a piece of tin or zinc, becomes rapidly colourless, and then assumes a deep indigo-blue colour.

TABLE X.

[Lustre sub-metallic, Infusible; or fusible on thinnest edges only.]

A — Yielding Sulphur-reaction with carb. soda on charcoal.

(*Zn reaction*).

SPHALERITE OR ZINC BLENDE: Zn 67, S 33. Reg.; H 3·5-4; G. 3·9-4·2. Brown, black, red, &c.; streak light-brown; lustre in most examples, non-metallic, but sub-metallic in many dark varieties.

(*Mn reaction*).

ALABANDINE: Mn 63·2, S 36·8. Reg.; H. 3·5-4; G 4; black, brownish, dark steel-grey. Streak greenish, Becomes greyish-green on ignition. Scarcely fusible, but slags upon surface and edges in prolonged heat. No sublimate in closed tube.

HAUERITE: Mn 46·2; S. 53·8. Reg., crystals small, parallel-planed hemihedrons; H 4; G 3·46. Dark red-brown, brownish black; streak brownish or brownish red. In closed tube turns green and gives sublimate of sulphur.

B.—Magnetic before or after ignition.

B¹.—ANHYDROUS.

MAGNETITE (Magnetic Iron Ore): Fe 72·41, O 27·59, = Fe O. 31, Fe^2O^3 69. Reg.; H 5·5-6·5; G 4·9-5·2. Iron black, with black streak. Strongly magnetic, often showing polarity. Diamagnetite (of Shepherd) in long rhombic prisms is probably pseudomorphous after Lievrite (Dana). Pseudomorphs in rhombohedrons, after Spathic iron ore, also occur.

FRANKLINITE: ZnO, MnO, FeO; Fe^2O^3, Mn^2O^3, in variable proportions, but giving the common formula RO, R^2O^3. Reg.; H. 6-6·5; G 5·0-5·1. Black, with brownish streak. Often strongly magnetic.

CHROMITE (Chromic Iron Ore): FeO, MgO, CrO; Al^2O^3, Cr^2O^3, Fe^2O^3 = RO, R^2O^3. Reg.; H 5·5; G 4·4-4·6. Black, brownish or greenish black; streak blackish brown to nearly black. Sometimes magnetic.

HÆMATITE (Red Iron Ore): Fe 70, O 30 (= Fe^2O^3). Hemi-Hex., H (ordinary examples) 5·5-6·0; G 5-5·3. Steely-red, bluish-red, with cherry-red streak.

ILMENITE (Titaniferous Iron Ore): $Fe^2O^3 Ti^2O^3$ in variable proportions. Hemi-Hex.; H 5·5-6; G. 4·5-5·3. Black, brownish-black; streak black to brownish-red. *See* Note to TABLE VIII.

B².—YIELDING WATER ON IGNITION IN BULB-TUBE,

TURGITE: Fe^2O^3 94·7, H^2O 5·3. H 5-5·5; G 3·55-4·7; black, reddish-brown, streak dull red. Hydrohematite is identical or closely related.

GŒTHITE: Fe^2O^3 90, H^2O 10. Rh.; H 5-5·5; G 3·8-4·2. Dark brown, streak brownish yellow. Lepidochrocite and Stilpnosiderite are merely varieties, usually containing 3 or 4 p. c. more aq, and thus passing into ordinary Brown Iron Ore.

LIMONITE or BROWN IRON ORE: Fe^2O^3 8-5·6, H^2O 14·4. Massive fibro-botryoidal, &c., often in pseudomorphs after cubical pyrites and other ferruginous species. H. commonly, 5-5·5, but often lower; G 3·5-4. Aspect sub-metallic in some varieties only. Brown, brownish black; streak brownish-yellow. *See*, also, TABLE XXIII.

C.—Not Magnetic after ignition.

C¹.—READILY DISSOLVED (IN POWDER) BY HOT HYDROCHLORIC ACID, WITH PRODUCTION OF CHLORINE FUMES.*

(B. B. strong Mn reaction).

BRAUNITE: Mn 69·2, O 30·8. A little BaO is often present as in most manganese ores, and many impure varieties are strongly siliceous. Tet.; H. 5·5-6·5; G 4·7-4·9. Brownish-black, with similar streak.

HAUSMANNITE: Mn 72, O 28, but BaO, SiO^2, &c., commonly present as impurities. Tet.; H 5-5·5; G. 4·7-4·9. Black, brownish black, with dark-brown streak. Braunite and Hausmannite are comparatively rare, closely related, species. The crystals are small Tetragonal octahedrons, often twinned.

PYROLUSITE: MnO^2. Black; soils; H 2-2·5. Aspect commonly metallic. Fibrous. *See* TABLE VIII.

MANGANITE: $Mn^2O^3 + H^2O$. Steel-grey, iron-black; H 3·5-4. Aspect commonly metallic: *See* TABLE VIII.

PSILOMELANE: Iron-black, dark steel-grey; H. 5-6. Gives aq in bulb-tube. Aspect commonly metallic beneath dark surface tarnish. *See* TABLE VIII.

---

* Recognized unmistakably by the odour. The student should become familiar with this by warming a little black oxide of manganese with hydrochloric acid.

CHALCOPHANITE: MnO, ZnO, H^2O. Hemi-Hex.; H. 2·5; G 3·9. Blue-black. BB. becomes reddish or copper-coloured.

(Strong Cu reaction).

CREDNERITE: CuO 43, Mn^2O^8 57, but generally impure from presence of BaO, SiO^2, &c. Iron-black, streak black. The hydrochloric acid solution is green or bluish, and becomes deep blue on addition of ammonia, Mn^2O^3 gradually precipitating.

C².—NO CHLORINE FUMES PRODUCED BY TREATMENT WITH HYDROCHLORIC ACID. Sp. Gr. OVER 2·0.

(Decomposed or attacked by hot sulphuric acid).[*]

COLUMBITE: FeO, MnO, Nb^2O^5, Ta^2O^5, &c. Rh.; H 6; G 5·37-6·5. Iron-black, brownish-black. Streak reddish or greyish-black. Commonly yields a little tin by blowpipe reduction.

SAMARSKITE: YO, FeO, CeO, U^2O^3, Nb^2O^5, Ta^2O^5, &c. Rh.; H 5·6; G 5·6-5·8; black; streak red-brown. Diff. fusible into steel-grey mass. Nohlite (with 4·6 aq) is regarded as an altered variety.

POLYCRASE: YO, CeO, ErO, &c., with TiO^2, Nb^2O^5, and small percentage of water. Rh.; H 5·6; G 5·5·15. Black; streak brownish.

ÆSCHYNITE: CeO, LaO, YO, &c., with TiO^2, Nb^2O^5, ThO^2, &c., and 1 or 2 p. c. aq. Rh.; H 5-5·5; G 5-5·25; black, dark-brown; streak brownish.

MENGITE: Fe^2O^3, ZrO^2, TiO^2, &c. Rh.; H 5-5·5; G 5·48. Black; streak dark-brown.

POLYMIGNITE: YO, CaO, FeO, ZrO^2, TiO^2, &c. Rh.; H 6·5. G 4·75-4·85. Black; streak blackish-brown.

PYROCHLORE: CaO, CeO, Na^2O, Fl, ThO^2, Nb^2O^5, TiO^2, &c. Reg.; H 5; G 4·18-4·37. Blackish or reddish-brown, with light brown streak. Fusible on edges into a yellowish slag. Generally yields a little aq in bulb tube.

PEROWSKITE: CaO 40·6, TiO^2 59·4. Reg., with cubical cleavage. H 5·5; G. 4-4·1. Iron-black, yellowish, with metallic adamantine lustre.

WARWICKITE: MgO, FeO, B^2O^3, TiO^2. Clino-Rh.; H. 3-4; G 3·2-3·5. Brown, black, reddish, with dark streak. When moistened with sulphuric acid, or glycerine, imparts green colour to flame.

PITCHBLENDE—Slightly attacked by sulphuric acid. *See* below.

---

[*] The solution diluted slightly and boiled with addition of hydrochloric acid and a piece of zinc or tin, assumes a blue, greenish, or violet colour (from presence of Ta, Nb, or Ti).

(Not attacked, or very slightly attacked, by sulphuric acid).

PITCHBLENDE (Pitch Uran Ore, Nasturan): UO, U^2O^3 (?) with various impurities. Reg. (?); H (usually) 5-6; G 6·5-8. Black, brownish-black, with black or dark brown streak. Commonly yields a little aq on ignition. Decomposed in powder, by nitric acid, forming a yellow solution. *See* TABLE XXIII.

CASSITERITE (Tinstone): Sn 78·6, O 21·4. Tet.; H 6-7; G 6·5-7·1. Black, brown, greyish, &c. Lustre, as a rule, non-metallic: *See* TABLE XXIV. BB., with reducing flux, yields metallic tin.

TANTALITE: FeO, MnO, Ta^2O^5, Nb^2O^5, &c. Rh.; H 6-6·5; G 6·3-8 (usually about 7). Iron-black; streak dark-brownish. Commonly gives BB with reducing flux a little tin. TAPIOLITE is apparently a Tetragonal Tantalite.

YTTROTANTALITE: YO, ErO, FeO, CaO, Ta^2O^5, WO^3, &c., with 4-6 p. c. aq, but the latter probably a product of alteration. Rh.; H 5-5·5; G (as regards the black sub-metallic varieties) 5·4-5·7; black, brownish-yellow. Becomes yellow and yields aq in bulb-tube. With reducing flux gives generally a little tin. HJELMITE is a related tantalate, containing SnO^2, WO^3, &c. G 5·82. Black.

FERGUSONITE: YO, ErO, CeO, FeO, &c., with Nb^2O^5 and Ta^2O^5, and 1-7 p. c. aq. Tet.; H. 5·5-6; G 5·6-5·9. Black, blackish-brown, with pale brown streak. Tyrite and Bragite are varieties.

EUXENITE: YO, CeO, UO, &c., with TiO^2, Nb^2O^5, and 2-3 p. c. aq. Rh.; H. 6·5; G 4·6-5; black, brownish-black; streak, red-brown. Burns brownish-yellow and yields aq by ignition in bulb-tube.

[NOTE.—The Nio-tantalates and Nio-titaniates of this and the preceding section are for the greater part very imperfectly known, and all are of rare occurrence. Several have probably little claim to rank as distinct species.]

RUTILE: Ti 61, O 39. Tet.; Crystals commonly prismatic, and often in geniculated twins; sometimes acicular. H 6-6·5; G 4·2-4·3; red, with metallic-adamantine lustre; more rarely black (Nigrine), or yellowish; streak pale brown.

ANATASE or OCTAHEDRITE: Ti 61, O 39. Tet., crystals commonly pyramidal, of small size. H 5·5-6; G 3·8-4; dark indigo-blue, greyish, brownish, with, in general, adamantine lustre.

BROOKITE: Ti 61, O 39. Rh.; H 5·5-6; G 4-4·25. Hair-brown, reddish, yellowish, black (Arkansite). Comparatively rare.

C³.—NOT ATTACKED BY ACIDS. SPECIFIC GRAVITY UNDER 2.

ANTHRACITE : Carbon, with small amounts of H, O, and N ; hygroscopic moisture, and inorganic matter or "ash" (1 to over 20 p. c.) being also present in most examples. H 3 (or 2·5-3·25) ; G 1·2-1·8 ; black, often iridescent in places ; streak greyish-black.

NOTE ON TABLE X.

Excluding the manganese ores, Pyrolusite and Manganite, the lustre of which is essentially metallic (see TABLE VIII), the more commonly occurring minerals of this table comprise the following species : (1) the iron ores, Magnetite, Franklinite, Chromite, Hæmatite, Ilmenite, and Limonite ; (2) The sulphide Sphalerite or Zinc Blende ; (3) The tin ore, Cassiterite ; (4) The two forms of Titanic anhydride, Rutile and Anatase ; and (5) the coal variety, Anthracite.

As regards the iron ores, Magnetite and Franklinite are strongly magnetic in their natural condition ; the others occasionally are feebly magnetic, but all attract the magnet strongly after ignition in the R. F. Magnetite is frequently in large masses, and also in regular octahedrons and rhombic dodecahedrons. Both colour and streak are black. Thin splinters may be fused at the extreme point. Franklinite is commonly in small rounded masses imbedded in crystalline limestone with red zinc ore, &c., less commonly in cubes and octahedrons, or in large masses. Its streak is reddish-brown. BB, with carb. soda it gives Mn and Zn reactions. Some examples are said to be slightly magnetic only. Chromite is almost always in granular masses of a black colour. Its sp. gr. is much lower than that of Magnetite and Franklinite ; and it forms with Borax a fine green glass, by which it is readily distinguished from the above species. The student must remember, however, that mixtures of these iron ores often occur.

Hæmatite is essentially distinguished by its cherry-red streak or powder. It is commonly in granular, slaty, or fibro-botryoidal masses. Its crystals generally present a strongly marked metallic lustre. They are mostly rhombohedral combinations with largely developed basal plane (*See* note to TABLE VIII.) Ilmenite is a titaniferous hæmatite, usually of dark colour and dark streak. Its crystals resemble those of hæmatite, but the interfacial angles are slightly different. It is best distinguished by the amethystine colour produced in its hydrochloric acid solution by boiling with tin or zinc.

Limonite or Brown Iron Ore is distinguished by its ochre-yellow streak, and by yielding water in the bulb-tube. It is commonly in dark brown masses of granular or fibrous structure. The surface is often iridescent. Frequently also it is found in coarse, brown cubes, and other pseudomorphous crystals, after iron pyrites. Light-brown examples also occur, but these present a silky or other non-metallic aspect. (*See* TABLE XXIII).

Zinc Blende is at once distinguished from other minerals of the Table—the very rare manganese sulphides excepted—by the sulphur reaction which it yields with carb. soda. Its powder warmed with hydrochloric acid also emits the odour of sulphureted hydrogen. Commonly in cleavable masses of a black-brown, dark-red or yellowish colour, or in groups of crystals (mostly tetrahedrons, or combinations of rhombic dodecahedron and tetrahedron) of the Regular System. A dark ferruginous variety (which becomes magnetic after ignition) has been named Marmatite; and a cadmiferous var. (mostly in dark sub-fibrous masses) is termed Przibramite. (*See* also the note to TABLE XVI.)

Cassiterite or Tinstone scarcely belongs to the present table, as in most examples the lustre is essentially non-metallic. Its great weight and hardness, tetragonal (often twinned) crystallization, and its property of yielding tin globules by reduction with mixture of carb. soda and borax, are its more distinctive characters.

Rutile and Anatase (two of the natural representatives of binoxide of Titanium, the comparatively rare Brookite being a third representative of that compound), have in most examples a non-metallic (adamantine) lustre, with a certain degree of translucency. But some examples are opaque. Rutile resembles Cassiterite (and also Zircon, TABLE XXIV.) in its crystallization. The crystals are commonly composed of two square prisms (forming a pseudo-8-sided prism) with pyramidal terminations. The prism-planes are striated vertically in most cases, and the basal plane (as in Zircon) is constantly wanting. Geniculated twins are common. The colour is generally dark brownish-red or blood-red, but light-brown and other tints also occur. Anatase occurs in small pyramidal crystals, usually composed of two or several square octahedrons, the more common one having the angle over a polar edge $= 97°50'$, and over a middle edge $= 136°36'$. Prism planes and basal plane are also occasionally present, and some crystals are tabular from predominance of the latter. The colour is usually indigo-blue, brown, or greyish-blue. Both Rutile and Anatase, when fused in fine powder with caustic potash (or with carb. soda and borax), are attacked or dissolved by hydrochloric acid, the diluted solution becoming of a deep amethystine tint when boiled with metallic tin.

Anthracite is at once distinguished from other minerals of the Table by its low specific sp. gr. (1·2-1·8). The lustre, moreover, is properly non-metallic. *See* TABLE XXV.

TABLE XI.

[Aspect non-metallic. Readily inflammable: * burning with sulphurous or alliaceous odour.]

A.—Burning with sulphurous odour.

(Streak, yellow).

NATIVE SULPHUR: Rh.; H 1·5-2·5; G 1·9-2·1; yellow, brownish, reddish-yellow. See Note, below.

(Streak, red or brown).

CINNABAR: Hg 68·2, S 13·8. Hemi-Hex.; H 2-2·5; G (normally) 8-9, but often lower in dark carbonaceous varieties. Red with red streak; but sometimes brown from admixture with carbonaceous matter.

IDRIALINE: A mixture of Cinnabar with earthy matter and C^3H^2. Brownish-black; streak brown or reddish. H 1-1·5; G 1·4-1·6.

KERMESITE: (Sb, S, O). Inflammable in some varieties only; mostly fibrous or acicular. G 4·5. See TABLES IX, XV. BB, copious antimonial fumes.

(Streak, black).

COVELLINE: Cu 66·46, S 33·54. Hex.; H 1·5-2; G 4-4·6. Dark coppery-blue, blackish-blue. BB, copper reaction.

B.—Burning with alliaceous (arsenical) odour.

(Colour, yellow).

ORPIMENT: As 6, S 39, Rh.; H 1·5-2; G 3·4-3·5. Bright yellow, commonly with metallic-pearly lustre; streak yellow. In thin pieces, flexible.

(Colour, red).

REALGAR: As 78, S 30. Clino-Rh. H 1·5-2; G 3·5-3·6. Red, streak orange-yellow.

NOTE ON TABLE XI.

The principal minerals of this Table are N. Sulphur, Orpiment, Realgar, and Cinnabar. The latter is distinguished more especially by its high sp. gr. and its red streak.

---

* To test this property, a small piece of the mineral may be taken up by the steel forceps and held for an instant against the edge of a Bunsen-flame or the flame of a common candle.

Native sulphur, when crystallized, is commonly, in acute rhombic-octahedrons of small size. It occurs generally in indistinct druses, massive or efflorescent on pyrites, &c. It melts into red-brown drops which become pale yellow on cooling. From Orpiment, which is equally inflammable, it is distinguished by its low sp. gr. and by the absence of arsenical odour during combustion.

Orpiment is occasionally in small prismatic crystals, but occurs generally in foliated or other examples. It dissolves entirely in caustic potash, and is re-precipitated from the solution by hydrochloric acid.

Realgar is distinguished from Cinnabar by its orange-yellow streak, as well as by its lower sp. gr., and the arsenical odour evolved on combustion. Its crystals are small Clino-Rhombic prisms with largely developed basal plane, but are generally in druses, or otherwise indistinct. Most commonly it occurs in granular or other masses. In caustic potash it leaves a brown residuum of sub-sulphide. Otherwise like Orpiment.

Cinnabar is the essential ore of mercury. Under normal conditions it presents a scarlet red colour (whence its old name of Native Vermilion) and unchanged streak, but the surface is usually brownish, and many examples are dark-brown from intermixed earthy or bituminous matter (Liver Ore, &c.) The crystals are combinations of rhombohedrons and hexagonal prism, the triangular basal plane being especially apparent. Tetartohedral forms have been recognized, but in general the crystals are small, and more or less indistinct. Cinnabar occurs more commonly in granular masses, and occasionally in thin coatings or incrustations. Metallic mercury is easily sublimed from it by ignition with dry carb. soda, iron filings or other reducing agents, in a small flask or test-tube. Scarcely attacked by caustic potash, or by nitric or hydrochloric acid. Soluble in aqua regia.

TABLE XII.

[Aspect non-metallic. Inflammable in candle flame, burning with bituminous or aromatic odour.]

A —Coaly, ligneous, or pitch-like aspect. Burning with bituminous odour.

BITUMINOUS COAL : C 74-96, H 0·5-5·5, O 3-20. Black, often iridescent ; streak, black. H 2-2·5 ; G 1·2-5.

LIGNITE or BROWN COAL: C 55-80 ; H 3-6 ; O 17-27. Dark-brown or black (jet) with brown streak. H 2-2·5 ; G 1·2-1.4. Massive, ligniform, sometimes foliated (Paper Coal), and earthy. Imparts a brown colour to caustic potash. "Torbarnite" is sometimes referred to this variety, but it is properly a mere bituminous shale.

BITUMEN or ASPHALT : C, H, O. Black, greenish-black. H 0·5-2·0 ; G 1·0-1·2. Semi-fluid or pasty in ordinary examples, also in stalactitic and other more or less brittle masses with conchoidal fracture. Passes into Petroleum.

ALBERTITE : C, H, N, O. Black, highly lustrous, brittle. H 2-2·5 ; G 1-1·1. Scarcely attacked by alcohol, but partially dissolved by oil of turpentine. STELLARITE and GRAHAMITE are related substances.

ELATERITE (Elastic Bitumen) : C, H, O. Dark-brown or black. Soft and flexible, resembling caoutchouc. Passes into ordinary bitumen. G 0·8-1·2. DOPPLERITE is a closely related substance.

B.—Resinous (or when dark coloured somewhat coaly) in aspect, but burning with aromatic (non-bituminous) odour.

PIAUZITE : dark-brown, with yellowish-brown streak. H 1·5-2 ; G 1·18-1·22. Soluble in ether and in caustic potash. Pyroretine is apparently related.

AMBER (Succinite, Bernstein) C, H, O (= C 79, H 10·5, O 10·5?). Yellow, brownish, reddish, greyish-white. Mostly in nodular masses. H 2-2·5 ; G 1·0-1·1. Electric by friction.

RETINITE, KRANTZITE, IXOLITE, SIEGBURGITE, PYROPISSITE, and other obscurely known amber-like substances, belong also to this group.

C.—Wax-like in aspect.

OZOKERITE (Neftgil): Essentially C 85·7, H 14·3 (2-3 per cent. O present in some examples). Green, brownish (by transmitted light, yellowish or red). Very soft, pasty; G 0·95. Emits *per se* an aromatic odour. Easily sol. in oil of turpentine. Scarcely or slowly sol. in ether and alcohol.

PARAFFINE—URPETHITE—HATCHETTINE—GEOCERITE—GEOMYCERITE—EUOSMITE: Greyish-white to brownish yellow, soft wax-like substances, more or less readily soluble in ether.

HARTITE.—A white or brownish crystalline, wax-like substance, soluble in ether. *See* under D, below.

D.—Crystalline in aspect.

FICHTELITE: C 87·13, H 12·87. In white, pearly, crystalline laminæ, soluble in ether. After fusion, becomes again crystalline on cooling. TEKORETINE (Clino-Rhombic) is identical.

SCHEERERITE (Könleinite): C and H. In white acicular or lamellar crystals (Clino-Rhombic). G 1-1·2. Dissolves readily in ether, but rapidly separates again.

HARTITE: C and H. In soft paraffine-like, white or brownish crystalline lamellæ, or small (anorthic) crystals. H 1·0-1·5; G slightly over 1·0. Largely soluble in ether. BOMBICCITE is a related crystalline (anorthic) compound, but is said to contain nearly 15 per cent. O. Easily soluble in ether and in alcohol.

NOTE ON TABLE XII.

The substances included in this Table are essentially hydro-carbon compounds, probably in great part (or wholly, according to the common view), of organic origin. The absolutely organic nature of asphalt and other bituminous substances, remains, however, yet to be proved. Many other compounds enumerated by chemists might have been referred to in the Table; but the composition of these hydro-carbons appears to be more or less variable, and their physical characters, in most instances, cannot be very rigorously defined. The more common representatives of the Table comprise—Bituminous Coal, Brown Coal, and Amber. The latter occurs mostly in nodular or irregular masses of a light or deep yellow colour, but is sometimes greyish-white or brownish, and frequently clouded. Some examples are quite transparent, others only translucent, and many are quite opaque. Leaves and insects are frequently enclosed in these nodules, and thus amber is usually regarded as a coniferous gum or resin of Cainozoic age. Fraudulent imitations of insect-holding amber are often imposed, however, on the unwary. Like other resinous bodies, amber is rendered strongly electrical by friction.

Bituminous coals generally leave, by ignition in closed vessels, a semi-fused agglutinated coke. These are commonly known as "caking coals." In brown coals, proper, the coke remains unfused. In all kinds of coal, sulphur (from pyrites, and occasionally from gypsum,) is present more or less ; and all coals contain a certain amount of intermixed earthy matter or "ash." This latter may vary from 2 or 3 to 10 or 15 per cent , but many coals pass into coal shales, when the amount of earthy matter (essentially a silicate of alumina) may exceed 50 per cent. All coals, moreover, contain hygroscopic moisture, varying (according to conditions of exposure, &c.,) from about 3 or 4, to over 10 or 12 per cent., or higher in many brown coals. See Appendix to "Blow-pipe Practice," page 76, On the Examination of Coals by the Blowpipe.

TABLE XIII..

[Non-metallic aspect. Readily sol., BB., in phosphor-salt. Effervescing in diluted hydrochloric acid. (N. B.—The acid in some cases must be gently heated.)]

A.—Yielding metallic globules, per se, or with carb. soda on charcoal.

A¹—ANHYDROUS SPECIES. NO WATER, OR FAINT TRACES ONLY, IN BULB-TUBE.

(No reaction of S or Cl).

CERUSSITE: PbO, 83·52, CO^2 16·48 = Pb 77·6. Rh.; H 3-3·5; G (normally) 6·4-6·6, but lower in impure earthy varieties. Colourless, or grey, nearly black, yellowish, &c.; streak white. IGLEASITE is a zinc-holding variety.

PLUMBO-CALCITE: = Plumbiferous var. of Calcite or Calc Spar. TARNOWITZITE = Plumbiferous var. of Arragonite. G about 2·8. Both give a lead sublimate on charcoal, but metallic globules are not readily obtained.

(S reaction).

LEADHILLITE: PbO, CO^2 72·56, PbO, SO^3 27·44 = Pb 75. Rh.; H 2·5-3; G 6·2-6·6. Yellowish-white, grey, brownish, &c., streak white. SUSANNITE is a supposed rhombohedral variety (G 6·55). MAXITE is probably an altered var., containing a small percentage of water.

CALEDONITE: PbO, CuO, SO^3 (CO^2 by alteration or admixture). Light-green. See Table XVI.)

(Cl reaction).

PHOSGENITE (Kerasine): PbO, CO^2 49, $PbCl^2$ 51, = Pb 73·8. Tet.; H 2·5-3; G 6-6·3. Yellowish-white, grey, yellow, green; streak white.

A²—YIELDING WATER ON IGNITION.

(Cu reaction).

MALACHITE: CuO 71·95, CO^2 19·90, H^2O 8·15. Clino-Rh., but rarely crystallized. H 1·0-4; G 3·7-4. Green, often zoned in different shades; streak light-green. Some varieties are calcareous. ATLASITE is a variety containing copper chloride.

AZURITE (Chessylite): CuO 69·2, CO^2 25·6, H^2O 5·2. Clino-Rh.; H 1-4; G 3·7-3·8. Blue, paler in the streak.

See also TIROLITE, Table XIV, many examples of which contain intimately intermixed carbonate of lime. Green or blue radiated masses, or earthy. BB, strong arsenical odour.

(Cu and Zn reactions).

AURICHALCITE : CuO 28, ZnO 46, $CO^2$16, H^2O 10 (?). Acicular or fibrous. H 2 ; G about 3·3. Green or bluish; streak paler. BURATITE is a calcareous variety.

(Bi reaction).

BISMUTITE : BiO, CO^2, H^2O. H 4-4·5 (?) G 6·8-6·9 (?). Yellow, grey, green; streak paler. A doubtful species, more or less variable in characters and composition.

B.—No metallic globules obtained by fusion with carb soda on charcoal.

B¹—ANHYDROUS SPECIES. NO WATER, OR TRACES ONLY, IN BULB TUBE.

(NOTE:—The presence of Ca, Ba, Sr, singly or together, in carbonates of this group, is very readily ascertained by a small, direct vision spectroscope. *See* Outline of Blowpipe Practice, pp. 55, 57.

† *Magnetic after ignition.*

SIDERITE (Spathic Iron Ore): FeO 62, $CO^2$38, = Fe 48·2 ; part of the FeO, however, often replaced by MgO, MnO, CaO. Hemi-Hex.; H 3·5-4·5 ; G 3·7-4·1 ; yellowish-grey, yellow, brown, olive-green, &c., streak paler. SPHEROSIDERITE is a fibrous-spherical variety from trap rocks ; CLAY-IRONSTONE, BLACK BAND, &c., are impure argillaceous or bituminous varieties from coal strata. SIDEROPLE-SITE, MESITINE and PISTOMESITE (G 3·3-3·6) are crystalline magnesian vars. ; and OLIGON SPAR, a variety containing 25·5 p. c. of MnO CO^2. In the typical rhombohedron, R : R = 107°, whilst in the Mg and Mn examples it varies from 107°3′ to about 107°18′. Crystals, however, commonly present curved planes.

ANKERITE : (CaO, MgO, MnO, FeO) CO^2. Hemi-Hex., with RR about 106°12′. White, yellowish, brownish; streak, in unweathered examples, white. H 3-4 ; G 2·9-3·3. Merges into Siderite, Calcite, and Dolomite.

See also dark-coloured varieties of MAGNESITE and DOLOMITE.

†† *Not magnetic on ignition, and no marked alkaline reaction.*

(Strong reaction of Mn).

RHODOCHROSITE (DIALLOGITE, MANGANESE SPAR): MnO 61·74, CO^2 38·26, but MnO often in part replaced by CaO and MgO. Hemi-Hex, with R : R (normally) 106°5'. H 3·5-4·5 ; G 3·3-3·6. Rose-red, pink-brownish when weathered; streak very pale red, reddish-white. Blackens on ignition. RŒPPERITE is a calcareo-magnesian variety.

(Co reaction).

SPHÆROCOBALTITE (Cobalt spar): CoO 63, CO^2 37. H 4 ; G 4·0-4·1. In spherical concretions, black externally, red within. A doubtful species.

(Zn reaction).

SMITHSONITE (Calamine, Zinc Spar): ZnO 64·8, $Co^2$35·2. Hemi-Hex., with R : R = 107°49'. H 5 ; G 4·4·5. Colourless, pale-greyish, greenish, brownish; streak, white. Many varieties contain FeO and MnO. HERRERITE is a cupreous variety.

†↓† *Alkaline reaction, after strong ignition.*

(Ba reaction: flame coloured pale-green).

WITHERITE : BaO 77·67, CO^2 22·33. Rh., with pseudo-hexagonal aspect. H 3-3·5 ; G 4·2-4·4. Colourless, pale-grey, yellowish; streak white. BB, entirely soluble in carb. soda.

ALSTONITE (Bromlite) : BaO, CO^2 66·33 + CaO, CO^2 33·67. Rh. ; H 4-4·5 ; G 3·6-3·8. Colourless, greyish; streak white. BB, only in part sol. in carb. soda.

BARYTO-CALCITE : Composition and general characters as in ALSTONITE; but crystallization Clino-Rhombic, with V : V 84°52'.

(Sr reaction: crimson flame-coloration).

STRONTIANITE : SrO 70·27, CO^2 29·73. Rh. (V : V = 117°19') ; H 3·5 ; G 3·6-3·8. Colourless, greenish, yellowish, &c.; streak white. Some varieties are more or less calcareous; others (Stromnite) contain baryta.

(Ca reaction: flame, after prolonged ignition of test-substance, coloured red).

CALC SPAR or CALCITE: CaO 56, $CO^2$44. Hemi-Hex, with rhombohedral cleavage (R : R 105°5' ; or varying from about 105 to 105° 18', part of the CaO being commonly replaced by MgO, FeO, &c). H (normally) 3, but often lower; G 2·6-2·8. Colourless or variously tinted; streak white. *See* Note below.

DOLOMITE (Bitter Spar): CaO, CO^2 54·35, MgO, CO^2 45·65, but often more or less ferruginous, &c. Hemi-Hex (R : R 106°15'—106° 20'); H 3·5-4 ; G 2·8-3·0. Colourless, yellowish, brownish, &c. The varieties containing FeO are commonly called Brown Spar. Through these there is a complete transition into Ankerite and Siderite. GURHOFIAN and KONITE are impure silicious varieties, with H = 4·5-5·5.

ARAGONITE : CaO 56 ; CO^2 44. Rh. (V : V 116°10'); H 3·5-4 ; G 2·7-3·0, normally 2·94. Colourless, light-yellow, brown-violet, reddish, greenish ; streak white. Commonly falls into powder on ignition. Some examples contain a small percentage of strontia. *Flos Ferri* is a coralloidal var., accompanying iron ore at certain localities. TARNOVITZITE is a highly plumbiferous variety.

(My reaction : No flame coloration, if pure ; reddened by ignition with cobalt-solution).

MAGNESITE : MgO 47·62, CO^2 52·38, but part of MgO commonly replaced by FeO, CaO, &c. Hemi-Hex. (R : R 107°16'—107°29'). H 3-4·5, or lower ; G 2·8-3·1. Colourless, snow-white, yellow, greyish, &c. ; streak white. GIOBERTITE is merely crystallized Magnesite.

B²—HYDROUS SPECIES, YIELDING WATER BY IGNITION IN CLOSED TUBE.

† *Soluble or partly sol. in water.*

NATRON: Na^2O 22, CO^2 15, H^2O 63. Clino-Rh. (V: V = 79°41'), but chiefly earthy and efflorescent. H 1-1·5 ; G 1·4-1·5. Normally colourless.

THERMONATRITE : Na^2O 50, CO^2 35·5, H^2O 14·5. Rhombic, mostly in rectan. tables ; H 1·5 ; G 1·5-1·6. Normally colourless.

TRONA: Na^2O 38, CO^2 40, H^2O 22. Clino-Rh.; H 2-3; G 2·1-2·2. Normally colourless. Commonly mixed with $NaCl$.

GAYLUSSITE: Na^2O CO^2 35·50; CaO, CO^2 34·08, H^2O 30·42. Clino-Rh (V : V 68·51'). H 2·5 ; G 19·4 ; colourless. Slowly, and only in part, soluble in water.

†† *Insoluble in water. Giving BB with borax an uncoloured or very lightly-tinted glass.*

GAYLUSSITE : Partly sol. *See above.*

HYDROMAGNESITE : MgO 44, CO^2 36·2, H^2O 19·8. Clino-Rh., or Rh. (V : V 87°—88°), but commonly massive or earthy ; white ;

H 1·3·5 ; G 2·14-2·18. Some of the earthy varieties give only 4 or 5 p. c. water on ignition. BAUDISSERITE is an impure silicious var. LANCASTERITE, according to Smith and Brush, is a mixture of Hydromagnesite and Brucite. Hydrodolomite, in white or yellowish spherical masses from Vesuvius, is a compound of Hydromagnesite with Calcite or Dolomite.

HYDROZINKITE (Zink Bloom): ZnO 75·24, CO^2 13·62, H^2O 11·14. In white or yellowish earthy or oolitic masses, or efflorescent on zinc ores. G 3·25.

DAWSONITE: A compound (or mixture produced by alteration?) of Al^2O^3, CaO, Na^2O, CO^2 and H^2O. In colourless, thin-bladed aggregations or coatings on compact trachyte, Montreal. H 3 ; G 2·4. HOVITE, in white earthy crusts, is apparently related in composition.

TENGERITE : YO, CO^2, H^2O. In white or yellowish earthy crusts on certain examples of Gadolinite.

LANTHANITE : LaO 52·6, CO^2 21·3, H^2O 26·1. Rh. (V : V 92°50′ —94°), generally tabular. H 2·5-3·0 ; G 2·67 ; greyish or yellowish-white, pale red. BB, with borax, a pink or pale violet bead, apparently from the presence of Didymium.

†‡† *Insoluble in water. Giving, with borax, a strongly-coloured glass.*

WISERITE : MnO, CO^2, H^2O. In yellowish or pale-red fibrous coatings on certain examples of Hausmannite and other manganese ores.

ZARATITE (Texasite) : NiO, CO^2, H^2O. In thin emerald-green coatings on nickle ores. Also on examples of Chromic Iron Ore from Texas, Penn.

REMINGTONITE : CoO, CO^2, H^2O. In pinkish, or greyish-blue coatings on cobalt ores.

LINDAKERITE (Calc-Uran Carbonate). In coatings and crusts on Pitchblende. Yellowish-green. Contains (according to Lindaker) UO 37·03, CaO 15·55, CO^2 24·18, H^2O 23·24. VOGLITE is a cupreous variety. LIEBIGITE is also a closely related compound, but with 45 p. c. aq. All occur in connection with pitchblende.

NOTE ON TABLE XIII.

The more important minerals of this Table comprise : (1) Calcite, Dolomite, Magnesite, Siderite, Rhodochrosite, and Smithsonite, of the group of

Rhombohedral Carbonates; (2) Aragonite, Witherite, Strontianite, and Cerussite, of the group of *Prismatic Carbonates;* and (3), the *Cupreous Carbonates,* Malachite and Azurite.

Calcite, in its crystalization, chiefly affects three series of forms : (i) Rhombohedrons, acute and obtuse ; (ii) Scalenohedrons ; and (iii) Hexagonal Prisms, the latter commonly terminated by the three planes of a rhombohedron, pentagonal in shape in some cases, rhombohedral in others. The basal plane, when present, is usually rough or dull. Some of the more common rhombohedrons comprise: $-\frac{1}{2}R$ (polar angle 135°); $-2R$ (polar angle 79°); and $4R$ (p. a. 66°). The most common scalenohedron has the following interfacial angles : over long polar edge 159°24′ ; over shorter polar edge 138°5′ ; over middle edge 64°54′. All crystals and lamellar examples cleave readily into a rhombohedron of about 105°5′ and 74°55′, but these angles vary to within about 30′ in consequence of isomorphous replacements, a small portion of the lime carbonate being almost constantly replaced by carbonate of MgO, FeO, or MnO. Transparent examples show strong double refraction in the direction of the longer diagonal of a rhombohedral face. Pseudomorphs, after Orthoclase, Fluor Spar, Barytine, Celestine, Gypsum, Gaylussite, &c., are not uncommon. Calcite occurs likewise in rock-masses, forming crystalline limestone (marble), ordinary limestone, oolitic limestone, chalk, &c., and in various stalactitic, tufaceous, and other conditions. Calcite, after simple ignition (without the aid of hydrochloric acid, although it is always advisable to add a drop of this), shews the red and green calcium lines in the spectroscope very distinctly.

Dolomite much resembles calcite in its general characters and rhombohedral crystallization, but it dissolves, as a rule, in cold acids with comparatively feeble effervescence. Both hardness and sp. gr. are also slightly higher. The most certain method of distinction is the determination of magnesia in the hydrochloric acid solution. For this purpose the diluted solution is first boiled with a drop or two of nitric acid, and ammonia is then added in slight excess. This will cause a slight flocculent precipitate if iron be present. Oxalate of ammonia is then added to precipitate the lime ; this is filtered off ; the filtrate tested with another drop of oxalate of ammonia to make sure that all the lime has been thrown down, and the magnesia is precipitated by some dissolved phosphor-salt. It can be collected, if necessary, and ignited with nitrate of cobalt for the production of the characteristic flesh-red tinge. Many so-called limestones when examined in this manner are found to be "dolomitic." Ferruginous varieties of Dolomite pass into Ankerite.

Magnesite is comparatively rare in crystals, but occurs commonly in more or less compact or granular masses, beds, or layers of a white, pale-grey or yellowish colour. The small rhombohedrons show over a polar edge the angle 107°16′ to 107°29′. The powder by ignition with a drop of cobalt solution, is distinctly reddened. The absence of lime can be proved by the spectroscope ; and the presence of magnesia by the cobalt test or by precipitation, as explained under Dolomite.

Siderite or Spathic Iron Ore occurs under various conditions : crystallized in metallic veins, &c. ; fibro-botryoidal ; in spherical concretions in basaltic rocks; pisolitic in Jurassic and other strata; massive; and lithoidal. The crystals are usually small rhombohedrons of a yellow colour (with R : R 107°, but frequently with curved faces), also acute rhombohedrons and scalenohedrons. The spheroidal basaltic variety is usually dark-green or yellowish-brown, with radio-fibrous structure. The pisolitic variety, dark-brown or grey, and opaque; and the lithoidal and massive examples dark-grey, brown or black, and also opaque. These latter kinds commonly occur in oval or nodular masses in coal strata, or in layers mixed with coaly matter. Under the name of Clay Iron-Stone, Black Band, &c., they furnish a large part of the iron of commerce, but are always very impure from admixture with clay, silica, &c. They are also more or less altered, as a rule, into brown iron ore. The nodules, when split open, are usually found to contain the impression of a fern-frond or other organic body.

Rhodochrosite or Manganese carbonate is of less frequent occurrence than the preceding carbonates. Its crystals are mostly small rhombohedrons (with usually curved faces) sometimes shewing a triangular basal plane (R : R 106° 51'—107°); but it occurs commonly in botryoidal, granular or lamellar masses, of a pink or rose-red colour, with dark-brown altered patches. As in Magnesite and Siderite, it effervesces feebly unless the acid be heated. Its red colour and intense manganese reaction, BB, with carb. soda, generally serve to distinguish it at once from other carbonates; but many examples of Magnesite, Siderite, &c., give a more or less strongly-marked manganese reaction. No very definite lines of demarcation, in fact, can be drawn between the rhombohedral carbonates generally.

Smithsonite, or zinc carbonate, occurs mostly in aggregations of minute rhombohedrons, or in botryoidal or incrusting examples of a white, brownish, grey, yellowish, or green colour. It is usually more or less vitreous and transparent; but is sometimes in opaque, grey or brown, earthy or porous masses. The streak is white, and the hardness just sufficient to scratch glass; or sufficient, at least, to scratch fluor spar very strongly. In powder, with a mixture of carb. soda and borax, it yields on charcoal a sublimate of ZnO,—bright-yellow and phosphorescent, hot; white, cold; and light-green after ignition with cobalt solution.

Aragonite—the typical representative of the group of Prismatic carbonates—is identical in composition with the rhombohedral calcite. It occurs frequently crystallized, and in fibrous, coralloidal, and other masses. The crystals belong to the Rhombic System, and are generally six-sided prisms, composed of four V planes with the two side planes of a brachy-prism $\breve{V}$, terminated by a brachy-dome $\breve{P}$, and by the planes of a rhombic octahedron P; but the latter form is often absent. V : V = 116°10'; V : $\breve{V}$ = 121°55'; $\breve{V}$: $\breve{P}$ = 125°47'. Twins and compound crystals are very common. Some of the latter, composed of three or more individual crystals, are strikingly pseudo-hexagonal in character, presenting the appearance of a simple six-sided prism with large base. The colour is white, yellow, brownish-violet, &c. All examples dissolve with strong effervescence in cold acids, and show, after moderate ignition, the characteristic red and green calcium lines in the spectroscope.

Witherite, carbonate of baryta, also presents in its crystallization a pseudo-hexagonal aspect. The crystals are, very generally, six-sided pyramids, but are regarded as compound crystals, made up of interpenetrating rhombic-octahedrons. Columnar, botryoidal, and massive examples are however its principal forms of occurrence. Its high sp. gr. (over 4·0), and the green colour which it imparts to the flame border, sufficiently distinguish it from other carbonates.

Strontianite, like Witherite, is entirely dissolved by fusion with carb. soda; and its sp. gr. is comparatively high (3·6-3·8). It is readily distinguished, however, by the intense crimson coloration which it communicates to the flame-border, and by the characteristic blue, orange, and red lines, of its spectrum. Its crystallization is identical with that of Arragonite, and is characterized by pseudo-hexagonal combinations and twin forms (V : V = 117°19′; V : $\ddot{\text{V}}$ = 121°20′30″; V : 2$\overset{\cup}{\text{P}}$ = 145°22′). Strontianite occurs more commonly, however, in columnar, fibrous, granular and other examples.

Cerussite, or lead carbonate, is also identical in crystallization with Aragonite, and is particularly characterized by its stellate and cruciform groups (V : V = 117°14′, $\ddot{\text{V}}$: 2$\overset{\cup}{\text{P}}$ = 145°20′). The lustre is strikingly adamantine. This character, with the high sp. gr. (6·5) of the species, its remarkable fragility, and its blowpipe reactions, sufficiently distinguish it.

The copper carbonates, Malachite and Azurite, yield water on ignition, and are otherwise distinguished by their deep green and blue colours, and their copper reactions. Malachite (although often, as a product of alteration, entirely coating octahedrons and dodecahedrons of red copper ore, Cu^2O) is very rarely crystallized, but occurs commonly in botryoidal, fibrous and massive examples, and as an earthy coating on copper ores generally. Azurite, the blue carbonate, is frequently in groups of small clino-rhombic crystals, more or less indistinct in form. It occurs also in columnar and other masses, and in earthy coatings on copper ores.

TABLE XIV.

[Aspect non-metallic. BB, on charcoal, arsenical fumes or odour.]

A —Entirely volatilizable, or leaving only a minute residuum.

(Streak white).

ARSENOLITE (Arsenious acid): As 75·8, O 24·2. Reg.; H 1-2; G 3·7; in white, crystalline or acicular groups and coatings, and in earthy crusts. CLAUDETITE (Dana) is a rhombic species, in small sub-pearly laminæ. G 3·85.

(Streak black).

NATIVE ARSENIC, weathered examples. In dull, black, earthy masses, often coating the metallic-grey or tin-white unaltered metal. See TABLE I.

B —Yielding, BB, metallic globules on charcoal. (A mixture of carb. soda and borax assists the reaction).

(BB, a silver globule).

PROUSTITE (Light-red Silver Ore). Ag 65·46, As 15·15, S 19·39. Hemi-Hex.; H 2-2·5; G 5·4-5·6; red, more or less translucent, with adamantine lustre; streak, red.

XANTHOCONE: Ag 64·08, As 14·83, S 21·09. Hemi-Hex., mostly tabular. H 2-2·5; G 5·0-5·2. Orange or brownish-yellow, translucent or transparent, with adamantine lustre. Streak orange-yellow.

RITTINGERITE: Normally AgAs (with 57·7 Ag), but S commonly present. Clino-Rh.; H 2·5-3; G 5-6·3. Iron-black, red by transmitted light; streak orange-yellow. Lustre in general strongly sub-metallic. See TABLE IX.

POLYBASITE, arsenical varieties. Ag (64-74), Sb, As, S; Rh.; H 2·5; G 6·0-6·2. Iron-black, red in thin pieces by transmitted light; streak, commonly dark-red. Lustre, usually metallic. See TABLE IV.

All the above arsenical silver ores fuse *per se* in the flame of a candle, without the aid of the blowpipe. Rittingerite and Polybasite are still imperfectly known.

(Cu reaction).

OLIVENITE: CuO 56·15, As^2O^5 40·66, H^2O 3·19. Rh. (V:V 92°30'); H 3; G 4·3-4·6; dark-green, brownish; streak, paler.

EUCHROITE: CuO 47·15, As^2O^5 34·15, H^2O 18·70. Rh. (V : V 117°20′); H 3-4; G 3·3-3·5; emerald-green, leek-green; streak, paler. CHLOROTILE is closely related.

ERINITE: CuO 60, As^2O^5 34·6, H^2O 5·4. Mostly in concentric-lamellar examples; H 4·5; G 4·0; emerald-green; streak, paler.

TIROLITE (Kupferschaum): CuO 50·32, As^2O^5 29·15, H^2O 20·53. Mostly in radio-fibrous mammillary examples. H 1-2; G 3·1. Green or greenish-blue; streak, paler. Most examples are intimately mixed with CaO, CO^2. The presence of Ca, readily shewn by spectroscope.

CLINOCLASE (Abichite, Aphanese, Strahlerz): CuO 62·65, As^2O^5 30·25, H^2O 7·10. Clino-Rh.; H 2·5-3; G 4·2-4·4; dark-green, bluish-green, blackish externally; streak, paler.

LIROKONITE (Linsenerz): CuO, As^2O^5, Al^2O^3, H^2O (25 per cent.). Clino-Rh.; H 2-2·5; G 2·8-2·95; light-blue, sometimes green; streak, paler.

CHALCOPHYLLITE (Copper Mica): CuO, As^2O^5, Al^2O^3, H^2O (23-32 per cent.). Hemi-Hex., tabular, micaceous. H 2; G 2·5; bright emerald-green; streak, paler.

ZEUNERITE: CuO 7·71, U^2O^3 55·95, H^2O 14. Tet., isomorphous with Chalcolite or Torbernite. H 3·5; G 5·76; orange or wax-yellow, with adamantine lustre; streak, paler.

ADAMITE: Cupreous varieties. Green. G 4·35; zinc sublimate with carb. soda on charcoal. *See* below.

(*Pb reaction*).

MIMETESITE: PbO, As^2O^5 90·7; $PbCl^2$ 9·3. Hex. (crystals often sub-spherical). H 3·5-4; G 7-7·3; yellow, green, greyish, colourless, with resino-adamantine lustre. KAMPYLITE and HEDYPHANE (G 5·5) are more or less calcareous and also phosphatic varieties. Some of the orange-yellow examples contain lead chromate. All give Cl reaction with phosphor-salt and CuO.

ARÆOXENE: PbO, ZnO, V^2O^5, As^2O^5. Radio-fibrous; H 3; G 5·8; brownish-red; streak, yellow.

CARMINITE (Karminspath): PbO 23·62, Fe^2O^3 29·14, As^2O^5 47·24. Acicular, mammillated. H 2·5; G 4·1; red; streak, reddish-yellow.

BEUDANTITE: PbO, Fe^2O^3, P^2O^5, As^2O^5, SO^3, H^2O. Hemi-Hex. (?) H 3·5; G 4·0. Olive-green; streak, yellowish. A doubtful species.

(Bi reaction).

RHAGITE: Bi^2O^3 79.5, As^2O^5 15.6, H^2O 4.9. Mostly botryoidal or in small spherical examples. H 4.5-5; G 6.82; light-green; streak, very pale green or white. In bulb-tube crumbles into yellow powder. Accompanies uran ores at Schneeberg.

WALPURGINITE: Bi^2O^3, U^2O^3, As^2O^5, H^2O (4-5). Clino-Rh.?; orange or wax yellow, with resino-adamantine lustre; streak, paler; H 3.5; G 5.76. Accompanies uran ores at Schneeberg. ATELESITE is apparently related.

C.—No metallic globules, BB, on charcoal.

(Zn reaction. Characteristic sublimate with carb. soda on charcoal).

ADAMITE: ZnO 56.6, As^2O^5 40.2, H^2O 3.2, but some green examples contain CuO, and red examples, CoO. Rh.; H 3.5; G 4.3-4.35. Normally, yellow; but often violet, red, or green; streak, paler.

KÖTTIGITE: Zinc-holding var. of ERYTHRINE. See below.

(Co reaction).

ERYTHRINE (Cobalt Bloom): CoO 37.56, As^2O^5 38.40, H^2O 24.04. Clino-Rh.; H 2.5; G 2.9-3.0. Red, purplish-red; streak, paler. Some earthy varieties contain intermixed arsenolite. KÖTTIGITE is a zinc-holding var.

ROSELITE: CaO, MgO, CoO, As^2O^5, H^2O (8.20). Rh. or Cl.-Rh. H 3-3.5; G 3.46; deep rose-red; pale-reddish or white streak. The presence of Ca easily shewn by the spectroscope.

CABRERITE: A cobaltiferous var. of ANNABERGITE. See below.

(Ni reaction).

ANNABERGITE (Nickel Green): NiO 37.25, As^2O^3 38.59, H^2O 24.16. Acicular, efflorescent; H 1-2.5; G 3; apple-green, greenish-white. CABRERITE is a variety containing CoO and MgO. Green and yellowish anhydrous nickel arseniates have also been recognized (G 4.9).

(Fe reaction, BB, magnetic slag or bead).

PHARMACOSIDERITE (Cube Ore). Fe^2O^3 40, As^2O^5 43.13, H^2O 16.87. Reg. *(See* Note at end of TABLE). H 2.5; G 2.9-3; dark-green, yellow, brownish,; streak, paler. Mostly in minute cubes tetrahedrally modified.

SCORODITE: Fe^2O^3 34·63, As^2O^5 49·78, H^2O 15·59. Rh.; H 3·5-4; G 3·1-3·3 ; dark-green, brownish, indigo-blue ; streak, paler.

ARSENIOSIDERITE : Fe^2O^3 39·4, CaO 13·8, As^2O^5 37·9, H^2O 8·9. Fibrous-botryoidal. H 1-2; G 3·9; brownish-yellow, with silky lustre.

SYMPLESITE : FeO, Fe^2O^3, As^2O^5, H^2O (25-28 per cent.). Clino-Rh., acicular; H 2·5.; G 2·9-3·0 ; pale blue, green, with pearly lustre.

PITTIZITE : Fe^2O^3, As^2O^5, SO^3, H^2O (12-29 per cent.). Amorphous, stalactitic. H 2·5-3; G 2·3-2·5; brownish-yellow, dark-brown; streak, paler.

CARMINITE ; BEUDANTITE : Contain PbO. See above.

(*MnO reaction*).

CHONDRO-ARSENITE : MnO, MgO, CaO, As^2O^5, H^2O (7·8 per cent.). In small granular concretions of a wax yellow colour. H 3·0.

BERZELITE : Gives Mn reaction in most examples; Ca-lines in spectroscope; no water. See below.

DURANGITE : Strong Na and F reactions. Orange-red. See below.

(U^2O^3 *reaction*).

TROEGERITE : U^2O^3 65·95, As^2O^5 17·55, H^2O 16·50. Clino-Rh., tabular; H 2-2·5 ; G 3·23-3·27; lemon yellow; streak, yellowish-white. Easily fusible.

URANOSPINNITE : U^2O^3 59·18, CaO 5·47, As^2O^5 19·37, H^2O 16·19. Tetr. ? scaly or thin tabular; H 2-2·5 ; G 3·45 ; yellowish-green.

(*MgO and CaO reactions. Ca-lines well shewn in spectroscope*).

BERZELITE (KUHNITE) : CaO, MgO, MnO, As^2O^5. Massive; H 5 (or 4·5); G 2·5-2·55 ; yellow, yellowish-white. Nearly infusible. No water evolved in bulb-tube.

PHARMACOLITE ; CaO 24·90, As^2O^5 51·10, H^2O 24. Clino-Rh., but mostly acicular, fibrous, earthy, &c.; H 1·5-2·5 ; G 2·73. Normally colourless or white. Easily fusible.

HAIDINGERITE: CaO 28·81, As^2O^5 56·87, H^2O 14·32. Rh.; G 2·9; otherwise like Pharmacolite, but of rare occurrence.

WAPPLERITE : CaO, MgO, As^2O^5, H^2O (18-20 per cent.). Clino-Rh.; H 1·5-2·5 ; G 2·5 ; colourless or white. Very easily fusible.

HOERNESITE is a related, but purely magnesian, arseniate (with H^2O 29 per cent.), recognized by Kengott in the kaiserlichen min. Cabinet of Vienna.

(*Na reaction*).

DURANGITE (J. G. Brush): Na^2O, Li^2O, Al^2O^3, Fe^2O^3, MnO, As^2O^5, F. Clino-Rh.; H 5; G 3·94-4·07; orange-red. Easily fusible. With sulphuric acid, fluorine reaction. Hitherto, only recognized as accompanying tin ore and colourless topaz in the Province of Durango, Mexico.

NOTE ON TABLE XIV.

This Table is composed essentially of arseniates. The exceptions comprise a few silver sulpharsenites in which the lustre is mostly non-metallic, and the naturally occurring arsenious acid or anhydride As^2O^3. The only minerals of the Table likely to come under ordinary observation, include: (1) The "Light-Red Silver Ore," Proustite; (2) The Cupreous Arseniates—Olivine, Clinoclase, Liroconite, Chalcophyllite and Tirolite; (3) The Cobaltic Arseniate, Erythrine; (4) The Ferruginous Species, Pharmacosiderite and Scorodite; (5) The Lime Arseniate, Pharmacolite; and (6) the Lead Chloro-Arseniate, Mimetesite.

Proustite or light-red silver ore, the arsenical silver blende of some nomenclatures, is readily recognized by its deep or bright red colour, red streak and adamantine lustre; as well as by the large silver-globule obtained from it by the blowpipe. It frequently accompanies Native Arsenic. It occurs both crystallized and massive. The crystals are generally small, and are not always readily made out in consequence of distortion by irregularity in the size of corresponding planes. Commonly, they consist of hexagonal prisms terminated by a rhombohedron (with $R : R = 107°50'$), or of scalenohedrons. Small fragments melt in the candle flame, without the aid of the blowpipe. Boiled with caustic potash, the powder becomes immediately black, and As^2S^3 is dissolved. This is thrown down, as a yellow flocculent precipitate, by a drop or two of hydrochloric acid.

The copper arseniates are green, or more rarely blue, in colour, and, as a rule, they detonate or deflagrate somewhat strongly when ignited on charcoal. Olivine and Clinoclase are usually dark-green or blackish-green (though sometimes brown or brown-yellow), and both occur frequently in small crystals, and in radiated-fibrous, reniform, and other uncrystallized examples. The Olivine crystals are rhombic, and the Clinoclase crystals clino-rhombic combinations. Clinoclase is almost constantly in radiated groupings, whence its old German name of *Strahlerz*. Olivine yields only 3·20 per cent. aq. Clinoclase 7 per cent. Liroconite is very usually of a light-blue colour, though sometimes green. It occurs mostly in very small clino-rhombic crystals which present in general an ortho-rhombic aspect, and sometimes resemble slightly distorted octahedrons. In the bulb-tube it yields (without decrepitation) a large quantity of water (25-26 per cent.). Chalcophyllite is rarely in distinct crystals, but generally in micaceous or thin tabular examples of a bright emerald-green colour, with metallic-pearly lustre on the broad surfaces of the laminæ. In the bulb-tube it decrepitates strongly and yields a large amount of water (23-32 per cent.). Tirolite or Tyrolite is unknown in crystals. Most commonly it occurs in bright green or blue radiated examples, or in reniform or fine scaly masses. Thin foliæ are flexible. The specimens hitherto examined contain 13-14 per cent. carbonate of lime, either in combination or as an intermixture. The presence of Ca is readily shewn by the spectroscope, especially if the copper be first reduced by fusion with carb. soda on charcoal, and the resulting slag be moistened with a drop or two of hydrochloric acid. The amount of water equals 20-21 per cent.

Erythrine, the cobaltic arseniate, is especially distinguished by its peach-blossom red colour, and by the deep-blue glass which it forms by fusion with borax. It occurs in small clino-rhombic crystals, but more commonly in bladed, acicular and efflorescent examples. The thin foliæ are flexible. Easily fusible. Water, 24 per cent.

The ferruginous arseniates, Pharmacosiderite and Scorodite, distinguished from the cupreous and other arseniates by the magnetic slag which they yield, BB, on charcoal, are distinguished individually by their crystallization. Pharmacosiderite is almost always in very minute cubes, truncated on alternate angles by the triangular planes of the tetrahedron. Its colour is dark-green, passing into brownish-yellow and brown, and the little crystals are usually in drusy aggregations. Scorodite when crystallized is commonly in small prisms terminated by an acute rhombic pyramid, but it occurs also frequently in fibrous and other examples. The colour is dark-green or indigo-blue, inclining to reddish-brown in some specimens. The hardness exceeds that of calcite, whilst Pharmacosiderite is slightly under calcite in hardness.

Pharmacolite, the ordinary lime arseniate, is comparatively unimportant. It occurs mostly as a white efflorescence, or in acicular crystals, on arsenical cobalt and iron ores.

Mimetesite, chloro-arseniate of lead, is readily distinguished from other minerals of the Table by its high sp. gr. (7·0-7·3), as well as by the lead globule which it yields, BB, on charcoal. It belongs by its crystallization and chemical formula to the Apatite group, and often passes into Pyromorphite, the corresponding lead phosphate. The crystals, hexagonal prisms, or combinations of prism and pyramid, are very commonly curved into almost globular shapes. The colour is generally yellow, more rarely grey, brown, or green, with resino-adamantine lustre. Fused in the platinum forceps, the bead crystallizes on cooling, but on charcoal it becomes reduced.

TABLE XV.

[Lustre non-metallic. BB, on charcoal, antimonial fumes and deposit.]

A —On charcoal, reducible to metallic antimony and rapidly volatilized.

(S reaction with carb. soda).

KERMESITE (Pyrostibite, Red Antimony Ore, Antimony Blende). Sb^2S^3 70, Sb^2O^3 30. Red, bluish or brownish red, with red streak and adamantine lustre. Rh. (or Clino Rh. ?), but mostly acicular or fibrous; H 1-1·5; G 4·5-4·6. Fusible in candle-flame.

(No S reaction).

VALENTINITE: Sb 83·56, O 16·44. Rh., mostly tabular or acicular. H 2-3; G 5·3-5·6; normally white, but sometimes pale reddish or brownish from admixtures. Becomes yellow on ignition, and melts very easily. In the bulb-tube, sublimes entirely, if pure. SENARMONTITE has the same composition (Sb^2O^3) and general characters, but is Regular in crystallization. The crystals are commonly octahedrons, often with curved planes.

CERVANTITE: Sb^2O^3 47·40, Sb^2O^5 52·60. Rh. ? acicular, encrusting; H 3·0-4·0 (or 5?); G 4·08. Yellow, yellowish-white. Infusible, but reducible on charcoal. Not volatile in the bulb-tube.

(No S reaction; aq in bulb-tube).

ANTIMONY OCHRE: Sb^2O^3, mixed more or less with Sb^2O^5, and yielding H^2O on ignition. Earthy, encrusting; G 3·8; yellow, yellowish-white. Reduced and volatilized on charcoal.

STIBLITE: Sb^2O^3, Sb^2O^5, H^2O (5·6 per cent.). Compact, pseudomorphous after antimony glance. Yellow, yellowish-white. Reduced and volatilized on charcoal, the reduction (as in all compounds of $Sb^2O^3 + Sb^2O^5$) assisted by addition of carb. soda.

B.—On charcoal partially vol , a metallic globule remaining.

(Ag reaction).

PYROSTILPNITE (Fire Blende in part): Ag (62 per cent.) Sb, S. Clino-Rh. ? tabular, foliated. H 2; G 4·2-4·3; orange-yellow, brownish-red; streak, red or yellow; lustre pearly-adamantine. BB, antimonial fumes and large silver-globule.

PYRARGYRITE; POLYBASITE: Ag, Sb, S. Iron-black, or deep-red in thin pieces by transmitted light. Streak, red. Lustre essentially metallic or sub-metallic. See TABLES IV. and IX.

(*Cu reaction*).

RIVOTITE: CuO, Sb^2O^5, mixed with carb. lime, &c. A doubtful species. Compact; yellowish-green; H 3·5-4·0; G 3·55-3·62 (Ducloux).

(*Pb reaction*).

BINDHEIMITE (Bleiniere): PbO, Sb^2O^5, H^2O (6 per cent.). A doubtful compound. Massive, earthy, &c.; H 1-4; G 3·9-4·7. Greyish-white, yellowish, brownish, green, &c. Often veined or clouded in different tints.

NADORITE: PbO, Sb^2O^3 + $PbCl^2$. Rh., tabular. H 3·0; G 7·02; yellowish or greyish-brown. Hitherto found only in calamine deposits in Algeria.

C.—On charcoal partially vol., an earthy mass remaining.

ROMEITE: CaO 19·5, Sb 63·8, O 16·7. In groups of small tetragonal octahedrons of a yellow or reddish colour. H 5·5 (?), G 4·67-4·71. The presence of Ca in the residuum, left on charcoal after roasting, is easily recognized by the spectroscope. Part of the CaO is commonly replaced by MnO and FeO.

NOTE ON TABLE XV.

The minerals of this Table consist chiefly of rare or obscurely known examples of antimonial oxides, alone, or combined with lead oxide, &c. None of these compounds are of mineralogical importance. The only species of ordinary occurrence referred to in the Table is the mineral Kermesite or Pyrostibite, a compound of $2Sb^2S^3$ with Sb^2O^3. This occurs commonly in association with Antimony Glance. It is usually in radiating-fibrous or tufted plumose masses of a deep bluish-red or brownish-red colour, with red streak and adamantine (more or less sub-metallic) lustre. In caustic potash, the powder assumes a yellow colour, and on boiling is rapidly dissolved. Fusible and volatilizable in the candle-flame without the aid of the blowpipe.

TABLE XVI.

[Lustre non-metallic. BB, with carb. soda strong sulphur-reaction.]

A.—Anhydrous species. No water (or traces only) in bulb-tube.

A^1.—REDUCIBLE TO METAL *PER SE* OR WITH CARB. SODA.

(*BB*, *a lead globule*).

ANGLESITE: PbO 73·6, SO^3 26·4. Rh. (V : V 103°44′); H 3 (or sometimes slightly lower); G 6·1-6·4 (commonly 6·3); colourless, grey, yellowish, &c.; streak, white. Sol. in caustic potash. SARDIANITE is a supposed clino-rhombic species of similar composition.

LANARKITE: PbO, SO^3 57·6 + PbO 42·4. Clino-Rh.; H 2·0-2·5; G 6·5-6·9; pale greenish-white, yellowish, grey. Flexible in thin pieces. By alteration, partially converted into carbonate, and then effervesces in acids.

LEADHILLITE: PbO, CO^2 + PbO, SO; effervesces in acids. *See* TABLE XIII.

(*Pb and Cu reactions. Flame coloured strongly green. With carb. soda, lead sublimate. With boracic acid, copper globule*).

CALEDONITE; PbO, CuO, SO^3 (CO^2 and H^{2}O by alteration ?) Rh. (or Clino-Rh. ?) V : V 95°. Light bluish-green; streak, greenish-white; H 2·5-3; G 6·4. Generally effervesces in acids.

A^2.—NOT REDUCIBLE TO METAL, BB. ATTACKED OR DISSOLVED IN POWDER BY HOT HYDROCHLORIC ACID, WITH EMISSION OF H^{2}S ODOUR.

(*With carb. soda, zinc sublimate on charcoal*).

SPHALERITE or ZINC BLENDE: Zn 67, S 33. Reg. (mostly inclined hemihedral); H 3·5-4; G 3·9-4·2. Brown, black (often red by transmitted light), green, yellow, rarely colourless; streak, mostly pale-brown. Many yellow examples are phosphorescent by surface-abrasion. Practically infusible. The lustre varies from adamantine to sub-metallic and metallic proper. *See* TABLES III. and X. MARMATITE and CHRISTOPHITE are dark, ferruginous varieties.

VOLTZINE: ZnS 82·7, ZnO 17·3. H 3·5-4·0; G 3·5-3·8. Brownish-red, yellow, greenish; streak, pale-brown. Practically infusible.

(*With carb. soda, red-brown cadmium-sublimate*).

GREENOCKITE: Ca 77·8, S 22·2. Hex., hemimorphic (crystals mostly small acute pyramids, with lower half entirely replaced by basal plane). H 3-3·5 ; G 4·8-4·9. Yellow, orange, brownish, with yellow streak and adamantine lustre. Infusible. On ignition, becomes deep-red whilst hot, but generally decrepitates.

(*With carb. soda, strong manganese reaction*).

ALABANDINE : Mn 63·2, S 36·8. Black, brownish-black, with greenish streak, and, in general, sub-metallic aspect. No sublimate in closed tube. *See* TABLE X.

HAUERITE : Mn 46·22, S 53·78 ($= MnS^2$). Reg., parallel-hemihedral, and thus resembling Iron Pyrites in crystallization. Dark red-brown, brownish-black, with brownish-red streak, and, in general, sub-metallic lustre. In closed tube, turns green, and gives sublimate of sulphur. *See* TABLE X.

A^3.—NOT REDUCIBLE TO METAL. NO ODOUR OF H²S EVOLVED BY TREATMENT WITH HYDROCHLORIC ACID.* TASTELESS, INSOLUBLE.

† *Entirely diss·lved, BB, by carb. soda.*

(*Flame coloured apple-green*).

BARYTINE (HEAVY SPAR) : BaO 65·7, SO^3 34·3, a portion of the BaO sometimes replaced by SrO or CaO. Rh. (V : V 101°40'). H 3·5 ; G 4·3-4·7 ; colourless, white, yellow, flesh-red, brown, &c., with white streak. BB, generally decrepitates. Fusible into a white caustic enamel, the flame coloured pale-green. BARYTO-CELESTINE (G 4·24) is a mixture or isomorphous union of BaO, SO^3 and SrO, SO^3. BARYTO-CALCITE (G 4·0-4·3) = BaO, SO^3 + CaO, SO^3. As regards the latter, see below.

(*Flame coloured crimson*).

CELESTINE : SrO 56·52, SO_3 43·48. Rh. (V : V 103°40' - 104°10'); H 3-3·5 ; G 3·9-4·0 ; colourless, pale-blue, indigo-blue, yellowish, &c., with white streak. BB, generally decrepitates. Fuses into a white caustic enamel, and imparts a crimson coloration to the flame.

†† *In part, only, dissolved, BB, by carb. soda.*

ANHYDRITE : CaO 41·18, SO^3 58·82. Rh. (V : V 100°30'); H 3-3·5 ; G 2·8-3. Mostly in colourless, white, bluish, or reddish lamellar masses, with pearly lustre on cleavage planes ; streak, white. BB, fusible into a white caustic enamel. Colours the flame-border red, after prolonged exposure.

---

* The presence of Ba, Sr, and Ca, in minerals of this section, is readily determined by the spectroscope. *See* Outline of Blowpipe Practice, page 55.

BARYTO-CALCITE: A compound of the sulphates of baryta and lime. Properly, a calcareous var. of Heavy Spar, but referred to here as being only partially soluble, BB, in carb. soda, the lime remaining unattacked. G 4·0-4·3. Imparts a pale-green tint to the flame-border; but the orange-red Ca-line comes out prominently in the spectroscope.

A⁴.—SOLUBLE. SALT OR BITTER TO THE TASTE.
(Ammonia reaction. BB, entirely vol.)

MASCAGNINE: Am 39·4, SO^3 60·6. Rh. (V : V 121°8′), but chiefly in white or yellowish crusts or mammillated masses on certain lavas. H 2-2·5; G 1·7-1·8. Taste, sharp and bitter.

(Flame coloured violet. Red K-line well defined in spectroscope).

GLASERITE (ARCANITE): K^2O 54, SO^3 46. Rh. (V : V 120°24′), but mostly in white earthy crusts. H 2-5; G 2·7. Taste, bitter. BB, generally decrepitates, melts easily, and crystallizes over the surface on cooling. APTHALOSE is a rhombohedral potash sulphate from Vesuvius.

(Flame coloured intensely yellow. Na-line, only, in spectroscope).

THENARDITE: Na^2O 43·66, SO^3 56·34. Rh., but mostly in drusy or earthy crusts and coatings of a white or greyish colour; H 2·5; G 2·67. Taste, saltish, alkaline. Easily fusible, and on charcoal reduced to sulphide, and absorbed.

(In spectroscope, green and orange-red Ca-lines, and yellow Na-line).

GLAUBERITE: Na^2O, SO^3 51, CaO, SO^3 49. Clino-Rh.; H 2·5-3·0; G 2·7-2·8. Taste, saltish and bitter. White, grey, yellowish, red, &c. Somewhat deliquescent. BB, decrepitates, fuses easily, and becomes reduced to sulphide. In carb. soda, the lime remains undissolved. In water, only partially soluble.

B.—Hydrous compounds. Yielding water by ignition in bulb-tube

B¹.—FORMING, BB, WITH BORAX A PRACTICALLY UNCOLOURED BEAD.

† *Readily soluble in water, and possessing a bitter or other taste.*

(On charcoal, BB, entirely absorbed. Flame coloured intensely yellow).

MIRABILITE (GLAUBER'S SALT): Na^2O 19·3, SO^3 24·8, H^2O 55·9. Clino-Rh., but mostly efflorescent, &c.; H 1-2; G 1·4-1·5. Taste, cooling and bitter.

(*The residuum, left BB on charcoal, assumes by ignition with cobalt solution a fine blue colour*).

ALUM (Potash Alum, Kalinite): K^2O 9·95, Al^2O^3 10·82, SO^3 33·75, H^2O 45·48. Reg., octahedral, but commonly in white or greyish crusts, &c. H (crystals) 2-2·5; G 1·7-1·9. Red K-line in spectroscope.

SODA-ALUM : Na^2O, Al^2O^3, SO^3, H^2O. In white or greyish crusts, &c. BB, strong yellow flame, and yellow Na-line in spectroscope.

AMMONIA ALUM (Tschermigite) : Am, Al^2O^3, SO^3, H^2O. Earthy crusts. BB, partially vol. with strong ammoniacal odour. If pure, no lines in spectroscope.

MAGNESIA ALUM (Pickeringite): MgO, Al^2O^3, SO^3, H^2O. In white or greyish crusts, &c. If pure, no lines in spectroscope.

ALUNOGENE (*Hair-salt* in part) : Al^2O^3 15·4, SO^3 36, H^2O 48·6. In efflorescences of minute acicular crystals on various coals, shales, &c. Normally colourless, but often gre nish or brownish from admixture with iron-vitriol or iron-alum.

(*The residuum, left BB on charcoal, assumes by ignition with cobalt solution a flesh-red colour*).

REUSSIN : A compound of soda sulphate (Mirabilite) with 30-31 per cent. of magnesia sulphate. In white feathery crusts, &c. Colours flame intensely yellow.

EPSOMITE: MgO 16·26, SO^3 32·52, H^2O 51·22. Rh. (V : V 90°38') but commonly efflorescent, &c.; colourless ; H (crystals) 2-2·5 ; G 1·7-1·8. After strong ignition, gives alkaline reaction. If pure, no lines in spectroscope.

BLŒDITE (Astrakanite): Na^2O 18·65, MgO 11·95, SO^3 47·90, H^2O 21·50. Clino-Rh., but mostly in lamellar masses, crusts, &c. H (crystals) 2·5-3·5 ; G 2·2-2·3. White, grey, pale-reddish, greenish, &c. Colours flame intensely yellow. LŒWITE is a related soda-magnesia sulphate, but apparently distinct in crystallization, and with only 14·66 per cent. water.

KAINITE: MgO, SO^3 48·3, KCl 30, H^2O 21·7. Clino-Rh. (tabular), but commonly in granular masses, &c. H 2·5 ; G 2·13. Yellowish-white, greyish. BB, with phosphor-salt and CuO, strong chlorine reaction (azure flame). Part of the KCl sometimes replaced by NaCl.

(*The residuum, left BB on charcoal, assumes by ignition with cobalt solution a light-green colour*).

GOSLARITE : ZnO 28·22, SO^3 27·88, H^2O 43·90. Rh. (V : V 91°5'). H (crystals) 2-2·5 ; G 2·0-2·1 ; colourless, greyish-white. BB, with carb. soda, gives zinc sublimate on charcoal.

(*The residuum, on charcoal, with cobalt solution becomes on ignition dark-grey. In spectroscope, Ca (or Ca and K) lines*).

POLYHALLITE ; SYNGESITE ; ETTRINGITE ; KIESERITE : Soluble in part only, or very slowly sol. Taste, very feeble. *See* below.

†† *Insol. or very slowly sol. in water. Taste, 0, or very feeble.*

(*BB, imparts a green colour to the flame-point, and gives P^2O^5 reaction (yellow precipitate) with amm. molybdate in the nitric acid solution*).

SVANBERGITE : Na^2O, CaO, Al^2O^3, SO^3, P^2O^5, H^2O 6 per cent. Hemi-Hex. (RR about 88° or 90°); H 4·5 ; G 2·57. Yellow, orange-red. Very rare, and imperfectly known.

(*BB, with Co-solution, a bright blue colour*).

ALUMINITE (Websterite) : Al^2O^3 29·77, SO^3 23·23, H^2O 47. In white or yellowish-white earthy or porous masses which adhere to the tongue ; H 1·0; G 1·7-1·8. BB, infusible ; evolves SO^2. FELSO-BANYITE, in small groups of rhombic (pseudo-hexagonal) tabular crystals, is related in composition, but yields 38·67 per cent. water.

ALUNITE (Alumstone) : K^2O 11·33, Al^2O^3 37·10, SO^3 38·56, H^2O 13·01. Hemi-Hex. (R : R 89°10'), but commonly in granular masses. H 4-5 ; G 2·6-2·8. White, pale-reddish, yellowish, &c. Infusible; generally decrepitates. Evolves, on strong ignition, SO^2. LŒWIGITE is identical or closely related, but yields 18-18·5 per cent. water.

(*BB, with Co-solution, a flesh-red colour*).

KIESERITE : MgO 29, SO^3 58, H^2O 13, but commonly yields more water, from hygroscopic absorption. Clino-Rh., but commonly in fine-granular masses. H 3 ; G 2·57. Very slowly soluble in water. BB, infusible ; gives off SO^2.

(*BB, with Co-solution, a dark-grey colour. In spectroscope, Ca (or Ca and K) lines*)[*].

GYPSUM (Selenite) : CaO 32·54, SO^3 46·51, H^2O 20·95. Clino-Rh. (*See* note at close of Table). H 1·5 ; G 2·2-2·4. Colourless, white,

---

[*] These spectrum-lines come out most distinctly when the ignited test-substance is moistened by hydrochloric acid. *See* Outline of Blowpipe Practice, pp. 55-59.

pale-reddish, bluish, yellowish, &c.; streak, white; aspect vitreo-pearly in crystallized and lamellar examples, silky in most fibrous varieties, sometimes earthy. In thin pieces, somewhat flexible. BB, becomes immediately opaque, and fuses into an opaque white bead. On prolonged ignition, reacts alkaline, and tinges the flame-border distinctly red.

POLYHALLITE : K^2O, SO^3 28·93, CaO, SO^3 45·17, MgO, SO^3 19·92, H^2O 5·98. Rh., but commonly fibrous, lamellar, &c. H 3-3·5 ; G 2·7-2·77. Red, flesh-red, greyish, colourless. Partly sol. in water. Taste, feebly bitter. BB, very easily fusible into an alkaline (hollow) bead. Some examples give Cl-reaction with phosphor-salt and CuO.

SYNGENITE : K^2O, SO^3, CaO, SO^3, with 5·5 per cent. water. Clino-Rh. (tabular); H 2·5 ; G 2·6 ; colourless. BB, easily fusible, but generally decrepitates. Partly soluble in water. Distinguished chemically from Polyhallite by absence of MgO. (*See* Outline of Blowpipe Practice, page 55).

ETTRINGITE: CaO, Al^2O^3, SO^3, H^2O (45·82 per cent.). In delicate, silky, hexagonal prisms on the lava of the Laacher See. G 1·75. BB, swells up, but remains unfused. Partly soluble in water.

B².—FORMING, BB, WITH BORAX A STRONGLY COLOURED BEAD.

† *Soluble or partly sol. in water, and possessing a metallic or other taste.*
(*Cu reaction*).

CHALCANTHITE or BLUE VITRIOL. CuO 31·85, SO^3 32·07, H^2O 36·08. Anorthic, but commonly in drusy or earthy crusts of a blue or greenish-blue colour ; streak bluish-white ; H (crystals) 2·5 ; G 2·2-2·3. Taste, strongly cupreous and unpleasant. Moistened and rubbed on a knife-blade, deposits metallic copper. PISANITE is a cupreous Iron-Vitriol or Melanterite. LETTSOMITE and WOODWARDITE are hydrated sulphates of CuO and Al^2O^3. The first occurs in druses of deep-blue hair-like crystals; the second in mammillated examples of similar colour.

(*Fe reaction*[*], *BB, a magnetic slag*).

MELANTERITE or GREEN VITRIOL : FeO 25·90, SO^3 28·78, H^2O 45·32. Clino-Rh., but commonly in crusts and coatings on iron ores, &c. Pale-green, blue-green, often ochreous on surface. H

---

[*] The solution yields with Ferricyanide of potassium, or with the ferrocyanide, a deep-blue precipitate.

(crystals) 2; G 1·8-1·9. Taste, inky, metallic. PISANITE is a cupreous variety. TAURISCITE a rhombic variety isomorphous with Epsomite.

COQUIMBITE: Fe^2O^3 28·47, SO^3 42·70, H^2O 28·83. Hex., but commonly in small granular masses. H 2-2·5; G 2-2·1; very pale green, bluish, greenish-white. Taste metallic, inky. The aqueous solution deposits Fe^2O^3 on boiling.

BOTRYOGENE: MgO, FeO, Fe^2O^3, SO^3, H^2O (28-30 per cent.). Clino-Rh., fibro-mamillated, &c. Red, orange or brownish-yellow; streak yellow; H (crystals) 2-2·5; G 2·1. Taste, feebly metallic. Partly soluble in water. RŒMERITE is closely allied or identical, but part of the FeO is replaced by ZnO.

IRON-ALUM (Feather Alum; Halotrichite, in part): Composition very variable, but essentially an alum, with FeO and Fe^2O^3 largely replacing the other bases. Greenish or brownish, in coatings and minute hair-like crystals. *See* under ALUNOGENE in B^1 †, above.

VOLTAITE: FeO, Fe^2O^3, Al^2O^3, SO^3, H^2O (15·3 per cent.). An altered Iron-Alum? Reg.; dark-green, black; streak, greenish-grey. H 2·5-3·0; G 2·8. Slowly soluble in water. Taste, feebly metallic.

Other ferruginous sulphates, Glockerite, Pissophane, Apatelite, Copiapite, Jarosite, &c., are insoluble or very slightly soluble in water. *See* below.

(Co reaction).

BIEBERITE (Cobalt Vitriol): CoO, SO^3, H^2O, but part of CoO often replaced by FeO or MgO. Isomorphous with Melanterite, but occurs only in efflorescent coatings of a pale rose-red colour. Easily distinguished by its blue borax-glass.

(Ni reaction).

MORENOSITE (Nickel Vitriol). NiO (MgO), SO^3, H^2O (40-45 per cent.). Isomorphous with Epsomite, but occurring only in efflorescences of hair-like crystals or in amorphous coatings. Green, greenish-white.

(Uranium reaction).

JOHANNITE: UO, SO^3, H^2O. Clino-Rh.; H 2-2·5; G 3·2. Grass-green; streak paler. Slowly soluble in water. Various other uranium sulphates (in some of which U^2O^3 is present) have been recognized (Medjidite, Zippeite, Voglianite, &c.), but the composition of these is more or less inconstant, and their characters are very imperfectly known.

(*Mn reaction*).

APJOHNITE (Manganese Alum): Essentially an alum with MnO replacing part of the other bases. In hair-like efflorescences of a pale reddish or brownish colour.

FAUSERITE (Manganese Vitriol): MnO, MgO, SO^3, H^2O (42·66 per cent.). Rh. (V:V 91° 18'); H 2-2·5; G 1·9. Pale reddish, yellowish-white.

† † *Insoluble (or practically insol.) in water. Taste, 0, or very slight.*
    (*Pb and Cu reactions. BB, on charcoal a yellow coating*).

LINARITE: PbO 55·7, CuO 19·8, SO^3 20, H^2O 4·5. Clino-Rh.; H 2·5-3; G 5·3-4·45. Azure-blue; streak pale-blue.

(*Cu reaction*).

BROCHANTITE: CuO 70·34, SO^3 17·71, H^2O 11·95. Rh.? (V:V 104° 32'); H 3·5-4; G 3·8-3·9. Emerald-green, dark-green; streak pale-green. KRISUVIGITE is identical. TANGITE and WARRINGTONITE are closely related, but yield 15·33 per cent. water. All form a deep-blue solution with ammonia.

LETTSOMITE (Velvet Copper Ore). CuO, Al^2O^3, Fe^2O^2 SO^3, H^2O (23·34 per cent.). In delicate hair-like crystals of a deep blue colour. WOODWARDITE, in blue mamillated masses, is identical.

(*Fe reaction: BB, a magnetic slag or crust*).

COPIAPITE: Fe^2O^3, SO^3, H^2O 24·5 per cent. In six-sided pearly, tables, and granular masses. H 1·5; G 2·14. Yellow.

STYPTICITE: Fe^2O^3, SO^3, H^2O 36 per cent. In greenish or yellowish-white fibrous crusts. FIBRO-FERRITE is apparently identical, but some so-called Fibro-ferrites are soluble in water. Owing to their inconstancy of composition, due to alteration and intermixture, no very strict definitions are possible as regards ferruginous sulphates generally.

GLOCKERITE: Fe^2O^3, SO^3, H^2O (20·7 per cent.). Stalactitic, botryoidal. Black, dark-brown, yellowish, dark-green; streak brownish-yellow. PISSOPHANE, in dark-green and brown botryoidal and earthy masses, is apparently a variety, or a closely related substance, but yields 40-41 per cent. water. VITRIOL-OCHRE is an earthy, ochre-yellow variety (H^2O 21 per cent.).

APATELITE: Fe^2O^3, SO^3, H^2O (4 per cent.). In small nodular earthy masses of a yellow colour and streak, from Auteuil, near Paris. Chiefly distinguished by its low amount of water.

JAROSITE: K^2O 9·38, Fe^2O^3 47·91, SO^3 31·93, H^2O 10·78. Hemi-Hex. (R : R about 89°), mostly tabular from predominance of the basal plane, also in scaly and fine granular examples; H 3-4; G 3·2-3·6. Dark-brown, brownish-yellow, red by transmitted light; streak ochre-yellow. Shews the red K-line in spectroscope.

DIADOCHITE: Fe^2O^3, SO^3, P^2O^5, H^2O 30·3 per cent. H 2·5-3; G 1·9-2·5. Brown, brownish-yellow; streak somewhat lighter. Mostly in mammillated, concentric-lamellar examples. BB, on charcoal, a magnetic bead. In the forceps, tinges the flame-point green.

PITTICITE (Iron Sinter) resembles Diadochite in general characters, but contains As^2O^5. The composition, however, varies greatly in different examples.

NOTE TO TABLE XVI.

This Table is composed, apart from a few sulphides of non-metallic aspect, entirely of sulphates.

Sphalerite or Zinc Blende is the only commonly-occurring sulphide referred to in the Table. This mineral presents in many of its varieties a more or less metallic lustre; but in others, the light-coloured varieties especially, the lustre is non-metallic and usually adamantine. Sphalerite is commonly in lamellar masses (of easy cleavage parallel to the planes of the rhombic dodecahedron), or otherwise in crystals of the Regular System. These consist chiefly of tetrahedrons, twinned octahedrons, rhombic-dodecahedrons, and a combination of rhombic-dodecahedron with a half-trapezohedron or pyramidal tetrahedron $\frac{3\cdot3}{2}$. Sub-fibrous and granular examples are also known, and some of these, more especially, are cadmiferous. Certain Blendes, likewise, contain thallium in minute quantity; and in almost all the dark Blendes small proportions of Fe and Mn are present. Many varieties also contain traces, and even workable amounts, of gold and silver. The more common colours are dark-brown and black, with light-brown streak, and these dark examples are often blood-red in thin pieces by transmitted light. Less common colours are dark-green and yellow: colourless examples are still more rare, and hitherto have been found only in New Jersey. Yellow varieties (especially) often shew strong phosphorescence when scratched or abraded in the dark. All varieties give a zinc-sublimate on charcoal if fused in powder with carb. soda and borax; and all emit the odour of sulphuretted hydrogen when warmed in powder with hydrochloric acid.

Natural sulphates fall into five more or less well defined groups. These comprise: (1) Anhydrous Prismatic Sulphates; (2) Gypsums; (3) Bitter-Salts and Vitriols; (4) Alums; and (5) Alumstones.

The anhydrous prismatic sulphates are principally represented by Anglesite, Barytine, Celestine, and Anhydrite. These have the common formula RO,

SO^3, and a common Rhombic crystallization, with V : V (the prism-angle in front) 100° 30'—104° 30', according to the species.

Anglesite (PbO, SO^3) occurs in small crystals, mostly colourless or greyish, with strong adamantine lustre; and also in small mamillated and lamellar examples, and in earthy masses, white, yellowish, &c., arising from decomposition of galena. The crystals are generally in drusy aggregations, and are very brittle. They are either tabular, from predominance of B or $\bar{V}$; prismatic, vertically, from predominance of V; prismatic, transversely, from extension of $\tfrac{1}{2}\bar{P}$ or $\tfrac{1}{4}\bar{P}$; or pyramidal from preponderance of P. It much resembles the lead carbonate cerussite, but is distinguished (when the two are not intermixed) by blowpipe and acid reactions.

Barytine or Heavy Spar, sulphate of baryta, is very widely distributed, and is especially abundant as a gangue or veinstone in lead, zinc, silver, and other metallic veins. It occurs most commonly in lamellar or fibrous masses, but is also very common in crystals. The latter are sometimes of comparatively large size, and are almost always sharply-defined and distinct. They belong to the Rhombic System, and present chiefly four leading types : (1) tabular, with V and B, or rectangular-tabular with $\tfrac{1}{2}\bar{P}$, $\overset{\prime\prime}{P}$, and B, as principal forms, B predominating; (2) transversely prismatic in a macro-diagonal direction, with V and $\tfrac{1}{2}\bar{P}$ as chief forms, the latter elongated; (3) transversely prismatic in a brachy-diagonal direction, with $\tfrac{1}{2}\bar{P}$ and $\overset{\prime\prime}{P}$ as chief forms, the latter elongated; and (4) pyramidal, from about equal predominance of the common front and side polars $\tfrac{1}{2}\bar{P}$ and $\overset{\prime\prime}{P}$. More common angles are as follows : V : V 101° 40'; $\tfrac{1}{2}\bar{P} : \tfrac{1}{2}\bar{P}$ over summit 102° 17'; B : $\tfrac{1}{2}\bar{P}$ 141° 8'; $\overset{\prime\prime}{P} : \overset{\prime\prime}{P}$ over summit 74° 36'; B : $\overset{\prime\prime}{P}$ 127° 18'. Barytine is commonly colourless, white, or yellow, but also frequently grey, reddish, bluish, &c., and in some stalactitic and radiospherical examples, deep-brown or greyish-black. BB, it melts into a bead which reacts alkaline after prolonged ignition, and it communicates to the flame-border the apple-green tint characteristic of barium compounds. In carb. soda, BB, it is rapidly and entirely dissolved. In acids, insoluble. In Bunsen flame, after sufficient ignition, it shews the green bands of the barium spectrum very distinctly.

Celestine, the strontia sulphate, differs remarkably from Barytine in its geological relations, occurring very rarely in metallic veins, but chiefly in cavities and fissures in stratified calcareous rocks. The finest crystals occur in connection with native sulphur in Sicily. These are colourless, but when in fibrous or lamellar masses celestine very commonly presents a pale-blue colour, whence its name. It is also white, pale-yellowish, &c. The crystals are Rhombic combinations, and are generally elongated in the direction of the brachy-diagonal. More common forms comprise B, V, $\overset{\prime\prime}{P}$, and $\tfrac{1}{2}\bar{P}$, with angles as follows : V : V about 104°, but varying from 103° 30' to 104° 30'; $\overset{\prime\prime}{P} : \overset{\prime\prime}{P}$ over summit 75° 52'; B : $\overset{\prime\prime}{P}$ 127° 56'; $\tfrac{1}{2}\bar{P} : \tfrac{1}{2}\bar{P}$ 62° 40'; B : $\tfrac{1}{2}\bar{P}$ 121° 20'. BB melts, colours the flame-border crimson, and reacts alkaline. Entirely dissolved,

BB, by carb. soda. Insoluble in acids. In Bunsen flame, after short ignition, shews very distinctly the blue, orange-red, and group of crimson lines, of the strontium spectrum. These lines come out still more prominently by crushing the ignited or fused bead (as obtained in a reducing flame on charcoal), and moistening the powder with hydrochloric acid.

Anhydrite, lime sulphate, is generally in lamellar, granular or columnar masses of a white colour, though occasionally greyish or bluish, and sometimes brick-red. Crystals are comparatively rare. They consist chiefly of combinations of $\bar{V}$ and $\overset{..}{V}$ with several brachydomes or side-polars, $\overset{..}{V}$ predominating and imparting to the crystals a rectangular, tabular aspect. Also of combinations of $\overset{..}{V}$ and P, with brachy diagonal elongation. BB, fuses easily into an alcaline reacting bead, which imparts a comparatively feeble but distinct red colour to the flame border. In carb. soda, BB, not dissolved. Slowly soluble in hydrochloric acid. In the Bunsen flame (especially if first ignited and then moistened with hydrochl. acid), it shews the green and red lines of the calcium spectrum very distinctly.

The Gypsum Group consists of hydrated sulphates, with lime, or limemagnesia, and alkalies, for base. It is chiefly represented by Gypsum and Polyhallite.

Gypsum, in analytical formula, $CaO, SO^3 + 2 H^2O$, is a widely distributed mineral. It occurs chiefly in Clino-Rhombic crystals and in lamellar, laminar, fibrous, columnar, and granular masses, either colourless, or of a white, reddish, yellowish or other tint, and occasionally red, brown, black, &c., from ochreous or carbonaceous admixtures. Small transparent pieces become immediately opaque if held at the edge of a candle flame, and all varieties may be scratched by the nail. The crystals are often of considerable size. The most common, perhaps, are combinations of the Vertical prism V with the Side-vertical or Clino-pinakoid $\acute{V}$, and the Hemi-pyramid P. The latter form occurs necessarily as a pair of inclined planes (often curved) at each extremity of the crystal. The $\acute{V}$ or side planes usually predominate, and thus give a somewhat flattened aspect to most crystals. Two of these crystals are frequently united in reversed positions, producing arrow-headed or lance-headed twins. Transparent examples of Gypsum are commonly known as Selenite. The lustre is partly pearly and partly vitreous, and in most fibrous examples, satin-like. The ignition-loss (water) is nearly 21 per cent. In the Bunsen flame, the red and green lines of the calcium spectrum come out very prominently, especially if the ignited test-matter be moistened with a drop of hydrochloric acid. Gypsum, although tasteless, and thus for practical purposes regarded as insoluble, is dissolved in fine powder by about 450 parts of water.

Polyhallite (see composition in Table) is comparatively unimportant. It occurs commonly in sub-fibrous or columnar masses of a pale reddish or greyish colour. In water it is partially dissolved, a residuum of lime sulphate remaining. Very easily fusible. Ignition-loss under 6 per cent., but examples are often mixed with clay, gypsum, &c.

The group of Bitter-Salts and Vitriols falls into three sections: a rhombic section, with the analytical formula $RO, SO^3 + 7H^2O$, represented by Epsomite, Goslarite, Morenosite; a clino-rhombic section, represented by Melanterite, Bieberite, &c., also with the formula $RO, SO^3 + 7H^2O$; and a triclinic or anorthic section, with the formula $RO, SO^3 + 5H^2O$, represented by Chalcanthite or Copper Vitriol. These compounds in their actual occurrence as minerals, however, are of comparatively little interest, as they occur chiefly in solution or in the condition of efflorescent coatings, &c., rarely in distinct crystallizations. All possess an intensely bitter or metallic taste, and give off sulphurous acid on strong or prolonged ignition. The water, evolved in the bulb tube, has thus an acid reaction.

The group of Alums, characterized by octahedral crystallization and the general formula $RO, SO^3 + R^2O^3, 3SO^3 + 24H^2O$, is represented primarily by ordinary or potassic alum, and subordinately by soda alum, magnesia alum, iron alum, &c. These compounds in their natural occurrence, present themselves merely in efflorescent crusts and coatings, and, as minerals, are of no special interest. All are soluble and sapid, and evolve SO^2 on strong ignition. The alum of commerce is essentially a manufactured product, derived chiefly from decomposing pyritous shales.

The Alumstones are insoluble aluminous sulphates, represented chiefly by Alunite and Aluminite or Websterite. Alunite is a rhombohedral potassic species, occurring essentially in connection with volcanic or trachytic rocks. It differs from most sulphates by its hardness, which, in granular varieties especially, often exceeds that of fluor spar. It is infusible, but becomes decomposed on strong ignition, and evolves SO^2. In the Bunsen flame (especially if the ignited test-matter be moistened by hydrochloric acid), it shews the red line of the K-spectrum very distinctly.*

Aluminite or Websterite is of little importance. It is a simple sulphate of alumina with 47 per cent. water, mostly in white or yellowish-white earthy or nodular masses, which adhere strongly to the tongue and are scratched by the finger-nail. BB, infusible, but evolves SO^2.

---

* See PART I., page 58, 59.

TABLE XVII.

[Lustre non-metallic. Easily soluble, BB, in borax or phosphor-salt. Nitric-acid solution (on warming) yielding yellow precipitate with amm. molybdate.]

A.—Fluo-Phosphates -Chloro-Phosphates. Giving,in powder, with sulphuric acid in glass tube, strong fluorine-reaction; or with phosphor-salt and copper oxide, BB, an azure flame-coloration.

A¹.—YIELDING METALLIC LEAD, BB, WITH CARB. SODA ON CHARCOAL.

PYROMORPHITE : PbO P^2O^5 89·7, $PbCl^2$ 10·3, but part of the PbO sometimes replaced by CaO, part of the P^2O^5 by As^2O^5, and part of the $PbCl^2$ by $CaFl^2$. Hex. ; H 3·5-4 ; G 6·9-7·0 ; green of various shades, light or dark brown, ash-grey, rarely yellow or colourless. BB, melts into a bead which crystallizes with broad surface-facets on cooling. *See* Note at close of present Table.

A².—INFUSIBLE, OR FUSIBLE ON EXTREME EDGES ONLY.

APATITE : var. 1, *Fluor-Apatite:* CaO, P^2O^5 92·27, $CaFl^2$ 7·73 : var. 2, *Chlor-Apatite:* CaO, P^2O^5 89·34, $CaCl^2$ 10·66. But in var. 1, a small amount (usually 0·20-0·60 per cent.) of $CaCl^2$ is commonly present; whilst in var. 2, the $CaCl^2$ is almost always largely replaced by $CaFl^2$, the latter usually averaging 4 or nearly 5 per cent. of the entire components of the apatite. Crystal-System Hex. ; H 5·0 ; G 2·9-3·3 ; green of various shades, greenish-white, light-red, reddish or chocolate brown, sometimes colourless. BB, practically infusible, or rounded only on the thinnest edges. Phosphorite, Francolite, Osteolite, Talc-apatite, Eupychroite, are merely varieties (in some cases more or less decomposed) of apatite proper. In these, as well as in many unaltered crystals, &c., intermixed carbonate of lime is often present. *See* Note at close of present Table.

WAGNERITE: MgO, P^2O^5, 81, $MgFl^2$ 19. Clino-Rh. ; H 5·5-5 ; G 3·0-3·15 ; yellow, yellowish-white. BB, fusible on thin edges only. Very rare. The Norwegian Kjerulfin is closely related, if not an altered variety.

A³.—VERY EASILY FUSIBLE.

(*Strong Mn reaction with carb. soda*).

TRIPLITE : (FeO, MnO) P^2O^5, R Fl^2. Clino-Rh.? ; H 5-5·5 ; G 3·6-3·9; dark-brown; streak yellowish-grey. Occurs only in cleavable masses of vitreo-resinous lustre. Easily fusible into a dark globule. With carb. soda, strong manganese-reaction. Zwieselite is closely related, but is apparently Rhombic in crystallization.

(*Red flame-coloration, and distinct Li-line in spectroscope*).

AMBLYGONITE: Al^2O^3, P^2O^5, (LiNa) Fl. Anorthic; H 6; G 3·0-3·12; greenish-white, greyish or bluish-green. Easily fusible into a white opaque bead, with red coloration of the flame. With cobalt-solution, after ignition, assumes a fine blue colour. HEBRONITE (Montebrasite) is closely allied, but yields water on ignition. Perhaps an altered amblygonite?

.*. The imperfectly known HERDERITE or ALLOGONITE (Rh., with pseudo-hexagonal aspect; yellowish-white; H 5; G 2·9-3), and some varieties of WAVELLITE (mostly in greenish-white or green radiated fibrous examples, *see* under D, below), are also fluorine-containing phosphates. These assume a fine blue colour after ignition, in powder, with nitrate of cobalt. KAKOXENE, in yellow silky tufts (*see* under C¹, below), shews also, in most examples, a slight fluorine-reaction. BB, a magnetic slag.

B.—No Fluorine reaction. No water evolved by ignition in bulb-tube.

B¹.—EASILY FUSIBLE.

(*Fusion-globule magnetic*).

TRIPHYLINE: Li^2O, Na^2O, K^2O, MnO, FeO, P^2O^5. Rh., but occurring only in cleavable masses of a greyish-green, light grey, or grey-blue colour. H 4·5; G 3·5-3·6. Colours the flame distinctly red, if moistened with hydrochloric acid, or fused with chloride of barium, and shews the red Li-line in spectroscope.*

B².—INFUSIBLE, OR FUSIBLE ON EXTREME EDGES ONLY.

XENOTIME: YO 62·13, P^2O^5 37·87, but with part of the YO always replaced by CeO. Tetr.; H 4·5; G 4·45-4·6; yellowish-brown, red-brown, pale-red. Scarcely attacked by boiling acid; but, on dilution with water, sufficient is dissolved to give a yellow coloration to a fragment of amm. molybdate dropped into the solution and gently warmed.

CRYPTOLITE: CeO, LnO, DiO, P^2O^5. Pale-yellow or reddish; G 4·6; in minute acicular crystals in certain Apatites. PHOSPHOCERITE,

---

* Some examples of Triphyline shew this crimson line very distinctly *per se*, but in general it is only obtained by moistening the mineral with hydrochloric acid, or mixing it in powder with chloride of barium. The latter reagent answers perfectly, and has the advantage of being conveniently carried in the blowpipe case.

in very minute, apparently Tetragonal, crystals in certain Swedish examples of cobaltine, is identical in composition. G 4·78; pale greenish-yellow.

MONAZITE: CeO, LnO, ThO, P^2O^5. Clino-Rh.; H 5-5·5; G 4·9-5·3; reddish-brown, yellowish-red, pale-red. Many examples give traces of tin by the reducing process: See page 17. EREMITE (MONAZITOID) and TURNESITE are varieties. In some of these a small percentage of Tantalic acid is present.

C.—Hydrous Phosphates. Water evolved on ignition in bulb-tube.

C¹.—MAGNETIC AFTER FUSION OR IGNITION, OR GIVING STRONG REACTION OF MANGANESE WITH CARB. SODA.

(This section includes a series of iron or manganese-phosphates, in most of which the composition is very uncertain, owing to changes in the oxidation of the base, or loss or gain of water. Many of these phosphates can scarcely rank as definite species. In the present Table they are arranged after the average percentage of water which they contain. Where the iron is in the condition of protoxide, the ignition-loss will necessarily be slightly lower (about 1 per cent.) than the actual percentage of water present in the mineral).

KAKOXENE: Fe^2O^3 47, P^2O^5 21, H^2O 32. In delicate tufts and fibro-mammillated examples of a yellow colour with silky lustre. G 2·4. BB, a dark magnetic slag.

VIVIANITE: FeO (rapidly changing into Fe^2O^3), P^2O^5, H^2O 28 per cent. Clino-Rh., but commonly in bladed and fibrous examples of a greenish-blue or deep indigo-blue colour; rarely colourless, and then containing FeO only; H 2; G 2·6-2·7. Flexible in thin pieces. BB becomes red, and fuses into a magnetic globule. LUDLAMITE, from Cornwall, is closely related, but has less water (17 per cent.).

STRENGITE: Fe^2O^3, P^2O^5, H^2O 19-20 per cent.; Rh., but chiefly in fibrous mamillated examples of a bluish-red or pink colour, rarely colourless; H 3-4; G 2·9. BB, easily fusible into a magnetic globule.

CHILDRENITE: MnO 10, Fe^2O^3 29, Al^2O^3 14, P^2O^5 29, H^2O 18. Rh.; H 4·5-5; G 3·2-3·3. Yellowish-white, yellow, blackish-brown. BB, intumesces, and forms a dark magnetic slag or semi-fused mass.

BERAUNITE: Fe^2O^3, P^2O^5, H^2O 16·5 per cent. In radiated and leafy examples of a red or red-brown colour and yellow streak; H 2; G 2·9-3; BB, fusible, magnetic.

166   BLOWPIPE PRACTICE.

HUREAULITE: MnO 41, FeO 8, P^2O^5 39, H^2O 12. Clino-Rh., mostly tabular; also coarse-fibrous, &c. H 3·5; G 3·2; yellowish-red, red-brown, more rarely violet or reddish-white. BB, easily fusible into a dark, feebly-magnetic globule.

HETEROSITE: FeO (changing into Fe^2O^3), MnO (changing into MnO^2), P^2O^5, H^2O 4·4 per cent. Massive; H 4·5-5; G 3·4-3·5; greenish or bluish-grey, violet, brown. BB, fusible, magnetic.

C^2.—WITH BORAX, BB, A GLASS COLOURED BY COPPER OR URANIUM OXIDE· STREAK LIGHT-GREEN OR YELLOW.

† *Water-percentage* 14-19.

LIME-URANITE (AUTUNITE): CaO 6·10, U^2O^3 62·75, P^2O^5 15·47, H^2O 15·68, but sometimes, and normally, nearly 19 per cent. of water present. Tet., or Rh. with marked tetragonal aspect, mostly tabular from predominance of basal plane, and thus passing into foliated examples. Yellow, greenish-yellow; H 1-2; G 3-3·2; BB intumesces slightly, and fuses into a dark bead with crystalline surface. In nitric acid forms a yellow solution. URANOSPHŒRITE is a related uranium phosphate, but with baryta in place of lime. Yellowish-green; G 3·5.

COPPER-URANITE (CHALKOLITE, TORBERNITE): CuO 8·43, U^2O^3 61·19, P^2O^5 15·08, H^2O 15·30. Tet., mostly tabular, passing into foliated micaceous examples. Emerald-green, paler in the streak, with metallic-pearly lustre; H 2-2·5; G 3·5-3·6. BB, fusible and reducible to metallic copper. Forms in nitric acid a yellowish-green solution.

CHALKOSIDERITE: CuO (8·15), Fe^2O^3, Al^2O^3, P^2O^5, H^2O 15 per cent. In small, light-green, anorthic crystals. G 3·1.

† † *Water-percentage* 8-11.

TAGILITE: CuO 61·85, P^2O^5 27·64, H^2O 10·51. Clino-Rh.? but mostly fibrous, mammillated, &c.; emerald-green; H 3; G 4·4·1. BB, fusible and reducible.*

EHLITE: CuO 67, P^2O^5 24, H^2O 9. Rh.? but mostly in foliated and bladed examples, with pearly lustre on cleavage surface; H 1·5-2; G 3·9-4·3. Decrepitates in bulb-tube. On charcoal, reduced.†

---

* The reduced copper-globule is surrounded by a black coating of unreduced phosphate· With carb. soda, perfect reduction ensues.

† The reduced copper-globule is surrounded by a black coating of unreduced phosphate. With carb. soda, perfect reduction ensues.

MINERAL TABLES:—XVII. 167

PHOSPHORCHALCITE: CuO 70·88, P^2O^5 21·10, H^2O 8·02. Clino-Rh., but mostly fibrous, mammillated, &c.; green, blackish-green; H 4·5-5; G 4·1-4·3; decrepitates and blackens on ignition; fuses to a dark bead with crystallized surface; on charcoal, reduced.* DIHYDRITE is closely related, but consists of CuO 69, P^2O^5 24·7, H^2O 6·3.

† † † *Water-percentage under 4.*

LIBETHENITE: CuO 66·5, P^2O^5 29·7, H^2O 3·8. Rh., crystals very small; dark-green, blackish-green; H 4; G 3·6-3·9. Decrepitates and blackens in bulb-tube. In forceps, melts to a dark bead with crystallized surface. On charcoal, forms a black globule surrounding reduced copper.

⁎ Some examples of Wavellite, Peganite, Fischerite, and Turquoise (*see* under C³, below), contain a small amount of CuO, and thus give a copper-reaction with borax.

C³.—IN POWDER, COLOURED BRIGHT-BLUE BY IGNITION WITH COBALT-SOLUTION.

† *Water-percentage* 20-40.

WAVELLITE: Al^2O^3 38·10, P^2O^5 35·16, H^2O 26·47, but traces of Fluorine often present. Rh. (crystals mostly small and indistinct), commonly in botryoidal radiated-fibrous examples of a pale green, greenish-white, or yellowish-white colour; H 3·5-4; G 2·3-2·5. BB, swells up, separates into fibres and becomes opaque-white, but does not fuse.

FISCHERITE—PEGANITE—VARISCITE: Hydrated aluminous phosphates closely related to Wavellite. Rhombic in crystallization, but commonly in radiated fibrous examples of a green or white colour. H 3-5; G about 2·5. BB, like Wavellite. Planerite, Striegisan, Richmondite, Evansite and Zepharovichite are probably altered examples. These minerals can only be distinguished by accurate chemical analysis. Many give a slight copper-reaction. The percentage of water is as follows: Variscite 23, Peganite 24, Wavellite 26·5, Zepharovichite 27, Fischerite 29, Richmondite 35, Evansite 40-42.

CALAITE or TURQUOISE: Al^2O^3 47, P^2O^5 32·5, H^2O 20·5. In light-blue and bluish-green amorphous masses; H 6; G 2·6-2·8. BB,

---

* The reduced copper-globule is surrounded by a black coating of unreduced phosphate. With carb. soda, perfect reduction ensues.

decrepitates, and often blackens, but remains unfused. Many examples shew traces of copper.

†† *Water-percentage under 13.*

LAZULITE: MgO, FeO, Al^2O^3, P^2O^5, H^2O (5-7 per cent.). Clino-Rh. (but scarcely differing from Rhombic in aspect and measurements). Blue, bluish-white; H 5-6; G 3-3·2. BB, exfoliates and crumbles, but does not fuse.

BERLINITE—TROLLEITE—AUGELITE: Hydrated aluminous phosphates of a blue or greenish-blue colour. Water percentage: 4, 6, and 12·5 respectively. Obscurely known or doubtful species.

C⁴.—IN POWDER, COLOURED PALE-RED, GREEN, OR DARK-GREY BY IGNITION WITH COBALT-SOLUTION.

† *With Co-solution, pale-red.*

LUNEBERGITE: MgO, P^2O^5, B^2O^3, H^4O (30·23 per cent.). In white, fibrous and earthy masses. H 1-1·5; G 2·05. Easily fusible, with green coloration of the flame-border. With sulphuric acid and alcohol, gives the green flame characteristic of B^2O^3.

STRUVITE: a hydrous phosphate of ammonia and magnesia. Rh. (hemimorphic). Colourless, yellowish, pale-brown; H 1·5-2; G 1·6-1·8. In peat-bogs, guano-deposits, &c. Evolves ammoniacal fumes on ignition.

†† *With Co-solution, light-green.*

(BB, on charcoal with carb. soda, a zinc sublimate).

HOPEITE: ZnO, P^2O^5, H^4O? Rh.; greyish-white; H 2·5-3; G 2·7-2·8. BB, fusible into a white bead. Some examples shew presence of cadmium.

††† *With Co-solution, dark-grey.*

BRUSHITE: CaO 32·6, P^2O^5 41·3, H^2O 26·1. Clino-Rh.; colourless, yellowish; H 1·5; G 2·2. Metabrushite and Isoclase are related products. In all, the presence of CaO is readily determined by the spectroscope.

CHURCHITE: CaO, CeO, DiO, P^2O^5, H^4O (15 per cent.). Clino-Rh. (?) radiated. Greyish-white, pale-red; H 2·5-3; G 3·1. Imperfectly known.

NOTE TO TABLE XVII.

This Table is represented by Phosphates, or by Phosphates combined with Fluorides or Chlorides. Its more important species may be referred broadly to the following groups: (1) Apatites; (2) Triplites; (3) Alumina Phosphates; (4) Iron and Copper Phosphates; (5) Uranium Phosphates.

The Apatite group is characterized by its Hexagonal crystallization, and by the common formula $3 (3 RO, P^2O^5) + R (Fl, Cl)^2$. It is represented by Apatite and Pyromorphite, and also by the related arseniate and vanadiate, Mimetesite and Vanadinite, the latter described, in a technical work of this kind, under other Tables. Apatite, often known commercially as "Phosphate," is largely employed in the manufacture of Superphosphate of lime, so extensively used as a fertilizer. It commonly presents itself in cleavable masses or hexagonal prisms of a light or deep green colour, but is frequently chocolate-brown, red, or almost colourless. Green and reddish tints are often intermingled. The edges of the crystals are frequently rounded. The more common crystals are simple six-sided prisms with large basal plane, or these with a slight pyramidal replacement on the basal edges; but Canadian crystals (when unbroken) shew complete pyramidal terminations, without any basal plane. As regards composition, Apatite includes two leading varieties: fluoride of calcium being present in one, and chloride and fluoride in the other. Both are readily dissolved, in powder, by nitric acid, and the diluted solution yields a yellow precipitate with amm. molybdate, especially on being warmed. Very carefully neutralized by ammonia, it gives also a yellow precipitate with nitrate of silver. Heated with a few drops of sulphuric acid, both varieties, as a rule, give a marked fluorine-reaction, the evolved fumes exerting a strongly corrosive action on glass. Before the blowpipe, Apatite is infusible, or is rounded only on the thinnest edges. The powder moistened with sulphuric acid tinges the flame border pale green, thus shewing the presence of phosphoric acid; and in the spectroscope the green and red Ca-lines are readily produced, but this latter reaction is best obtained by moistening the powder with hydrochloric acid.

Pyromorphite is essentially a chloro-phosphate of lead. It is commonly in groups of small crystals of a dark or light green, brown, or grey colour. The so-called yellow varieties are mostly Mimetesite, or mixtures, at least, of phosphate and arseniate. The crystals are chiefly simple six-sided prisms, frequently barrel-shaped by curvature. The name Pyromorphite refers to the peculiar blowpipe reaction presented by the mineral. *Per se* (if free from arseniate), it is not reduced, but melts easily into a light-yellowish or greyish bead which crystallizes over the surface on cooling. Pyromorphite is easily soluble in nitric acid.

The group of Triplites is principally represented by Triplite, Triphyline, and Amblygonite, practically anhydrous phosphates or fluo-phosphates of easy fusibility. Triplite is mostly in dark-brown cleavable masses, giving marked reactions of manganese and fluorine. Triphyline is also in cleavable masses, but of a light colour, essentially pearl-grey, greyish-blue, or greyish-green.

It gives no marked fluorine reaction, but if moistened with hydrochloric acid, or mixed in powder with chloride of barium, it shews in the spectroscope the crimson Li-line very prominently. Amblygonite (*see* the Table) is a rare mineral. It gives both Fl and Li reactions.

The group of hydrated alumina-phosphates is chiefly represented by Wavellite and Kalaite, the latter more generally known as the Turquoise. Wavellite occurs rarely in distinct crystals, but is generally in botryoidal and radiated-fibrous examples of a green or greenish-white colour, and is found more especially in argillaceous slates. It is soluble in acids, and also in a strong solution of caustic potash. Before the blowpipe, it exfoliates, becomes opaque white, and tinges the flame pale-green, but does not fuse. Most specimens give with sulphuric acid a slight fluorine-reaction. When pure, the water-percentage = $26\frac{1}{2}$. Kalaite or Turquoise occurs chiefly in small nodular or flattened masses of a bright blue, bluish-white, or bluish-green colour. These scratch glass slightly; but many so-called turquoises are merely pieces of fossil bone coloured by copper oxide. In these, the hardness rarely exceeds 3; and they give off in most cases a marked ammoniacal odour on ignition. In the true turquoise the water percentage = $20\frac{1}{4}$.

Vivianite, Phosphorchalcite, and Libethenite are the chief representatives of the group of Iron and Copper Phosphates, characterized by their peculiar blue and green colours. Many of these are isomorphous with arseniates of corresponding formulæ. Vivianite, normally, is colourless, but the FeO, present in it, becomes rapidly converted into Fe^2O^3, and the mineral assumes a blue or bluish-green colour, with pale blue or greenish streak. It is commonly in flat-fibrous or bladed masses. It reddens on ignition, and melts into a dark-grey magnetic bead. Easily soluble in acids. Blackened in a hot solution of caustic potash. Water percentage = 28. Phosphorchalcite occurs commonly in groups of small clino-rhombic crystals and in fibrous examples of a blackish-green or emerald-green colour, paler in the streak. It blackens in the bulb-tube, and evolves about 8 per cent. water. Before the blowpipe it commonly decrepitates, and then melts into a black globule containing in its centre reduced copper. If the dark globule be fused with a small cutting of metallic lead it crystallizes on cooling. Libethenite closely resembles it in general characters and in its blowpipe reactions, but is rhombic in crystallization, and yields only 3·77 per cent. water. Its colour also, as a rule, is much less bright. It is isomorphous with the arseniate Olivenite.

The group of Uranium Phosphates includes only the Autunite or Lime-Uranite, and the Chalkolite (Torbernite) or Copper-Uranite. The lime-uranite is distinguished by its pale yellow or yellowish-green tint, and the copper-uranite by its splendid emerald-green colour. Both occur commonly in small lamellar or micaceous examples, and in groups of small tabular crystals. These latter are rhombic in the lime-uranite (but with strongly tetragonal aspect), and tetragonal in copper-uranite. Both species fuse more or less easily, and the latter gives reduced copper. Both species dissolve readily in nitric acid, and are decomposed by caustic potash with abstraction of their phosphoric acid. Water percentage 15-16.

TABLE XVIII.

[Lustre non-metallic. Easily dissolved BB by borax or phosphor-salt. Green coloration of flame by treatment with sulphuric acid and alcohol.]

A.—Anhydrous Species. No water evolved (or merely traces) by ignition in bulb-tube.

BORACITE: MgO 27, B^2O^3 62·5, $MgCl^2$ 10·5. Reg. (*see* Note at end of Table); H 7; G 2·9-3; colourless, pale greenish, reddish, &c.; streak white. Mostly in small crystals imbedded in anhydrite or gypsum. BB fusible with intumescence, tinging the flame green. With CuO and phosphor-salt, gives chlorine-reaction. Slowly dissolved by hydrochloric acid. RHODIZITE (in small crystals on some Siberian tourmalines) is regarded as a lime boracite. H 8; G 3·3.

LUDWIGITE: MgO, FeO, Fe^2O^3, B^2O^3. In fibrous or prismatic masses of a dark-green or greenish-indigo colour; H 5; G 4; BB, fusible slowly into a dark magnetic bead. The only examples hitherto recognized occur with magnetic iron ore in the Bannat.

B.—Hydrous Species, yielding water on ignition.

B¹.—DISTINCTLY SOLUBLE AND SAPID.

SASSOLINE (Boracic Acid): B^2O^3 56·45, H^2O 43·55. Clino-Rh. or Anorthic (?), but essentially in small pearly-white scales and tabular examples, sometimes stained by ferruginous matter. H 1; G 1·4-1·5; bitter-acid taste, soapy to the touch. BB tinges the flame green, and melts with intumescence into a hard clear glass.

LARDERELLITE (Hydrated Borate of Ammonia): In small rhombic or rectangular plates and scales of a white colour. Scarcely soluble, except in hot water, and thus almost tasteless. *See* below, under B².

BORAX or TINKAL: Na^2O 16·2, B^2O^3 36·7, H^2O 47·1. Clino-Rh.; H 1·5-2·5; G 1·7-1·8; colourless, or stained brown, yellowish, &c., by impurities. Taste, slightly alcaline. BB, intumesces and melts easily, but (as regards natural or crude varieties) the glass is dark or more or less coloured. Moistened with sulphuric acid, or with glycerine, it tinges the flame green.

B².—PRACTICALLY INSOLUBLE AND WITHOUT TASTE; OR DECOMPOSED BY BOILING WATER ONLY.

† *No marked Mn or Fe reaction.*

STASSFURTITE (Massive and slightly altered BORACITE ?): In fine-granular or fibrous masses of a white or yellowish-white colour.

Yields 0·5-1 per cent. water on ignition: composition otherwise as in Boracite. H 4·5·5; G 2·9-3·0. Readily fusible.

SZAILBELYITE: MgO, B^2O^3, H^2O (7-12·5 per cent.). In small globular masses of radiated-fibrous structure and white colour; H 3·5; G 2·7. Easily fusible.

HYDROBORACITE: CaO, MgO, B^2O^3, H^2O (26 per cent.). In crystalline, radiated-fibrous or leafy masses of a white or pale reddish tint. H 2; G 1·9-2. Very easily fusible. Shews red and green Ca-lines in spectroscope if moistened with HC acid.

BOROCALCITE: CaO, B^2O^3, H^2O (35·5 per cent.). Clino-Rh. 1 Mostly in snow-white acicular crystals and incrustations. Very easily fusible. BECHILITE is closely related, but yields less water (25·75 per cent.). Both shew Ca-lines in spectroscope when moistened with HC acid. PRICEITE, a milk-white chalky borate of lime, with 20·3 per cent. water, from Oregon, is probably identical, the amount of water in these earthy borates being very inconstant.

ULEXITE (BORONATROCALCITE): Na^2O 6·80, CaO 12·21, B^2O^3 45·66, H^2O 35·33. In white, mamillated and fibrous masses. G 1·8. Very easily fusible with yellow coloration of the flame. Decomposed, in powder, by boiling water. TINKALZITE and CRYPTOMORPHITE are closely related substances.

LARDERELLITE: Ammonia 12·7, B^2O^3 68·6, H^2O 18·7. In white shining scales or small crystalline plates resembling Sassoline. Soluble in hot water. Yielding ammoniacal fumes on ignition. Fusible with strong intumescence.

†† *BB, marked reaction of Iron or Manganese.*

SUSSEXITE: MnO, MgO, B^2O^3, H^2O (9 per cent.). In white, or pale-reddish, silky-fibrous masses. H 2·5-3; G 3·42. Very easily fusible, with green flame-coloration.

LAGONITE: F^2O^3, B^2O^3, H^2O (12·73 per cent.). In yellow, ochreous masses from the boracic-acid lagoons of Tuscany.

NOTE TO TABLE XVIII.

This Table, apart from Boracic Acid, is composed exclusively of Borates, distinguished readily from other compounds by the peculiar yellowish-green coloration which they impart (when moistened with sulphuric acid) to the flame of alcohol. Many of these minerals are still imperfectly known, and are apparently of somewhat inconstant composition, more especially as regards the hydrous species. Boracite and Tinkal (crude Borax) are the principal representatives of the Table.

Boracite [2 (MgO, B^2O^3) + Mg Cl^2] occurs essentially in small hemihedrally-modified crystals of the Regular System, remarkable for their high degree of hardness, which equals that of ordinary quartz. Hence they scratch glass very distinctly. They are generally colourless, but sometimes present a pale grey, greenish or yellowish tint, and are always associated with anhydrite, gypsum, or rock salt. The most simple consist of the cube truncated on the alternate angles, and thus presenting a combination of cube and tetrahedron. Very commonly the cube-edges are also truncated by the planes of the rhombic dodecahedron; and the latter form predominates in some crystals. As in most other hemihedrally-modified minerals, Boracite is pyro-electric. The substance known (from its locality) as Stassfurtite appears to possess essentially the same composition, except that it yields a small amount of water on ignition. This substance is thus commonly regarded as massive Boracite, but its hardness is comparatively low, usually under 5. It occurs mostly in granular or subfibrous masses of a chalky-white colour. Tinkal or crude Borax (Na^2O, $2\ B^2O^3$ + $10\ H^2O$) is a product of certain salt lakes, and is mostly in the form of small granular or crystalline masses of a greyish or brownish-white colour. Moistened with sulphuric acid, or simply with glycerine, it imparts a distinct green coloration to the flame. *Per se*, it colours the flame intensely yellow, and melts with great intumescence into a more or less clear bead. In the bulb-tube it evolves $47 \cdot 2$ per cent. water. Its crystallization is Clino-Rhombic, and the ordinary borax crystals have a remarkable resemblance, even in their angle values, to those of Augite.

TABLE XIX.

[Lustre non-metallic. Easily dissolved, BB, by borax or phosphor-salt. Giving with the latter reagent and CuO an intensely azure-blue or green flame (Cl., Br., or I reaction)].

A.—Soluble in Water. Sapid.

A¹.—NO WATER (OR MERELY TRACES) IN BULB-TUBE.

† *Entirely dissolved BB by carb. soda.*

ROCK SALT (Halite): Sodium 39·31, Chlorine 60·69. Reg., with cubical cleavage; H 2; G 2·1-2·2; colourless, white, grey, greenish, red, violet, &c.; streak white; taste, strongly saline, sometimes bitterish from presence of chloride of magnesium and other impurities. BB, generally decrepitates, colours the flame strongly yellow, melts, and in prolonged heat sublimes.

SYLVINE: K 52·35, Cl 47·65, but generally contains NaCl. Reg.; H 2; G 1·9-2; colourless, greyish, reddish, &c.; taste, like that of rock salt. BB, easily fusible, colouring the flame violet if pure. In spectroscope, even if impure from NaCl, &c., it shews the red K-line very distinctly (*See* Part I., p. 58–59).

SAL AMMONIAC (Chloride of Ammonium): Reg., but commonly in crusts and earthy coatings; H 1·0-2; G 1·5-1·6; white, brownish, yellowish. Taste, pungent, saline. Entirely volatilizable without fusion. Ignited with caustic potash, gives off ammoniacal fumes.

† † *BB with carb. soda only partially attacked, an undissolved mass remaining.*

CHLOROCALCITE (Chloride of Calcium). In white crusts on some Vesuvian lavas, often associated with thin scales and crystals of Iron Glance. Shews red and green Ca-lines in spectroscope very distinctly.

TACHHYDRITE: $CaCl^2$ 21, $MgCl^2$ 37, H^2O 42. In rounded cleavable masses of a yellow colour associated with Carnallite, Anhydrite, &c. Very deliquescent. Easily fusible. Ca-lines in spectroscope.

CARNALLITE KCl 26·8, $Mg\ Cl^2$ 34·2, H^2O 39. Rh., with pseudo-hexagonal aspect, but commonly in fine granular examples of a white colour, or sometimes red from intermixed Fe^2O^3 or scales of Iron Glance. Deliquescent. Easily fusible. Shews red K-line in spectroscope, together with Na-line, part of the KCl being generally replaced by NaCl. Found essentially in salt deposits. Some examples are said to contain traces of Thallium, also Cæsium and Rubidium.

KREMERSITE: Am, K, Fe, Cl, H^{2}O. In small octahedrons, of a red colour, on some Vesuvian lavas. Easily soluble. Deliquescent.

B.—Yielding reduced metal, BB, with carb. soda on charcoal.

† *BB a silver globule.*

KERARGYRITE (Horn Silver Ore): Ag 75·3, Cl 24·7. Reg., but commonly in granular masses and coatings of a grey, greenish, or violet-brown colour, and waxy aspect. H 1-1·5; G 5·6; very sectile. BB melts with bubbling, and on charcoal is easily reduced.

BROMARGYRITE: Ag 57·4, Br 42·6. Yellow, yellowish-green. Sectile, and otherwise like Kerargyrite, but giving green and blue flame by fusion with phosphor-salt and CuO. Fused with bisulphate of potash in a small test-tube, it forms a blood-red globule which becomes yellow when cold, and turns green on exposure to sunlight. Chloride of silver (Kerargyrite) under this treatment, gives an orange-red bead which becomes yellowish-white on cooling, and dark-grey on exposure. Iodide of silver forms an amethyst-red globule which turns dingy-yellow on cooling, and does not undergo further change on exposure. *See* Appendix to Part I, page 90. Microbromite, Embolite, and Megabromite are isomorphous chloro-bromides, containing, respectively, 69·8, 66 9, and 64·2 per cent. silver. These (and intermediate varieties) resemble Kerargyrite in general characters.

IODARGYRITE; Ag 46, I 54, yellow, sectile, like Kerargyrite in general characters, but giving emerald-green flame with phosphor-salt and CuO. Fused with bisulphate of potash in closed tube, it forms a dark amethystine globule, which turns greyish yellow on cooling. Slight fumes of Iodine are also evolved. Torconalite is a rare compound of Ag I and Hg I. It yields a sublimate of mercury by fusion with reducing agents (dry carb. soda, neutral ox. potash, &c.) in closed tube.

†† *BB, copper globule and reactions.*

NANTOKITE: Cu 64, Cl 36. White or colourless, normally, but often greenish externally from partial conversion into the oxy-chloride Atacamite. In small granular and disseminated masses with cubical cleavage. H 2; G 3·9. Fusible, and in chief part volatilizable, BB, colouring the flame intensely blue. Soluble in ammonia and in acids.

ATACAMITE: $CuCl^2$ CuO, H^2O ($12\frac{1}{2} - 22\frac{1}{2}$ per cent). Rh., but mostly in small grains of a deep-green colour. H 3-3·5; G 3·7-3·77. Fusible and reducible, colouring the flame blue. Soluble in ammonia and acids. Atelite (from Vesuvius) and Tallingite (from Cornwall) are closely related compounds, the latter blue in colour. Atlasite (from Chili) is apparently Atacamite converted in chief part into green carbonate.

PERCYLITE: $CuCl^2$, $PbCl^2$, CuO, PbO, H^2O. Reg (crystals very small); sky-blue. Fusible and reducible, with strong coloration of flame and lead sublimate on charcoal. Hitherto only found with alluvial gold in Mexico.

††† *Pb or Bi globule and yellow coating BB with carb. soda on charcoal.*

COTUNNITE: Pb 74·5, Cl 25·5. Rh. (crystals acicular); H 1·5-2; G 5·24; white, with adamantine lustre. Fusible and volatilizable.

MATLOCKITE: $PbCl^2$ 55·5, PbO 44·5. Tet. Yellowish or greenish; H 2·5; G 7·2. BB, decrepitates and fuses. With carb. soda, easily reducible. Very rare. Hitherto only found in Derbyshire with lead carbonate and fluor spar.

MENDIPITE: $PbCl^2$ 38·4 PbO 61·6, but commonly in part altered to carbonate. In small sub-columnar or fibrous masses; Rh.?; H 2·5-3; G 7·0-7·1. BB decrepitates, fuses, and on charcoal is reduced. A rare mineral.

PHOSGENITE (Kerasine; Corneous lead ore): $PbCl^2$ 51, PbO, CO^2 49. Tet.; yellowish-white, yellow, greenish, grey; H 2·5-3; G 6·0-6·3. Easily fusible into a yellowish bead with somewhat crystalline surface. With carb. soda, lead globules and yellow sublimate. In acids, soluble with effervescence. Very rare. Schwartzembergite (from the Atacama Desert) is a related compound containing Iodide of lead. Colour, yellow.

DAUBREITE: $BiCl^2$ 22·5, Bi^2O^3 72·6, H^2O 3·8, with small amount of Fe^2O^3, &c. From Bolivia. Characters undescribed.

C.—No reduced metal BB on charcoal, but mercurial sublimate with carb. soda in closed tube.

CALOMEL: Hg 85, Cl 15. Tet.; yellowish-white, grey; H 1-2; very sectile; G 6·4-7; BB entirely vol. Blackens in caustic potash; soluble in nitro-hydrochloric acid, but not in nitric acid alone.

COCCINITE (?): $Hg I^2$. Scarlet-red; Tetragonal. Doubtful as a naturally-occurring species.

NOTE ON TABLE XIX.

This Table consists entirely of Chlorides and Oxy-Chlorides. Other chlorine compounds combined with phosphates, &c., will be found in preceding Tables. The only important species, or those of tolerably frequent occurence, belonging to the present Table, consist of Rock Salt, Kerargyrite or Corneous Silver Ore, and Atacamite.

Rock Salt or Chloride of Sodium is widely distributed in the form of beds, in strata of various geological periods, and, in solution, in sea-water and numerous mineral springs. It occurs also as a product of sublimation in many volcanic regions. Normally, it is colourless and transparent; but is very generally of a red, greenish, grey, violet or other colour, from intermixed impurities. Its crystals belong to the Regular System, and consist chiefly of simple cubes, or of aggregations of small cubes presenting a hopper-shaped aspect. Other forms (the octahedron, &c.), are comparatively rare. The cleavage is cubical, and strongly marked. Lamellar, granular, and sub-fibrous examples are also abundant. These are very frequently associated with gypsum and gypsiferous clay. Although normally anhydrous, rock salt (more especially in its less pure varieties) absorbs moisture from the atmosphere, and runs gradually into deliquescence. It dissolves in somewhat less than 3 parts of water, and it possesses the peculiarity of being about equally soluble in hot and cold water. Most examples decrepitate very strongly on ignition. From other chlorides it is readily distinguished by its saline taste and cubical crystallization and cleavage, combined with its intensely yellow flame-coloration.

Kerargyrite, often known as "Horn Silver" or "Corneous Silver Ore," is readily distinguished by the large globule of silver obtained from it by the blowpipe, and by its waxy aspect, sectility, and shining streak. It gives also reduced silver if moistened and placed in contact with a piece of zinc. It occurs mostly in compact masses or thin layers of a pearl-grey, greenish or blueish colour, turning brown on exposure. Unattacked by nitric acid, it dissolves more or less readily in ammonia.

Atacamite is a hydrated compound of chloride and oxide of copper, but of somewhat unstable composition. In some examples the water equals 12-13 per cent., and in others it is as high as $22\frac{1}{2}$ per cent. The mineral by its green colour and general aspect resembles certain cupreous arseniates and phosphates, but from these it is distinguished by the azure-blue coloration which it communicates to the blowpipe flame, as well as by the precipitate formed in its nitric acid solution by nitrate of silver. As seen in mineral collections, it is generally in the form of a blackish-green or deep emerald-green sand. Its crystals are small, vertically-striated prisms, and rectangular octahedrons, of the Rhombic System. $V:V = 112° 18'$. Cleavage brachydiagonal.

TABLE XX.

[Lustre non-metallic. Readily soluble BB in borax or phosphor-salt. Warmed, in powder, with sulphuric acid, evolve glass-corroding fumes.]

A.—Fusible.

† *Anhydrous, or yielding merely traces of moisture on ignition in bulb-tube.*

FLUOR SPAR (Fluorite) : Ca 51·3, F 48·7. Reg., essentially cubical (*see* Note at end of Table), cleavage octahedral ; H 4 ; G 3·1-3·2; colourless, violet, yellow, pale-green, deep bluish-green, rose-red, &c., with white streak. In most cases phosphorescent when heated. BB, generally decrepitates, fuses into a white enamel, which tinges the flame-border distinctly red, and reacts alkaline, after prolonged ignition. Ratofkite is a mixture of fluor spar and marl, of a dull greyish-blue colour.

CRYOLITE : Na 32·8, Al 13, F 54·2. Anorthic, but mostly in lamellar masses with nearly rectangular cleavage; H 2·5-3 ; G 2·9-3·0; white, or sometimes slightly yellowish or reddish ; streak white ; brittle. Melts in candle-flame into a white enamel. BB on charcoal, leaves a white crust which becomes blue on cooling after ignition with cobalt solution. Soluble in boiling solution of caustic potash. Shews strong Na-line in spectroscope. Chiolite (Tetragonal), Nipholite, Arksutite, and Fluellite, are related compounds of similar aspect. In Arksutite part of the Na is replaced by Ca.

SELLAITE : Mg 38·7, F 61·3. Tet.; colourless. H 5 ; G 2·97. Easily fusible into a white enamel. Becomes pale-red by ignition with cobalt solution. Very rare. Accompanies anhydrite at the Gerbulaz glacier in Savoy.

LEUCOPHANE : CaO, BeO, SiO^2, NaF. Rh., but commonly in lamellar, cleavable masses. H 3·5-4 , G 2·9-3. Greenish-grey, yellow. Phosphorescent when heated or broken. BB, very easily fusible. MELINOPHANE (Meliphanite) is a closely related species of a yellow colour, but Tetragonal (?) in crystallization.

† † *Yielding water by ignition in bulb-tube.*

PACHNOLITE : Na 10·35, Ca 17·99, Al 12·28, F 51·28, H^2O 8·10. Clino-Rh. (?). In minute twin-crystals in cavities of Cryolite. Colourless, strongly shining. BB, crumbles and fuses into a white enamel. In bulb-tube falls into powder and yields 8 per cent. water. In spectroscope shews Na-line, and green and red Ca-lines. Thomsenolite is closely related or identical.

PROSOPITE : Ca, Al, Si, F, H²O. Anorthic ; colourless ; H 4-4·5 ; G 2·9. Often earthy from decomposition. Sometimes altered into fluor spar. An imperfectly known species accompanying Iron Glance at Altenberg, Saxony. When transparent and crystalline, yields 14·84 per cent water.

B.—Infusible.

FLUOCERITE : Ce, F. ; Hex. (crystals, small, tabular), but mostly in granular examples of a pale-red or yellowish colour. H 4-5 ; G 4·7. Whitens or becomes yellow on ignition. BB, infusible. Hydrofluocerite is closely related (if not an altered fluocerite) but yields on ignition about 5 per cent. water.

YTTROCERITE : Ca, Ce, Y, La, Di, Er, F, H²O. In crystalline-granular masses of a light greyish-violet or blue-grey colour, with imperfect (tetragonal) cleavage. BB, infusible.

PARISITE; HARMATITE: Fluorides combined with carbonates; hence effervescing in acids. *See* Table XIII.

NOTE ON TABLE XX.

This Table consists essentially of Fluorides. Other Fluor-compounds combined with phosphates (Apatite, Triplite, &c.) will be found in Table XVII. Fluo-silicates (apart from Leucophane, placed here on account of its ready solution, BB, in phosphor-salt and borax) belong to one of the succeeding Tables: XXIV-XXVII.

Fluor Spar is the only commonly-occurring or generally distributed mineral belonging to the present Table. It occurs very commonly with ores of lead, zinc, and silver, more especially in mineral veins ; but is also found in cavities and fissures in limestone and other stratified rocks. It usually forms groups of distinct crystals, but sometimes presents itself in columnar, sub-fibrous, lamellar, and compact examples. The crystals as a rule consist of simple cubes, or of cubes slightly bevelled on the edges by the planes of a tetrakis-hexahedron (mostly ∞ 3). In many examples the cube-faces present a four-fold series of striæ, meeting in a point at or near the centre of each face. These striæ, lines of growth in the formation of the crystal, indicate the edges of a suppressed tetrakis-hexahedron, so to say. Fluor spar is often colourless, but more frequently it presents an amethystine, pale-green, yellow, or deep blue-green colour, and occasionally a rose-red or pearl-grey tint. The cube edges by transmitted light often show a shade of colour more or less distinct from that of the faces; and columnar or fibrous examples are frequently zoned in different tints. In all varieties the streak is white. Hardness between that of calcite and apatite, or equal to 4 of the ordinary scale. Sp. gr. 3·15-3·2.

Most examples when moderately heated exhibit a green or bluish phosphorescence; but, if a fragment be heated rapidly, decrepitation almost invariably ensues. By fusion, BB, a white enamel is produced. This tinges the flame red, and reacts alkaline after sufficient ignition. The red and green Ca-lines show prominently in the spectroscope, if a small splinter be held for a few minutes in the outer edge of a Bunsen-flame.

TABLE XXI.

[Lustre non-metallic. Readily dissolved BB by borax or phosphor-salt. Warmed in a test-tube with sulphuric acid, evolve orange-red or brownish nitrous fumes.]

A.—Anhydrous Species. Entirely soluble BB in carb. soda.

NITRE (Saltpetre); K^2O 46·53, N^2O^5 53·47. Rh. (V : V 118° 49'); H 2 ; G 1·9-2·1; normally colourless. Easily soluble in water; taste, saltish, cooling. BB fusible with intumescence, colouring the flame-border clear-violet. On charcoal, deflagrates and is absorbed.

NITRATINE (Chile Saltpetre, Soda Nitre): Na^2O 36·47, N^2O^5 63·53. Hemi-Hex. (R : R about 106°) H 1·5-2 ; G 2·1-2·2 ; normally colourless, but often brownish or reddish from impurities. Easily soluble ; taste, saltish, cooling. Deliquescent. Colours flame intensely yellow; otherwise like potash-nitre.

B —Hydrated Species. In carb. soda, BB, only partially soluble.

NITROCALCITE : CaO 30·76, N^2O^5 58·80, H^2O 10·44. In white or greyish earthy efflorescences on the walls of limestone caverns, cellars, &c. Soluble ; deflagrating by ignition on charcoal, leaving a white, earthy, alkaline-reacting crust.

NITROMAGNESITE : Mg O 24·10, N^2O^5 65·10, H^2O 10·80. Occurs with, and closely resembles, Nitrocalcite ; but the white crust, left BB on charcoal, exhibits a pink tinge after ignition with cobalt-solution.

NOTE TO TABLE XXI.

This short Table comprises the three or four representatives of the group of Nitrates hitherto recognized as minerals. All are soluble and sapid. By ignition with organic bodies, they detonate more or less violently ; and when warmed with sulphuric acid, or fused with bisulphate of potash, they evolve reddish or brownish nitrous fumes. The bases (magnesia excepted) are readily recognized by the spectroscope. Soda nitre (often erroneously called "cubical nitre") is distinguished also from ordinary or potash nitre by its crystallization in small rhombohedrons, its deliquescence, and its property of communicating a deep yellow coloration to the Bunsen or blowpipe flame. In the spectroscope, many examples shew the red K-line as well as the Na-line, and the presence of lime is also sometimes revealed (*see* Part I., page 55).

[182]

TABLE XXII.

[Lustre non-metallic. Easily dissolved BB by borax or phosphor-salt. Forming by fusion with carb. soda and nitre an alkaline salt partly soluble in water, the solution assuming a blue, brown, or green colour by boiling with hydrochloric acid and a piece of tin or zinc.

A.—Anhydrous Species. Yielding no water (or merely traces of moisture) by ignition in bulb-tube.

A¹.—GIVING LEAD GLOBULES OR OTHER FUSIBLE METAL, BB, WITH CARB. SODA OR ALONE.

† *With Borax, BB, a bright-green glass.*

(*Streak, strongly-coloured.*)

CROCOISITE (Crocoite): PbO 69, CrO^3 31. Clino-Rh. (*see* Note at close of Table). H 2·5-3; G 5·9-6; red; streak orange-yellow. BB. generally decrepitates; fusible and reducible, under slight detonation, on charcoal. Produces chlorine fumes with hydrochloric acid. Forms a brown or yellow solution with caustic potash.

PHŒNICITE: PbO 77, CrO^3 23. Rh. (crystals tabular, indistinct), mostly bladed or fibrous, accompanying Crocoisite. Red; streak, red; H 3-3·5; G 5·75. Fusible and reducible.

VAUQUELINITE: PbO 61·48, CuO 10·95, CrO 27·57. Clino-Rh. (crystals very small, indistinct), commonly in coatings and botryoidal: H 2·5-3; G 5·5-5·8. Dark-green, greenish-black; streak green. BB, intumesces slightly; fusible and reducible. With borax in R. F. (especially on addition of tin) forms a brick-red opaque bead from presence of copper. Laxmannite is a variety in which both CrO^3 and P^2O^5 are present; but this is probably the case in most varieties of Vauquelinite.

DECHENITE: PbO 54·95, V^2O^5 45·05. Mostly in small botryoidal masses or groups of minute indistinct crystals; H 3·5; G 5·82; reddish-yellow, brown; streak, yellow or orange. Fusible and reducible.

EUSYNCHITE: essentially a lead and zinc vanadate, resembling Dechenite in colour and general aspect.

DESCLOIZITE: essentially a lead vanadate of a dark-green or greenish-black colour, with bands of yellow or brown.

PUCHERITE: Bi^2O^3 71·74, V^2O^5 28·26, but often showing traces of P^2O^5 and As^2O^5. Rh. (crystals very small); red, brown; H 4;

G 6·25. BB, decrepitates, and yields reduced metal, with yellow ring on charcoal. Soluble in hydrochloric acid, with development of chlorine fumes, the red or yellow solution yielding a precipitate on dilution.

(*Streak white or indistinctly coloured.*)

VANADINITE: PbO 70·83, V^2O^5 19·35. $PbCl^2$ 9·82. Hex. (isomorphous with species of the Apatite group); H 3; G 6·8-7·2. Yellow, reddish, brownish. BB, decrepitates, throws off sparks, and gives reduced lead. With phos. salt and CuO, gives azure flame.

† † *With Borax, BB, no green coloration; but green or blue glass with phosphor-salt in RF.*

WULFENITE: PbO 61·4, MoO^3 38·6 Tet.; H 3; G 6-7; yellow, yellowish-grey, red (the latter colour due apparently to presence of lead chromate), rarely colourless. BB, decrepitates, melts and gives reduced lead.

STOLZITE: PbO 49, WO^3 51. Tet. (*see* Note at end of Table); H 3; G 7·9-8·1; grey, also green, reddish, and brown. BB, melts easily into a bead which crystallizes on cooling. On charcoal in RF, reduced.

A².—NO REDUCED LEAD BB ON CHARCOAL.

† *BB, no magnetic globule.*

SCHEELITE: CaO 19·45, WO^3 80·55. Tet. (*see* Note at end of Table); H 4·5-5; G 5·9-6·2; colourless, greyish, pale-yellow, sometimes red, brown, or greenish; streak white. BB, fusible on the edges, or in thin splinters only.

TUNGSTIC OCHRE: W 79·3, O 20·7. In earthy coatings of a yellow or greenish colour. BC, infusible, blackens. Insoluble in acids; soluble in ammonia.

MOLYBDIC OCHRE: Mo 65·7, O 34·3 In earthy, yellow crusts and coatings. BB easily fusible. On charcoal, absorbed (if pure). Easily soluble in hydrochloric acid.

† † *BB, magnetic globule.*

WOLFRAM: MnO, FeO, WO^3. Dark-brown, reddish-brown, with dark streak. In Clino-Rhombic crystals and lamellar masses, which present in most cases a sub-metallic lustre. H 5·5-5; G 7·1-7·55. BB fusible to a magnetic globule with crystalline surface. With carb. soda, strong manganese-reaction. *See* Table IX.

B—Hydrous Species. Yielding water by ignition in bulb-tube.
(*Cu reaction.*)

VOLBORTHITE: CaO, CuO, V^2O^5, H^2O (5 per cent.). Hex.; green, greenish-yellow; streak, yellow; H 3; G 3·5. BB, blackens, and fuses on charcoal into a dark slag containing reduced copper.

(*Cu and Pb reactions.*)

MOTTRAMITE: CuO, PbO, V^2O^5, H^2O (3·7 per cent.). In dark crystalline coatings with yellow streak; H 3; G 5·9. On sandstone from Cheshire. PSITTACINITE (from Montana) is a related compound in green sub-crystalline and botryoidal coatings, with $8\frac{1}{2}$ per cent H^2O.

NOTE TO TABLE XXII.

This Table is composed essentially of Chromates, Vanadates, Tungstates, and Molybdates. The two first may generally be distinguished from other compounds by the clear emerald-green glass which they form BB with borax in a reducing flame. The colour comes out in its full purity as the glass cools. If fused in a platinum spoon with carb. soda and nitre a partially-soluble salt results. This, in the case of Vanadates, becomes blue when warmed with a few drops of hydrochloric acid. Chromates, thus treated, give a green solution. See also the reactions of the latter described in PART I. of this work, page 49. Tungstates (in the absence of colouring oxides) form BB with phosphor-salt in the RF a fine blue glass, whilst with borax the glass is of a yellowish or brownish colour. Molybdates give with phosphor-salt in the RF a fine green glass. See also the distinctive reactions of these bodies with hydrochloric acid and zinc, as given in PART I, pages 46, 47, 62.

With the exception of Wolfram (a species which commonly presents a sub-metallic aspect, and thus belongs more especially to Table IX.) no mineral of this Table can be regarded as of common occurrence. Attention, however, may be directed to the following: the chromate Crocoisite, the molybdate Wulfenite, and the tungstates Wolfram, Stolzite, and Scheelite.

Crocoisite is readily distinguished by its fine red colour and orange-yellow streak, and by the emerald-green glass which it forms BB with borax*. It occurs commonly in groups of small or acicular crystals, and in granular masses and coatings. The crystals are Clino-Rhombic combinations; most commonly, vertically-striated prisms terminated by the two planes of an acute hemi-pyramid; or the same prism terminated by a very acute front-polar or hemidome, thus closely resembling an acute rhombohedron.

---

* Deceptive specimens are occasionally made by placing a piece of quartz in a crystallizing solution of bichromate of potash.

Wulfenite (molybdate of lead) occurs in small Tetragonal crystals, mostly of a yellow or yellowish-grey colour, but orange-red in some chromium or vanadium-containing varieties. The crystals are either tabular or more or less flattened parallel with the base, or are otherwise small pyramidal combinations. As pointed out by Von Kobell, a beautiful azure-blue coloration originates if the finely-powdered mineral be warmed with concentrated sulphuric acid in a porcelain capsule, and some alcohol be then added.

Stolzite (tungstate of lead) and Scheelite (tungstate of lime) crystallize also in the Tetragonal System, but the latter often occurs in crystals of half an inch or more in length, usually a simple square-based pyramid, measuring 130° 33′ over the base or middle edge. Stolzite has a very high sp. gr., 7·9-8·1, and is usually grey or brownish in colour, more rarely green or red. Scheelite has a sp. gr. of 5·9-6·2, and is commonly grey or greyish-yellow, though occasionally also brown, red, or green. Both, when warmed with nitric acid, leave a yellow residuum of WO_3, soluble in caustic alkalies.

Wolfram is readily distinguished from the other minerals of the Table by its dark-brown or red-brown colour and streak; and by the magnetic globule which it yields before the blowpipe. With carb. soda, also, it gives a strong reaction of manganese. Its crystals, as a rule, are of comparatively large size. As regards their general character, see Note to Table IX.

TABLE XXIII.

[Lustre non-metallic. Easily dissolved BB by borax or phosphor-salt, but not yielding any reaction of the preceding Tables.]

A.—Streak or Powder distinctly coloured.

A¹.—MAGNETIC, OR BECOMING SO AFTER STRONG IGNITION.

† *Anhydrous species.*

MAGNETITE (Magnetic Iron Ore): FeO 31, Fe^2O^3 69. Black, with black streak. In octahedrons and other crystals of the Regular System, and in lamellar and granular masses, rarely earthy. H 5·5-6·5 : G 4·9-5·2. Lustre, commonly sub-metallic. *See* Table IX.

MAGNOFERRITE : MgO, Fe^2O^3. In small black octahedrons, as a product of sublimation of Vesuvian fumeroles. Streak, dark-brown ; strongly magnetic ; G 4·65. Accompanies thin, tabular crystals of Iron Glance.

JACOBSITE : MgO, MnO, Mn^2O^3, Fe^2O^3. Reg. ; granular ; black ; streak, reddish-black ; H 5·5-6·0 ; G 4·74-4·77 ; strongly magnetic ; practically infusible. Strong Mn reaction BB with carb. soda. In crystalline limestone from Sweden.

FRANKLINITE : ZnO, MnO, Fe^2O^3. Reg., but commonly in small rounded masses. Black ; streak, brown or brownish-black. More or less magnetic in most examples. H 6-6·5 ; G 5·0-5·1. Lustre mostly sub-metallic : *See* Table IX. BB, in powder, with carb. soda and borax on charcoal gives a sublimate of ZnO. With carb. soda, also, strong Mn reaction.

CHROMITE : FeO, MgO, Al^2O^3, Fe^2O^3, Cr^2O^3. Reg., but commonly in granular masses. Brownish-black ; streak, dark-brown. Sometimes magnetic. Infusible. With borax, BB, fine green glass. Lustre, commonly sub-metallic. *See* Table IX.

ILMENITE (Titaniferous Iron Ore): Fe^2O^3, Ti^2O^3, but FeO also present in some varieties. Hemi-Hex.; iron-black, mostly with submetallic lustre. H 5-6 ; G 4·3-5·2, commonly about 4·9. The hydrochloric acid solution, diluted, and boiled with a piece of tin or zinc, becomes at first colourless and then violet. *See* Table IX.

RED IRON ORE (Hæmatite, Red Ochre, &c.). Fe^2O^3, with 70 per cent. Fe. Hemi-Hexagonal; but when of non-metallic aspect, mostly in fibrous-botryoidal, lamellar, or earthy examples. Red, brownish or bluish-red, with cherry-red streak. H 5-6, or lower (1·5-3) in earthy and sub-earthy varieties; G 4·8-5·3. BB, blackens and becomes magnetic. Fusible only in fine splinters. *See* also Table IX.

† † *Hydrous species. Yield water by ignition in bulb-tube.*

BROWN IRON ORE (= Gœthite, Limonite, Stilpnosiderite, Lepidokrokite, Yellow Ochre, &c. These, although commonly ranked as distinct species, cannot properly be regarded otherwise than as varieties of Brown Iron Ore, only differing from one another by their percentage of water, a character by no means absolutely constant): $Fe^2O^3 + m\ H^2O$, with Fe 60-63, and H^2O 10-15 per cent. Rh. (Gœthite), but mostly in fibrous-botryoidal, massive, and ochreous examples. Dark-brown, light-brown, brownish-yellow, with yellowish-brown or dull yellow streak; H 3·5-5·5 (but lower in ochreous and earthy varieties); G 3·2-4·2, commonly about 3.8-4·0. BB, yields water, blackens, and becomes magnetic. Fusible in thinnest splinters only. TURGITE is a closely related compound, but has a red streak, and yields only 5·5·5 per cent. water. G 3·5-4·5.

OXALITE (Humboldtine): FeO 42·10, Oxalic Acid 42·10, H^2O 15·80. In hair-like crystals, fibrous and earthy examples, of a yellow colour; H 2; G 2·1-2·25. BB, blackens, becomes magnetic, and then becomes converted into red iron-oxide. If a particle be a fused into a bead of borax, coloured blue by copper oxide, the latter becomes rapidly reduced to Cu^2O, and the glass becomes opaque red, or shews red streaks, on cooling. By this character, Oxalite is readily distinguished from yellow-ochre.

A².—YIELDING, BB, WITH CARB. SODA ON CHARCOAL, A DISTINCT SUBLIMATE AND METALLIC GLOBULES.

MINIUM (Mennige, Red Lead): Pb 90·7, O 9·3. Earthy; or pseudomorphous after galena or cerussite; red; streak, orange-yellow; H 2 (or less); G 4·6-4·8. BB, darkens, and fuses easily; on charcoal reduced. In HCl acid becomes transformed into white $PbCl^2$, with evolution of chlorine fumes. Partly soluble in dilute nitric acid, leaving residuum of puce-coloured PbO^2. Insoluble in caustic potash.

MASSICOT (Litharge; Bleiglätte): Normally, Pb 92·8, O 7·2, but always impure from presence of Fe^2O^3, &c. Fine scaly, earthy; sulphur-yellow, orange-yellow; paler in the streak; G 7·8-8·0. BB, easily fusible and reducible. Soluble in hot solution of caustic potash, and reprecipitating partly in crystalline scales.

BISMUTH OCHRE : Bi 89·7, O 10·3. In yellow, grey, or greenish crusts on Native Bismuth, &c. ; G about 4·5. Fusible into a yellow crystalline bead ; on charcoal reducible.

ZINCITE (Spartalite) : Gives zinc sublimate with carb. soda, but no metallic globules. See below.

A^s. —WITH CARB. SODA ON PLATINUM WIRE, DISTINCT MANGANESE REACTION.

† *Anhydrous Species.*

(*BB, zinc sublimate on charcoal. Streak, orange-yellow.*)

ZINCITE (Spartalite) : Normally, Zn 80·3, O 19·7, but always contains a certain percentage of Mn^2O^3. Hex., but commonly in lamellar or granular examples, often partly coated by white zinc-carbonate ; red ; streak yellow ; H 4 ; G 5·5-5·7. Infusible. Soluble in acids.

(*BB, no sublimate ; no copper reaction. Colour and streak, black or dark-brown.*)

BRAUNITE ; HAUSMANNITE : In small crystals (mostly Tetragonal octahedrons) or granular examples of dark-brown or iron-black colour and sub-metallic lustre   G 4·7-4·9. See Table X.

PYROLUSITE : MnO^2. Iron-black, very soft, mostly in fibrous masses of essentially sub-metallic lustre. ; infusible.. See Table X.

(*BB, with borax in RF, an opaque-red cupreous bead.*)

CREDNERITE : CuO 43, Mn^2O^3 57. In iron-black, cleavable masses of essentially sub-metallic lustre. See Table X.

†† *Hydrous Species.*

MANGANITE : Mn^2O^3, H^2O. In dark steel-grey or iron-black crystals and other examples of essentially sub-metallic (or metallic) lustre.   H 3·5-4 ; G 4·4. See Table X.

PSILOMELANE : MnO, MnO^2, HO^2, in somewhat variable proportions, with part of MnO replaced by BaO, K^2O, &c. (*See* Note on page 119). In black, granular or sub-fibrous masses, with brownish-black streak and more or less dull, earthy aspect; H 5-6 ; G 4·0-4·4. Infusible, or fusible on the edges in some examples. Occasionally of sub-metallic lustre. See Table X.

WAD : MnO, Mn^2O^3, H^2O, in variable proportions, part of the MnO always replaced by BaO, CaO, or K^2O. Properly, a mere mixture or decomposition product. In brown or black, earthy, scaly, stalactitic, or botryoidal examples, occasionally inclining to sub metallic in lustre, H 1·0-3·0 ; G 2·2-2·7. Practically infusible. Grogroilite is a mixture of similar character.

PYROCHROITE (weathered examples): MnO, H^2O, mixed with carb. lime, &c. Brown or black, in small druses in magnetic iron ore. An imperfectly known substance. Normally, white and pearly: see under § B, below.

(*BB, with borax, strong copper reaction.*)

LAMPADITE (Kupfermanganerz): CuO, MnO, BaO, CaO, Fe^2O^3, MnO^2, H^2O. Properly, a mixture or product of decomposition. Amorphous; black or brown; H 2·0-3·5; G 3·0-3·3. Infusible; soluble in HCl acid, with development of chlorine fumes. Kupferschwärze and Pelokonite are related mixtures.

ASBOLAN: CuO, CoO, K^2O, BaO, Fe^2O^3, MnO^2, H^2O. Resembles Lampadite or Wad in general characters, but contains cobalt oxide. Rabdionite is a similar cobalt-holding mixture.

A⁴.—GIVING COPPER REACTION, BUT NO MARKED REACTION OF MANGANESE.

CUPRITE (Red Copper Ore, Ruby Copper): Cu 88·8, O 11·2 (= the suboxide Cu^2O). Reg., commonly in small octahedrons or rhombic dodecahedrons often coated with malachite, also massive, &c. Red, bluish-red, with lustre frequently inclining to sub-metallic; streak, red; H 3·5-4; G 5·7-6·0. BB, tinges the flame green, blackens, melts, and on charcoal is reduced. Soluble in hydrochloric and in nitric acid, also in ammonia. See Note to Table IX. Tile Ore is a more or less earthy variety, mixed with Fe^2O^3, &c.

HYDROCUPRITE: Cu^2O + aq. A doubtful species, in orange-yellow coatings on magnetic iron ore from Pennsylvania. Recognized by Genth.

MELACONITE (Black Copper Ore): Cu 79·85, O 20·15 (= CuO). In black earthy coatings on certain copper ores, also massive and in pseudomorphous cubo-octahedrons. H 1·0-3·0; G 6·2-6·3. BB, fusible and reducible. See TENORITE (the same compound, but with metallic or sub-metallic lustre, from Vesuvius), in Table IX.

A⁵.—COLOURED STREAK OR POWDER, BUT NO REACTIONS OF FE, PB, MN, OR CU AS IN THE PRECEDING SECTIONS.

(*BB, with borax, strong Co-reaction.*)

HETEROGENITE: CoO, Co^2O^3, H^2O (21 per cent.?) mixed with quartz, brown iron ore, &c. Black or dark-brown; massive, botryoidal, earthy. A product of decomposition resembling Asbolan (*see above*) but giving no copper reaction.

(*BB, with borax, strong Ni reaction.*)

BUNSENITE : Ni 78·6, O 21·4. Reg. (minute octahedrons); H 5·5 ; G 6·4 ; brownish-green, yellowish-green. Infusible. With carb. soda on charcoal reducible to magnetic grains.

(*Uranium reaction. Soluble in nitric acid, the diluted solution giving with ammonia a yellow precipitate.*)

† *In bulb-tube no water, or traces only.*

PITCHBLENDE : UO, U^2O^3, more or less impure from presence of Fe, Pb, As, &c. In black or greenish-black granular masses or disseminated grains; H, commonly, about 5, but varying from 4 to 6 ; G 5·0-8·0. Infusible.

† † *In bulb-tube more or less water.*

CORACITE : Impure variety of Pitchblende from Lake Superior. Black ; streak, grey or greenish-grey ; H 4-5 ; G 2·4-5·0. Commonly mixed with CaO CO^2, SiO^2, &c.

GUMMITE : U^2O^3, H^2O, mixed with CaO, MgO, Fe^2O^3, P^2O^5, SiO^2, &c. In small granular masses, strings and scattered grains; H 2·5-3·5 ; G 3·9-4·3 ; yellow or yellowish-red ; streak, yellow. Infusible. Eliasite (red-brown, with yellow streak) is identical or closely related.

URAN OCHRE : U^2O^3, H^2O, but always more or less impure, and commonly mixed with uranium sulphate. In earthy or fine-fibrous crusts of a yellow colour, on examples of Pitchblende.

B.—Streak-powder uncoloured.

B¹.—REMAINING WHITE ON IGNITION IN BULB-TUBE.

† *Anhydrous Species.*

PERICLASE : Mg 60, O 40. Reg. (in minute octahedrons, cubo-octns., or cubes); cleavage, cubical ; H 6·0 ; G 3·65-3·75 ; dark-green ; vitreous ; infusible. Hitherto only found at Monte Somma in ejected limestone masses.

† † *Hydrous Species.*

BRUCITE : MgO 69, H^2O 31, but often partially converted into carbonate. Hemi-Hex. (R : R 82° 22′, but crystals mostly tabular from predominance of basal plane). Commonly, however, in scaly, foliated, and sub-fibrous masses. H 2 ; G 2·3-2·4 ; white, greenish-white ; lustre pearly on B plane. Infusible. Nemalite is an asbesti-form, fibrous variety, white or pale-bluish in colour.

VÖLKNERITE (Hydrotalcite): MgO, Al^2O^3, Fe^2O^3, CO^2, H^2O. A mixture of Brucite with alumina-hydrate, &c., or a product of decomposition. White; foliated, or in tabular hexagonal crystals; H 2; G 2·0-2·1. BB, exfoliates, but remains unfused.

B².—BLACKENING ON IGNITION IN BULB-TUBE.

(*After strong ignition, assume a pink colour by treatment BB with cobalt solution*).

BRUCITE: MgO, H^2O. Occasional examples: *see* above. Alkaline reaction after ignition.

(*With carb. soda, BB, strong manganese-reaction*).

PYROCHROITE: MnO 79·8, H^2O 20·2. In white, foliated masses, forming strings in certain examples of magnetic iron ore, but weathering brownish-black from conversion of the MnO into higher degree of oxidation. BB, blackens; infusible.

(*Ca-lines in spectroscope, and alkaline reaction, after ignition*).

WHEWELLITE: CaO 38·36, C^2O^3 49·31, H^2O 12·33. Clino-Rh.; in small (commonly twinned) crystals on certain examples of Calcite. Colourless; lustre vitreo-adamantine; H 2·5-2·0; G 1·83; infusible; by gentle ignition converted into CaO, CO^2.

NOTE ON TABLE XXIII.

This Table is composed essentially of Oxides. The more commonly occurring species, belonging to it, may be grouped in four series, as follows:—(1), Iron Ores and related compounds; (2), Manganese Oxides; (3), Red Zinc and Copper Oxides; and (4), the magnesia hydrate, Brucite.

The Iron Ore group comprises, chiefly, (*i*) the anhydrous species of Regular crystallization, Magnetite, Franklinite, and Chromite (with common formula RO, R^2O^3); (*ii*) the anhydrous Hemi-Hexagonal species, Hæmatite and Ilmenite (with common formula R^2O^3); and (*iii*) the hydrous species, conveniently ranked together under the common name of Brown Iron Ore (with common formula = $R^2O^3 + m\ H^2O$). All the species of this group become magnetic after ignition or semi-fusion, and several are magnetic in their normal condition. In most cases the finely powdered ore dissolves without much difficulty in hot hydrochloric acid, but Chromite, Ilmenite, and titaniferous-holding Magnetite are exceptions. The two latter in the form of very fine powder generally yield to slow digestion (in a small, covered beaker on a sand bath, the acid being kept just at the boiling point), but Chromite (unless mixed with magnetite) is very slightly attacked. It may be decomposed however (sufficiently for determinative purposes) by gentle fusion, in fine powder, with a mixture of carb. soda, borax, and nitre. By this treatment an alkaline chromate, soluble in water, is formed. The solution, decanted from the insoluble residuum, may then be evaporated to dryness, and the resulting deposit fused with borax for the production of a chrome-green glass. The presence of chromium may also be shown by the deep green coloration produced by addition of sulphuric acid and alcohol : *see* PART I., page 49.

Comparatively few examples of Magnetite are referrible to the present Table, as in most specimens of that mineral the lustre is unmistakably metallic or sub-metallic (*see* TABLES VIII. and X.). Some examples, however, are obscurely metallic in aspect. These are black in colour, with black streak, and strongly magnetic. Commonly in granular or lamellar masses, with G averaging 5·0. When crystallized, in octahedrons and rhombic dodecahedrons.

Franklinite and Chromite much resemble examples of Magnetite with obscurely metallic lustre. They are mostly in black, granular masses, with normally dark brown or red brown streak, but the latter is often black from presence of magnetite, or greenish from intermixed chloritic or pyroxenic rock-matter. Franklinite is often strongly magnetic (probably from presence of Fe^3O^4). Chromite is only occasionally magnetic, and its specific gravity falls below 4·6, averaging usually 4·3 or 4·4. Franklinite with carb. soda, BB, forms a turquoise enamel (Mn reaction), and gives on charcoal (if treated in powder with carb. soda and borax) a sublimate of ZnO. Chromite with borax gives (on cooling) a fine green glass. See also its reactions described above.

Ilmenite resembles the above minerals by its black colour and brownish or black streak, as well as by its frequent occurrence in granular or scaly granular masses; but its crystals are rhombohedral combinations closely resembling those of Hæmatite (R: R 85°31′). It is most readily distinguished by the deep amethystine colour which results when its hydrochloric acid solution (somewhat diluted) is boiled for a few minutes with a piece of tin.

Hæmatite occurs under several more or less distinct conditions ; but in most cases it presents a metallic or well-marked sub-metallic aspect, and is thus referred to in preceding Tables (*see* Notes to TABLES VIII. and X.). The examples belonging more especially to the present Table commonly come under the designation of Red Iron Ore, of which Reddle or Red Ochre is an earthy variety. In these, the streak is always distinctly red, and the colour either brick-red, brownish-red, or bluish-red, the lustre in the latter case merging into sub-metallic. The harder examples are very frequently in fibro-botryoidal masses. BB, in the RF, all blacken and become magnetic.

Brown Iron Ore includes several so-called species or sub-species, compounds of Fe^2O^3 with variable amounts of water. All yield a yellow or yellowish-brown streak ; and all become red by ignition with free access of air (especially in powder), the water being driven off. Ordinary varieties assume a bright red colour on ignition, but varieties which contain much manganese give a dull-red or chocolate-red powder. Before the blowpipe in a reducing flame, all become black and magnetic, and fine splinters exhibit fusion. Practically, these compounds may be referred to three series :—(i) a series, typified by Gœthite, in which the water averages 10 per cent., the formula being Fe^2O^3,

H^2O ; (ii) a second series, typified by Limonite, the formula of which may be written Fe^2O^3, 3 H^2O, with 14 to 15 per cent. water ; and (iii), a series of Bog ores and Ochres containing 20 per cent. or more water, and having part of the iron in the condition of FeO combined with humic or other organic acid. No very strict lines of demarcation can be drawn, however, between these varieties. Gœthite, although frequently in fibrous and other examples, occurs occasionally in thin-scaly and acicular crystals of the Rhombic System. The other Brown Ores are unknown in true crystals, although cubes and other pseudomorphs derived from Iron Pyrites are not uncommon. They occur chiefly in fibro-botryoidal, granular, and earthy masses. Many of the fibrous examples present a silky lustre, and some are comparatively light in colour. Many brown ores, also, shew a variegated surface-tarnish.

The group of Manganese Oxides—referrible as regards some examples to the present Table—includes the comparatively rare species Braunite and Hausmannite, characterized chiefly by occurring in small Tetragonal crystals of a brownish-black colour and more or less sub-metallic aspect (see Table X.); certain examples of Pyrolusite and Manganite, occurring mostly in dark fibrous masses or crystal groups, usually of metallic or well-marked sub-metallic lustre (see Tables VIII. and X.); and the amorphous Psilomelane, with the earthy, ochreous mixtures known as Wad. The two latter alone belong properly to this Table ; and Psilomelane in many of its examples presents a more or less metallic aspect (see Table VIII.). These manganese oxides, if warmed in powder with hydrochloric acid, cause the evolution of chlorine fumes, a character by which they are readily distinguished from bodies of similar aspect. The green-blue enamel which they form, BB, with carb. soda, is also highly distinctive. Ignited by a Bunsen-flame and examined by the spectroscope, nearly all examples shew green Ba-lines, and Psilomelane and many Wads shew in addition the red K-line, and occasionally the crimson Li-line. (See Foot Note, page 119).

Pyrolusite and Wad are of low hardness, and thus soil more or less distinctly. Wad yields water on ignition ; Pyrolusite is anhydrous. The other manganese oxides of natural occurrence range in hardness from about 4·0 (Manganite) to 5·5 or 6·0. Psilomelane and Manganite yield water on ignition : Braunite and Hausmannite are anhydrous; but, as already remarked, these latter species scarcely require notice in the present Table, as their lustre in ordinary examples is at least sub-metallic.

The group of red zinc and copper oxides includes merely Zincite and Cuprite. Zincite or Red Zinc Ore, ZnO (with part replaced by MnO), is chiefly distinguished by its red colour, orange streak, and infusibility. With carb. soda and borax, BB, on charcoal, it gives a characteristic zinc sublimate, and also a strong reaction of manganese. It occurs chiefly in cleavable and scaly-granular masses, usually associated with Franklinite. The crystallization is Hexagonal, with basal cleavage, but crystals are rarely met with.

Cuprite (Red Copper Ore or Ruby Copper) occurs commonly in octahedrons (often with sunk faces) and in rhombic dodecahedrons and other forms and combinations of the Regular System, frequently converted into green carbonate on the surface. It is also found in acicular groups and in lamellar and other masses; and in a dull, sub-earthy condition (mixed with Fe^2O^3, &c.) forming the so-called "Tile Ore." Its more distinctive characters are its red colour and streak, and its easy reduction, BB on charcoal, to metallic copper. It dissolves with effervescence and production of coloured nitrous fumes in nitric acid, forming (as in the case of copper compounds generally) a green solution which becomes intensely blue on addition of ammonia.

Brucite, MgO, H^2O, is easily distinguished from the other commonly occurring minerals of this Table by its white streak, softness, pearly aspect, and its magnesia-reaction, BB, with nitrate of cobalt. On ignition it evolves 30 to 31 per cent. water, and reacts alkaline.

TABLE XXIV.

[Lustre non-metallic. BB, slowly attacked or only in part dissolved by borax or phosphor-salt. Infusible, or fusible on thinnest edges only. Hardness sufficient to scratch ordinary window-glass distinctly.*]

A.—Insoluble (in powder) in hydrochloric acid.

A¹.—SPECIFIC GRAVITY OVER 5·0.

† *With carb. soda and a little borax, BB, yielding metallic tin.*

CASSITERITE (Tinstone):—Sn 78·62, O 21·38, but most examples contain traces of Fe^2O^3, Mn^2O^3, &c. Tetragonal (crystals often twinned), see Note at end of Table; also massive and in rolled pebbles (= stream-tin, wood-tin) often with sub-fibrous structure. Brown, black, grey, reddish, &c., rarely colourless; H 6·0-7·0; G 6·7-7·0. Infusible, but reducible on charcoal (especially if fused with carb. soda, cyanide of potassium, or neutral oxalate of potash).

†† *With carb. soda, BB, forming a slaggy mass or remaining undissolved. Streak more or less distinctly coloured.*

[This subsection includes only some comparatively rare species (essentially Tantalates, Niobates, Nio-titanates) in which the lustre on the fractured surface is distinctly sub-metallic, at least in typical examples. These species belong properly, therefore, to Table X. When they occur in a fragmentary form, or are indistinctly crystallized, their correct determination is not easily effected. In most examples, traces of tin are obtained by the reduction process with carb. soda and borax; and by fusion in fine powder with bisulphate of potash, all are more or less decomposed, the fused mass becoming blue when warmed with a few drops of hydrochloric acid and a piece of tin or zinc.]

(*BB, unchanged*).

TANTALITE: FeO, Ta^2O^5, &c.; Rhombic; black; H 6·0-6·5; G 6·3-8·0, usually 7·0-7·5.

---

* Minerals in which the normal degree of hardness scarcely exceeds 5·0 do not scratch glass very distinctly; and if slightly weathered or altered they may not scratch glass at all. To avoid risk of error, therefore, infusible silicates of this character are placed both in the present Table and in Table XXV. In trying the hardness of a mineral by a piece of glass, the glass should be laid flat on a table, and the mineral drawn with rather strong pressure sharply across it—care, of course, being taken that no particles of quartz are attached to the substance. Several species placed in this Table are not absolutely infusible when tested in the form of a very fine splinter, although melting even then at the extreme point only, and requiring practice on the part of the operator to effect this; but, to avoid uncertainty in cases of this kind, the species in question are referred to, again, in either Table XXVI. or Table XXVII.—the first containing fusible anhydrous silicates, and the latter, hydrated species.

COLUMBITE (Dianite): FeO, MnO, Nb^2O^5, Ta^2O^5, &c.; Rhombic; black, generally somewhat iridescent; H 6·0; G 5·37-6·5. In fine powder partially attacked by hot sulphuric acid.

MENGITE: YO, CeO, ZrO^2, TiO^2, &c.; Rh.; black; H 5·5; G 5·48. Decomposed by hot sulphuric acid.

(*BB, becoming yellow or pale-greyish and yielding a little water in the bulb-tube*)

YTTROTANTALITE: YO, ErO, FeO, Ta^2O^5, WO^3, &c.; black, brownish, yellow, often spotted; H 5·0-5·5; G 5·4-5·8.

FERGUSONITE; POLYCRASE; EUXENITE; ÆSCHYNITE:—*See* TABLE X., pages 126, 127.

(*BB, partially fused or attacked on the surface or edges*).

SAMARSKITE: YO, FeO, CeO, U^2O^3, Nb^2O^5, Ta^2O^5, &c.; Rhombic; black; streak red-brown; H 5·0-6·0; G 5·6-5·8. Decomposed in powder by hot sulphuric acid.

A². —SPECIFIC GRAVITY 3·3-5·0.

† *With carb. soda, BB, forming a slag only, or remaining undissolved.*

(*H* = 10. *In fine powder slowly combustible*).

DIAMOND (Crystallized Carbon):—Reg., crystal-faces often curved (*see* Note at end of Table). Colourless, pale yellowish or variously tinted, sometimes black; lustre strongly adamantine; H 10; G 3·5-3·55, but in the black "carbonado" variety sometimes slightly lower. BB, in fragments, unaltered *per se* and not attacked by the fluxes, but in fine powder slowly combustible.

(*BB, with Co-solution, Al^2O^3 reaction*).

CORUNDUM (Sapphire, Ruby, Adamantine Spar, Emery): Al 53·2, O 46·8 (= Al^2O^3). Hexagonal (*see* Note at end of Table). H 9·0; G 3·8-4·2, usually 3·9-4·0. Pink, blue, red, brownish, colourless, dark-grey—the latter in the opaque variety Emery; many crystals colourless at one extremity, and blue or reddish at the other. BB, quite infusible; the powder fused with bisulphate of potash forms a salt soluble in water. Ammonia throws down gelatinous Al^2O^3 (generally somewhat brownish from accompanying Fe^2O^3) from the solution.

DIASPORE: Al^2O^3 85, H^2O 15. Rhombic, but often in foliated or scaly masses; H 6-6·5; G 3·3-3·5. Colourless, white, brown, violet, greenish, &c. In bulb-tube generally decrepitates, gives off water, and falls into scaly particles. BB, like Corundum.

TOPAZ: Al^2O^3, SiO^2, Fl. Rhombic (*see* Note at end of Table); H 8·0; G 3·5-3·57; yellow of various shades, pale bluish-green, reddish-white, colourless; cleavage very perfect, parallel with basal plane. Infusible, but becomes colourless and loses polish on strong ignition. BB, with fused phosphor-salt in open tube, gives fluorine reaction. Pycnite and Physalite are columnar, opaque or semi-opaque reddish-white or straw-yellow varieties.

CHRYSOBERYL (Cymophane): BeO 19·8, Al^2O^3 80·2; Rhombic (*see* Note at end of Table); H 8·0-8·5; G 3·65-3·85; green of various shades, greenish-white (often shewing a floating opalescence), and in many examples pale-red by transmitted light. BB, like Corundum.

SPINEL: Normally, MgO 28, Al^2O^3 72, but part of the MgO commonly replaced by FeO, and part of the Al^2O^3 by Fe^2O^3. Reg. (crystals mostly small octahedrons, often twinned: *see* Note at end of Table). H 8·0; G 3·5-4·1, usually about 3·55-3·6. Red, blue, green, of various shades; reddish-white, black, rarely colourless. BB infusible, but many red varieties appear green whilst hot. Decomposed in powder by fusion with bisulphate of potash.

SAPPHIRINE: Essentially composed of MgO, FeO, Al^2O^3, SiO^2. Occurs in small granular masses in mica-slate from Greenland; light-blue, bluish or greenish-grey; H 7·5; G 3·42-3·47. Infusible.

CYANITE (Disthene): Al^2O^3 62·10, SiO^2 36·90. Anorthic, but chiefly in bladed or flat-fibrous masses; H 7·0 on edges of crystals or laminæ, 5-5·5 on flat surfaces; G 3·48-3·68. Bluish-white, light-blue, grey, pale-green, reddish-white, tile-red. Infusible.

(*Zn reaction by fusion in powder with mixture of carb. soda and borax on charcoal*).

GAHNITE (Automolite): ZnO 38·7, Al^2O^3 61·3. but small amounts of MgO, MnO, FeO and Fe^2O^3 also frequently present. Reg. (crystals mostly small octahedrons, commonly twinned as in Spinel); H 7·5-8·0; G 4·0-4·6; dark green, greenish-black. Dysluite is a manganese-holding variety; Kreittonite a ferruginous variety. BB, infusible; the powder fused with equal parts of carb. soda and borax gives a zinc sublimate.

(*Fe reaction**).

PLEONASTE (Ceylanite): Black or dark-green variety of SPINEL, *see* above, containing as a rule too much iron to give a distinct

---

* A small particle, or some of the powder, added to a bead of borax coloured by copper-oxide, quickly reduces part of the CuO to red Cu^2O.

Al^2O^3 reaction with Co-solution. HERCINITE (mostly in small dull black granular masses) is still more ferruginous, practically all the MgO being replaced by FeO. In these dark varieties the sp. gr. is usually about 3·9 or 4·0.

STAUROLITE (Staurotide): Composed essentially of FeO, MgO, Al^2O^3, SiO^2. Rhombic: (crystals often cruciform twins, essentially rhombic prisms, with V : V near 129°, truncated on acute edges); H 7·0-7·5; G 3·4-3·8; brownish-red, dark-brown. BB (as regards true Staurolite) infusible. In powder attacked by sulphuric acid See Note at end of Table.

(*Chrome reaction: BB, with borax, emerald-green glass*).

UWAROWITE (Ouvarovite, Chrome Garnet): CaO, Al^2O^3, Cr^2O^3, SiO^2. Reg. (crystals, small rhombic-dodecahedrons); H 7·5; G 3·4-3·53; bright green. Infusible.

(*Sp. gr. 4·0-4·7*).

ZIRCON (Hyacinth): ZrO^2 67, SiO^2 33. Tetrag. (crystals, commonly, eight-sided prisms with pyramidal termination); H 7·5; G, usually about 4·4; yellowish-brown, grey, light-brown, red, rarely greenish or colourless. Infusible. Slowly attacked by sulphuric acid. *See* Note at end of Table. •Auerbachite, Ostranite, and Malakon (Tachyaphalite), are probably slightly altered varieties, the latter yielding 3 per cent. water. H about 6·5; G 3·9-4·1.

(*Yielding water in bulb-tube*).

OERSTEDITE: MgO, ZrO^2, TiO^2, SiO^2, H^2O (5·6 per cent.). Tetrag.; H 5·5-6·0; G 3·63; red-brown, brownish-yellow, with adamantine lustre. A rare, imperfectly-known species, allied to and resembling Zircon.

MALAKON: An altered Zircon : see above.

† † *With carb. soda, BB, dissolving more or less readily or forming a fused glass.*

(*Titanium reaction*).

RUTILE : TiO^2. Tetragonal (crystals essentially prismatic, often geniculated twins, sometimes acicular); H 6·0-6·5; G 4·2-4·3; red (with strong adamantine, often sub-metallic, lustre), black (Nigrine, mostly in rolled pebbles), yellowish-brown; streak, pale-brown. BB, unchanged. Fused in fine powder with carb. soda (or better with caustic soda or potash) forms a salt soluble in hydrochloric acid, the solution, slightly diluted and boiled with a piece of tin or zinc, assuming a violet colour.

ANATASE or OCTAHEDRITE: TiO^2. Tetrag. (crystals, small square-based octahedrons or pyramids); H 5·5-6·0; G 3·8-4·0; indigo-blue, brownish, yellowish-grey, with adamantine often sub-metallic, lustre. BB, like Rutile.

BROOKITE: TiO^2. Rhombic ? (V : V 99° 50'; V : $\overline{V}$ 139° 55'; P : P in front 115° 43', at side 101° 35', crystals mostly tabular); H 5·5-6·0; G 4·0-4·25; light-brown, yellowish, reddish, black in Arkansite variety; lustre adamantine to sub-metallic. BB, like Rutile.

A^3.—SPECIFIC GRAVITY UNDER 3·3.

† *With carb. soda, BB, forming a slag or semi-fused mass.*

TOURMALINE (Light-coloured, red, green, and other infusible varieties): Essentially composed of MgO, Al^2O^3, B^2O^3, SiO^2, with small amounts of Na^2O, Li^2O, Fl, &c. Hemi-Hexag. (crystals mostly nine-sided prisms, longitudinally striated, with differently modified summits: R : R about 133° 10', -½ R 152° - 2 R 103° 3'); green, brown, red (Rubellite), blue (Indicolite), colourless;* H 7·0-7·5: G 2·9-3·2; pyro-electric. Infusible, or slightly attacked BB on thin edges, as regards the varieties belonging to this Table. The powder ignited in a platinum spoon and boiled with a few drops of sulphuric acid, communicates a green tinge to the flame of alcohol or to the point of the blowpipe-flame. In many varieties, also, the ignited powder moistened with hydrochloric acid shews the red Li-line in the spectroscope.

ANDALUSITE (Chiastolite): Al^2O^3 63, SiO^2 37. Rhombic (V : V 90° 50' - 91° 4'); H, normally, 7·0-7·5, but often lower from partial alteration; G 3·10-3·20; greyish-white, pearl-grey; pale violet, red, reddish-white, greenish. BB, infusible; with Co-solution, after ignition, assumes a fine blue colour. CHIASTOLITE is a variety in narrow straw-like crystals, or occasionally in thick prisms, imbedded in clay slate, mica slate, &c., and presenting on the transverse section a dark cross or black lozenge-shaped figure arising from a symmetrical arrangement of the rock-substance in the centre and at the angles of the hollow prismatic crystal.

---

* The black opaque varieties known as Schorl, and many brown varieties are easily fusible. See TABLE XXVI.

SILLIMANITE: Al^2O^3 36·9, SiO^2 63·1. Rhombic in crystn., but commonly in fibrous or bladed examples; H, normally, 6-7; G 3·2-3·3; pale brown, yellowish-grey, greenish. BB, like Andalusite and Cyanite, these three minerals being identical in composition and closely related in other respects. Fibrolite, Bucholzite, Xenolite, Monrolite, and Wœrthite, are varieties.

IOLITE (Dichroite, Cordierite): MgO, FeO, Al^2O^3, SiO^2, with usually traces of MnO, and frequently (from alteration) a small amount of H^2O. Rhombic (mostly in short stout crystals of pseudo-hexagonal aspect, with V : V 119° 10′), but commonly in granular examples; H, normally, 7·0-7·5; G 2·5-2·7; blue, smoky-grey; brownish or yellowish in certain directions by transmitted light. BB, fusible with difficulty on thin edges; with Co-solution becomes bluish-grey or pale-blue.

† † *With carb. soda BB forming a fused glass or bead.**
(Cleavage-planes more or less distinct).

EUCLASE: Essential composition: BeO, Al^2O^3, SiO^2 (41-43 per cent.), with a small percentage of water only driven off by intense and prolonged heat, and therefore not detected in ordinary blowpipe operations. Clino-Rhombic: (crystals much resembling the common augite crystals,† small and brilliant); H 7·5; G 3·0-3·1; colourless, pale-green, bluish-white. BB, in fine splinters, becomes opaque, blisters slightly, and becomes rounded at the extreme point. With carb. soda in proper proportion, forms an opaque pearl. A very rare species.

BERYL (Emerald): BeO 14·14, Al^2O^3 19·05, SiO^2 66·84, with traces of Fe^2O^3, and in the bright-green varieties (Emerald) a small amount of Cr^2O^3. Hexagonal (crystals mostly six-sided prisms with large basal plane); H 7·5-8·0; G 2·66-2·76; pale green, greenish-white, emerald-green, occasionally pale yellow, bluish, or quite colourless. BB, in fine splinters, becomes opaque white, and melts with difficulty at the extreme point.

PHENAKITE: BeO 45·78, SiO^2 54·22. Hex. or Hemi-Hex.; H 7·5-8·0; G 2·9-3·0; colourless, pale yellowish. BB, infusible. With

---

* The flux should be added little by little. With too much, or too small a quantity, imperfect results are obtained.

† By atomic constitution, and also by crystallization, Euclase is regarded as related to Datolite; but the actual composition and geological relations of these minerals are very different.

small amount of carb. soda melts to a white bead; with larger quantity forms a slag. A very rare species.

ENSTATITE: MgO 40, SiO^2 60, but with part of the MgO replaced by small amount of FeO. Rhombic (V : V 91° 44' – 93°); mostly in greenish-white, grey, or green cleavable masses; H 5·5 to nearly 6·0; G 3·10-3·29. BB, fusible on thinnest edges only. Bronzite, commonly regarded as identical, is here kept distinct on account of its inferior hardness. See TABLE XXV.

ORTHOCLASE (Potash Feldspar): K^{2}O 16·9, Al^{2}O^3 18·4, SiO^2 64·7. Clino-Rh. (crystals often twinned: see Note to TABLE XXVI.); commonly in cleavable masses (the adjacent cleavage-planes meeting at 90°) of a white, red, greyish, or light-green colour; H 6·0; G 2·5-2·6. BB, fusible with difficulty or on the edges only, but a fine splinter is readily vitrified at the point. Red K-line clearly visible in spectroscope if the powder be ignited and then moistened with hydrochloric acid, or fused with carb. soda.

ALBITE (Soda Feldspar): Essentially, Na^{2}O 11·8, Al^{2}O^3 19·6, SiO^2 68·6. Anorthic (crystals often twinned: see Note to TABLE XXVI.); commonly in white, red, or other-coloured cleavable masses, with adjacent cleavage-planes meeting at 93° 36' and 86° 24', one of these planes being generally striated. H 6·0; G 2·58-2·64. BB, (in fine splinters) difficultly fusible, tinging the flame-border strongly yellow.*

TRITOMITE: Normally, pure SiO^2, but differing from quartz (although belonging to the same System of Crystallization) by the character of its crystals, the indications of cleavage which it shews in one direction, its somewhat lower sp. gr., and its solubility in a saturated, boiling solution of carb. soda. H 7·0; G 2·28-2·33; colourless, opaque white. The crystals are mostly tabular from predominance of the basal plane (practically unknown in quartz), or in fan-shaped or other twins.

ASMANNITE: Normally SiO^2, but differing essentially from quartz and tritomite by its Rhombic crystallization, closely identical with that of Brookite. H 5·5; G 2·245-2·247. Recognized by Maskelyne

---

* These feldspars are referred to in the present Table, because they are commonly regarded as infusible by students who have had but little practice with the blowpipe, or who persist in testing fragments of too large a bulk. They are described again in their proper place, with Anorthite and other distinctly fusible feldspars, in TABLE XXVI.

(in small cleavable grains with indications of rhombic crystallization) in the meteoric iron of Breitenbach in Bohemia.

(*No observable cleavage planes*).

QUARTZ (Rock crystal, Amethyst, Calcedony, Agate, &c.): Normally, pure silica; Si 46·67, O 53·33, but often coloured by traces of Fe^2O^3, Mn^2O^3, &c. Hexagonal or Hemi-Hexagonal (*see* Note at end of Table); crystals, commonly six-sided prisms striated transversely and terminated by a six-sided pyramid; often massive, botryoidal, granular; H 7·0; G 2·5-2·8 (clear examples and crystals commonly about 2·65); colourless, white, violet, smoky-brown, pink, red, green, grey, black, &c., the colours of massive examples often in stripes or spots: *see* Note at close of Table. BB unchanged. With carb. soda fusible. with effervescence (due to expulsion of CO^2) into a clear glass.

OPAL (Hyalite, &c.): SiO^2, with from 2 to 20 per cent. H^2O: the latter usually 3 – 10 per cent. Opaque and strongly coloured varieties also contain intermixed Fe^2O^3 and other impurities. Uncrystalline, and thus normally without action on polarized light. In nodular, botryoidal, and other massive examples; H (normally) 5·5-6·5; G 1·5-2·5, commonly 1·9-2·2; colourless, bluish-white, yellowish-red; with internal play of colours or iridescence (Noble Opal, Girasol, Fire-Opal); also colourless, forming vitreous coatings or botryoidal masses on lava (Hyalite); or white, yellow, brown, red, bluish-grey, &c., often in stripes or patches in the same specimen, and with more or less waxy or sub-resinous lustre (Common Opal, Semi-Opal, Wood Opal, &c.). BB usually decrepitates; in the bulb-tube yields a little water; otherwise like quartz. In powder, soluble in hot solution of caustic potash. Jasper Opal is an opaque red, dull-yellow or brown variety, mixed with a considerable amount of Fe^2O^3 or Fe^2O^3, H^2O. Menilite is a light-brown or bluish-grey variety in flat nodular pieces. Pearl-sinter, Siliceous Sinter, Geyserite, &c., are stalactitic, encrusting or porous varieties, deposited by many hot springs. Tripoli, Polishing Earth, Randanite, are forms of amorphous silica, made up of minute tests or coverings of diatoms.

B.—Readily decomposed or dissolved (in powder) by hot hydrochloric acid.*

B¹.—YIELDING NO WATER (OR TRACES ONLY) BY IGNITION IN BULB-TUBE.

† *Decomposed, without gelatinization, by hydrochloric acid.*

LEUCITE: K^2O 21·53, Al^2O^3 23·50, SiO^2 54·97, but part of the K^2O commonly replaced by Na^2O. Tetrag., but crystals closely resembling a trapezohedron of the Regular System. H 5·5-6·0; G 2·45-2·50; white, light-grey, yellowish or reddish-white. Only found in crystals or small rounded masses in certain lavas. Infusible; with Co-solution, BB, assumes a bright blue colour. In fine powder, decomposed by hydrochloric acid, with separation of granular silica. Shews red K-line distinctly in spectroscope when ignited and fused with carb. soda or moistened with hydrochloric acid.

POLLUX: Cs^2O, Na^2O, Al^2O^3, SiO^2, with about $2\frac{1}{2}$ per cent. H^2O, the latter easily escaping detection in the examination of small fragments. Reg. (crystals very minute combinations of cube and trapezohedron 2-2). Commonly in small camphor-like colourless masses. H 5·5-6·5; G 2·8-2·9. Fusible only on thin edges. The powder heated with fluoride of ammonium and then moistened with hydrochloric acid shews in the spectroscope the two characteristic Cæsium lines. These are bright blue and close together, one being almost in the position of the blue Sr-line. A rare species, hitherto only found in the Island of Elba.

† † *Decomposed, with separation of gelatinous silica, by hydrochloric acid.*

(*Zn reaction: characteristic ring-deposit on charcoal by fusion of test-substance with carb. soda*).

WILLEMITE: ZnO 73, SiO^2 27. Hemi-Hex. (crystals commonly six-sided prisms terminated by an obtuse rhombohedron of 128° 30', but very small, and often with rounded edges); H 5·5; G 3·9-4·2; white, brownish, red, green, &c. Infusible, or attacked, BB, on thinnest edges only.

(*Zn and Mn reactions*).

TROOSTITE: Like Willemite in composition but with part of the

---

* Reduce a small fragment (5 or 6 grains, or less) of the test-substance to powder; place this (by means of a folded slip of glazed paper) at the bottom of a clean test-tube; twist a rolled-up piece of soft paper round the top of the tube to serve as a handle, the ends of the paper being twisted together; cover the powder to the depth of about half-an-inch with strong hydrochloric acid, and boil gently (letting the flame touch the side of the tube near the top of the acid) for two or three minutes.

ZnO replaced by MnO and FeO. Hemi-Hex. (crystals comparatively large, mostly six-sided prisms with rhombohedral terminations). Commonly opaque or semi-opaque, yellowish-grey, greenish or brown. BB with carb. soda forms a turquoise-enamel. Otherwise like Willemite. Properly, a manganese variety of the latter species.

(*Fl reaction with sulphuric acid*).

CHONDRODITE: MgO, FeO, SiO^2 (33-37 per cent.), MgFl^2. Clino-Rh., but commonly in small granular masses of a yellow, yellowish-white, reddish, brown, or green colour imbedded in cryst. limestone; H 6·0-6·5; G 3·0-3·25. BB infusible, or rounded only on thinnest edges. CLINO-HUMITE is closely related.

HUMITE: a Chondrodite of Rhombic crystallization. In small crystals with numerous pyramidal planes, and generally a well-developed basal plane, chiefly from Monte Somma, but recognized also by E. Dana (with Chondrodite and Clino-Humite) from Brewster, N.Y.

(*No Zn or Fl reaction. G 3·0 to 3·5*).

CHRYSOLITE or OLIVINE (Peridot): Average composition, MgO 49, FeO 10, SiO^2 41; but in some varieties the FeO is higher, and MnO and TiO^2 are occasionally present. Rhombic, but often in small granular masses in basalt, &c. H 6·5-7·0; G 3·2-3·5. Green of various shades, yellow, brownish, rarely yellowish-red. BB, infusible, except as regards some very ferruginous varieties (Hyalosiderite, &c.) which yield a magnetic slag or globule: see TABLE XXVI. Forsterite (Boltonite) is identical in composition, crystallization and other characters. Hortonolite and Glingite are ferruginous varieties.

MONTICELLITE (Batrachite): Average composition, CaO 35, MgO 22, FeO 5·5, SiO^2 37·5. Rh.; H 5·5; G 3·12; colourless, greyish, pale greenish or yellowish-grey. BB, rounded on thinnest edges only. Ignited and then moistened with HCl acid, shews in spectroscope momentary red and green Ca-lines.

GEHLENITE: Essential composition, CaO, Al^{2}O^3, SiO^2 with small amounts of MgO, FeO, Fe^{2}O^3, and H^{2}O; Tetrag. (crystals chiefly simple square prisms); H 5·5-6·0; G 2·98-3·10; pale greenish-grey, green, brownish. BB, rounded on thin edges. In spectroscope (after ignition and moistening with HCl) shews Ca-lines very distinctly.

(*G 4 or higher; colour, black*).

GADOLINITE: YO, CeO, BeO, FeO, SiO^2, with traces of H^{2}O, and occasionally small amounts of ErO, CaO, &c. Rhombic or Clino-Rh., but chiefly in small granular masses without distinct cleavage. Black, greenish-black; streak greenish-grey; H 6·5-7·0; G 4·0-4·3. BB, many varieties emit a peculiar glow, and most examples swell up slightly and become greenish-grey, but none exhibit fusion, properly so-called.

B².—YIELDING WATER ON IGNITION.*

(*BB, strong Cu-reaction with borax, or when moistened with hydrochloric acid*).

DIOPTASE: CuO 50·44, SiO^2 38·12, H^2O 11·44. Hemi-Hexagonal (crystals chiefly combinations of hexag. prism and rhombohedron, with angle of 95° 28' over polar edges of the latter); cleavage rhombohedral, with R:R 125° 54'; bright emerald-green, with paler streak; H 5·0-5·5; G 3·27-3·35. BB, generally decrepitates, blackens, but remains unfused. With carb. soda, on charcoal, gives metallic copper. Gelatinizes in hydrochloric acid. A rare species. The amorphous copper silicate, Chrysocolla, has normally a low degree of hardness, and is decomposed by hydrochloric acid without gelatinization. *See* TABLE XXV.

(*BB, with carb. soda on charcoal, zinc reaction*).

CALAMINE: ZnO 67·5, SiO^2 25, H^2O 7·5. Rhombic (crystals hemimorphic, *i.e.*, with different terminations, but generally small, and somewhat indistinct); H 5·0; G 3·3-3·5; colourless, or variously tinted. The crystals pyro-electric. Frequently in botryoidal and other massive examples. BB, infusible; commonly decrepitates. Decomposed with gelatinization by hydrochloric acid.

(*No reactions of Cu or Zn. G 4·9 to 5·0*).

CERITE: CeO, SiO^2, H^2O (6·12 per cent.), but with part of the CeO constantly replaced by LaO, DiO, CaO, &c. Hexag. (?); mostly in massive examples of a red, reddish-grey, or brownish colour; H 5·5; G 4·9-5·0. Gelatinizes in hydrochloric acid. The solution (if not too acid) gives with oxalic acid a white precipitate which becomes converted into tile-red Ce^2O^3 by ignition in the platinum spoon (Von Kobell).

(*G under 3·0*).

POLLUX: Yields on ignition a very small amount of water. Mostly in small colourless camphor-like masses. *See* under B¹, above.

---

* The minerals of this section belong properly to Table XXV., as they scratch glass more or less indistinctly, but to avoid risk of error in their determination they are referred to also here.

NOTE ON TABLE XXIV.

This Table includes a series of hard, infusible or very difficultly fusible minerals of vitreous or other non-metallic lustre; with, in addition, a few species in which the lustre is occasionally sub-metallic. These latter are comparatively rare, and they belong normally to Table X.

The following are the only species of importance, or of ordinary occurrence, which possess sufficient hardness to scratch glass distinctly :—(1) The Diamond; (2) a group of closely allied Tetragonal species, comprising: Cassiterite, Rutile, Anatase, Zircon; (3) the purely or essentially aluminous species, Corundum, Chrysoberyl, Spinel, Gahnite; (4) the purely siliceous species, Quartz and Opal; and, (5), the silicates, Topaz, Beryl, Cyanite, Andalusite, Staurolite, Chrysolite, Chondrodite, Tourmaline, Iolite, Leucite, Orthoclase, Albite.

The Diamond is distinguished essentially by its extreme hardness, its peculiar adamantine lustre, and, in ordinary examples, by its crystallization. The latter is Regular, but the crystals have almost invariably curved planes. The principal forms comprise the tetrahedron and octahedron, and the adamantoid $3 \cdot 1 \cdot \frac{3}{2}$, the last often distorted both by curvature of faces and by elongation. The cleavage is octahedral. In the Bunsen flame on platinum foil, diamond dust burns slowly away, but small splinters remain unchanged.

The Tetragonal species, Cassiterite, Rutile and Anatase, have the common formula RO^2; and with these, from its close correspondence in crystallization with Rutile, the Zircon may be placed. Cassiterite, SnO^2, is readily distinguished by its high sp. gr. (6·7-7·0), and by yielding reduced tin, BB, with carb. soda or other reducing flux on charcoal. The crystals are commonly short eight-sided prisms, terminated by the planes of the two corresponding square pyramids (without basal plane); and they are very frequently in geniculated twins. P:P over polar edge = $121° 40'$; $\bar{P}:\bar{P} = 133° 30'$. In mineral veins, Cassiterite is very generally associated with Wolfram and Quartz, the latter forming the gangue or veinstone. The variety known as "stream tin" occurs in small rolled pebbles and grains in alluvial deposits. "Wood tin" is also an uncrystallized variety of light or dark brown colour and concentric-radiated structure. Rutile, TiO^2, distinguished in ordinary examples by its red or brown colour and adamantine lustre, closely resembles Cassiterite in crystallization, and especially in its geniculated twin-forms: P:P = $123° 8'$; vertical planes, in general, longitudinally striated. Rutile occurs also occasionally in acicular radiating crystals, traversing quartz; and in small dark pebbles (Nigrine). Anatase or Octahedrite, another form of TiO^2, is mostly in small pyramidal crystals of a greyish-brown or peculiar blue colour, with adamantine, more or less sub-metallic lustre. The crystals commonly shew a consecutive series of several pyramids, but are sometimes tabular from extension of the basal plane. The angle over middle edge in P = $136° 36'$ (over polar edge $97° 51'$); in $\frac{1}{2}$ P, $79° 54'$; in $\frac{1}{3}$ P, $53° 22'$; in $\frac{1}{4}$ P, $39° 30'$. Both Anatase and Rutile, and the Rhombic species Brookite, after fusion in fine powder with carb. soda, are dissolved by hydrochloric acid. The solution assumes a deep violet colour if slightly diluted and boiled with metallic tin.

Zircon, ZrO^2, SiO^2, occurs occasionally in small granular masses, but most commonly in simple crystals of the Tetragonal System. These are frequently small square prisms terminated by a square pyramid measuring 123° 20′ over polar edges, and 84° 20′ over middle edges. The basal plane is always absent. Other common crystals are eight-sided, from combination of the two square prisms, and in many a second pyramid is subordinately present. Some crystals, again, shew the planes of one or more octagonal pyramids, 3 P 3, 4 P 4, 5 P 5, but these planes are usually quite narrow or of small size. Zircon is mostly red or red-brown in colour, but sometimes pale yellowish-grey, orange-yellow, greenish, or colourless. Its hardness (7·5), and its high sp. gr. which averages 4·4 or 4·5, and always exceeds 4·0, are salient characters. BB loses colour, but is quite infusible. The powder is slowly taken up by borax, the saturated glass becoming opaque when flamed.

Corundum, Al^2O^3, is distinguished by its great hardness (9·0), its high sp. gr. (3·8-4·2), hexagonal or hemi-hexagonal crystallization, and complete infusibility; and by the fine blue colour imparted to it by treatment, BB, with cobalt solution. It occurs under three more or less distinct conditions: (1) in small transparent or sub-transparent crystals of a blue, pink, red, or other colour, or sometimes colourless, forming the sapphire, ruby, &c., of jewellers, according to the colour; (2) in coarser translucent or opaque crystals and cleavable masses of a greyish-green, red, brown or other tint, forming the variety known as Adamantine Spar; and (3) in fine-granular masses of a grey or dark bluish-grey or black colour, commercially known as Emery. The latter variety is sometimes mixed with grains of magnetic iron ore. The Corundum crystals are mostly small, pyramidal combinations, or six-sided prisms with narrow pyramidal planes and large basal face, and are frequently ill-formed. The cleavage is basal, and also rhombohedral, with R: R 86° 4′. Many crystals are parti-coloured, blue and white, &c.; and in some (asteria sapphire), a six-rayed opalescence is visible. The cleavage faces often shew a delicate striation. For blowpipe reactions, see the Table.

Chrysoberyl, BeO, Al^2O^3 (or perhaps Be^2O^3, Al^2O^3), is a comparatively rare species of a green or greenish-white colour, sometimes reddish by transmitted light, and often shewing a pale-bluish opalescence—whence the name Cymophane, by which this species is also known. The crystals are Rhombic combinations, and are frequently in pseudo-hexagonal stellate groups*—both simple and compound crystals being generally more or less tabular from extension of the front vertical form or macro-pinakoid $\overline{V}$. The hardness of chrysoberyl (8·5) nearly equals that of corundum; and its comparatively high sp. gr. (3·7-3·8) is also distinctive.

---

* Compound stellate and hexagonal groupings are common among crystals of the Rhombic System (Chrysoberyl, Marcasite, Discrasite, Aragonite, Cerusite, &c.), and are occasionally seen in Clino-Rhombic and Regular crystals (the latter in Camphor, &c.), but are apparently unknown among minerals and chemical products of recognized Hexagonal crystallization. The beautiful snow-crystals so common in Canadian winters are thus most probably not truly hexagonal, but compound Rhombic forms. See a brief communication by the writer in the Canadian Journal, 1860.

Spinel, normally MgO, Al^2O^3, is readily distinguished, in most examples, by its occurrence in small octahedrons, commonly twinned, as well as by its great hardness (8·0), and its high specific gravity (3·5-4·1). The colour is usually some shade of red, but colourless and other-coloured varieties are also known. After fusion in fine powder with bisulphate of potash it is partially soluble in water. Ammonia throws down flocculent Al^2O^3 from the solution.

Gahnite is properly a zinciferous spinel, commonly in small octahedrons, both simple and twinned, of black or dark-green colour, with greenish-grey streak. Combinations of the cube with the rhombic dodecahedron and several trapezohedrons, are also known. The simple octahedral crystals resemble generally those of magnetic iron ore, but from this species Gahnite is distinguished by its want of magnetism, its pale streak, lower sp. gr. and greater hardness, as well as by the zinc sublimate which it yields when fused, in powder, with a mixture of about equal parts of carb. soda and borax, on charcoal.

Quartz, SiO^2, is distinguished readily from the preceding minerals by its much lower sp. gr., as this never exceeds 2·7 or 2·8. Also by fusing readily with carb. soda, and forming with that reagent a clear glass. Its want of distinct cleavage is also characteristic. When crystallized, it is almost invariably in six-sided prisms, streaked across and terminated by the planes of a regular hexagonal pyramid, the basal plane being always absent. The pyramid-planes are often very irregular in size and shape. The principal angles are as follows : over polar edge, 133° 44′; over point of crystal, 76° 26′; on adjacent prism-plane, 141° 47′. If the pyramid be regarded as consisting of two complementary rhombohedrons, R on R equals 94° 15′; and in many crystals only three terminal planes of this kind are present ; or the six planes differ alternately in size, so as to form two sets of three. Many crystals also show a small plane (⅓ [2 P 2] in Naumann's notation), usually rhombic or rhomboidal in shape and often striated, on alternate angles of the prism-pyramid. Although normally colourless, Quartz very commonly presents various shades of violet, pink, red, yellow, green, brown, &c., and some rock-varieties are dark-grey or black. The crystallized examples comprise Rock-crystal, Amethyst, Cairngorm, Smoky Quartz, &c. Massive, crystalline, or sub-crystalline varieties include Common Quartz, Rose Quartz, Prase, some kinds of Jasper, &c. (many of these containing intermixed iron-oxide, chlorite, actynolite, or other foreign matters); whilst the nodular, stalactitic, and amygdaloidal examples, composed largely of amorphous silica, comprise Calcedony, Carnelian, Cat's-Eye, Chrysoprase (coloured apple-green by NiO), Agate, Flint, Blood-stone, and other varieties.

Opal consists of amorphous silica, and most, if not all, examples yield a certain amount of water on ignition. It occurs only in nodular, amygdaloidal or botryoidal masses, or in small veins, essentially in trappean or volcanic rocks. Its sp. gr. rarely exceeds 2·0 or 2·2, and its degree of hardness is always below that of ordinary quartz. In powder, it is dissolved more or less readily by a hot solution of caustic potash or soda. The noble opal is beautifully iridescent ; but ordinary varieties, comprising the so-called semi-opals, milk-opals, wood-opals, &c., much resemble calcedonic varieties of quartz, and are usually opaque-white, brown, red, yellow, or grey in colour. Hyalite is a transparent glassy variety in small botryoidal masses on lava. As regards these and other varieties of Opal (*see* the Table), the more distinctive characters are as follows : low sp. gr. (1·5-2·5); amorphous structure; infusibility ; presence of water; solubility (or partial solubility if mixed with quartz) in caustic potash.

Topaz is apparently an aluminous silicate combined with a fluoride. It contains $17\frac{1}{2}$ per cent. of fluorine, but gives a very feeble indication of that substance with sulphuric acid, owing to its general insolubility. If fused, however, with some previously fused phosphor-salt in a piece of open tube—the flame being directed into the tube upon the assay—the glass becomes corroded. Topaz occurs commonly in crystals, more rarely in small rolled pebbles (distinguished from quartz pebbles by their ready cleavage and higher sp. gr.), and occasionally in opaque, granular or columnar masses (Pycnite, Physalite) of a reddish-white or yellowish colour. In all, the hardness exceeds that of quartz, and the sp. gr. is comparatively high (3·5-3·6). The crystals belong to the Rhombic System, and are invariably prismatic in aspect, with $V : V$ 124° 17', and $V2 : V2$ 93° 11'. They are of three general types: (1), the Brazilian type, essentially of a wine-yellow colour, presenting several vertical prisms (some of which, however, are merely denoted by vertical striæ), terminated by four planes of a rhombic pyramid measuring 141° over front polar edge, and 101° 40' over side edge, the basal plane wanting ; (2) the Siberian type, essentially of a pale blucish-green colour, resembling that of ordinary beryls, and consisting of vertical forms with two more or less largely developed side-polars or brachydomes, $2\overset{\cup}{P}$, measuring 92° 42' over the summit—the basal plane being either absent or of comparatively small size, and other planes, if present, being also but slightly developed ; and (3) the Saxon type, of very pale-yellow colour, or nearly colourless, characterized essentially by its largely-developed basal plane, with polar planes (of several forms) subordinately present. These definitions hold good in the main, but crystals of intermediate type occasionally occur.

Beryl, a silicate of alumina and glucina, occurs only in crystals or crystalline columnar aggregations. The species consists of two leading varieties, comprising the Beryl proper and the Emerald. In both, the crystals as a rule are simple hexagonal prisms with largely-developed basal plane ; but in some the basal edges or angles (or both) are replaced by a border of narrow pyramidal planes ; and occasionally the vertical edges are replaced by the prism V2. In Beryl the vertical planes are generally longitudinally striated by an oscillation between the two prisms, and crystals are thus often rounded or rendered more or less cylindrical. The colour is usually greenish-white or some pale shade of green, greenish-blue or yellow, and crystals often occur of large size. In the Emerald the prism-planes are generally smooth, and the colour is emerald-green, derived from the presence of a very small amount of sesquioxide of Chromium. In both varieties the hardness exceeds that of quartz, and the average sp. gr. equals 2·7. In the blowpipe-flame, fine splinters lose their colour, become opaque, and vitrify at the extreme point; but, practically, the mineral may be regarded as infusible.

Cyanite or Kyanite, also known as Disthene, occurs commonly in long, bladed or broadly-fibrous aggregations of a mixed blue and white colour, but occasionally of a red, grey or other tint, and also in examples of narrow-fibrous structure. The flat surfaces are readily scratched by a knife, whilst the edges scratch glass strongly. The Crystal-System is Anorthic, but crystals as a rule are imperfectly formed. They consist of long narrow prisms, with indistinct terminal planes in most examples. The blue-white colour, bladed structure, perfect infusibility, and assumption of a blue colour by ignition with nitrate of cobalt, are the leading distinctive characters. In examples from St. Gothard, frequently seen in collections, Cyanite crystals in mica slate are closely conjoined with long narrow prisms of dark-red Staurolite.

Andalusite is identical with Cyanite in composition (Al^2O^3, SiO^2), but presents a very different aspect, and crystallizes in the Rhombic System. It is generally in granular masses, or in rectangular prisms, of a peach-blossom red or greyish colour. The crystals are often large and coarsely formed. In blowpipe characters, it resembles Cyanite.

Staurolite may in general be recognized easily by its dark-brown, brownish-black, or dark-red colour, its very common cruciform crystallization, and its infusibility. Simple crystals however are also of frequent occurrence. These consist invariably of an obtuse rhombic prism, with V : V 129° 20', truncated (sometimes deeply, sometimes very slightly) on the side or acute vertical edges, so as to form a six-sided prism, and carrying generally, in addition, a front-polar form, $\overline{P}$, mostly of small size. The basal plane has generally a rough surface, and many crystals are rough and dull throughout. These crystals occur very commonly in cruciform twins, in some of which the crystals cross each other at right angles, and in others obliquely. The prism-angle V : V appears to vary from about 128° 40' to 129° 30'. B : $\overline{P}$ averages 124° 30' to 125° 30'.

Chrysolite proper is commonly of a pale-yellow or yellowish-green colour; but in the variety known as Olivine, the colour is dark-green, or brownish-yellow, or occasionally red, and this variety occurs chiefly in small granular, more or less transparent masses, imbedded in basalt and lava. Chrysolite, proper, occurs in small crystals and crystalline grains, and is normally a pure silicate of magnesia, whilst in Olivine, much of the MgO is replaced by FeO. The crystals are Rhombic, and are mostly combinations of the vertical forms $\overline{\overset{\cup}{V}}$, V, $\overset{\cup}{V2}$, and $\overset{\cup}{V}$; the polar forms $\overline{P}$, P, and $2\overset{\cup}{P}$; and the base B, the latter sometimes failing. V : V = 130° 2' ; $\overset{\cup}{V2}$: $\overset{\cup}{V2}$, 94° 2' ; P : P over front edge, 139° 54' ; over side edge, 85° 16' ; $\overline{P}$: B, 128° 17' ; $2\overset{\cup}{P}$: $2\overset{\cup}{P}$, over B or summit, 80° 53'. Both varieties are decomposed, in powder, by hydrochloric acid, and also by sulphuric acid, the silica separating in (usually) a gelatinous condition. Normal examples are infusible, but very ferruginous varieties (hyalosiderite, &c.) often vitrify on thin edges. The leading characters of Olivine are its peculiar greenish-yellow or green colour, its occurrence in traps and lavas, its general infusibility, and its gelatinization in acids.

Chondrodite, essentially a magnesian fluo-silicate, occurs commonly in the form of small granular masses, chiefly of a yellow colour, imbedded in crystalline limestone; but green, yellowish-red, and other-coloured varieties are also known. It occurs also, though less commonly, in small crystals with numerous planes, belonging to the Clino-Rhombic System. Humite and Clino-Humite (chiefly from Vesuvius) are closely similar in composition and general physical characters, but the first is Rhombic in crystallization, and the latter presents different angular values. All give a marked fluorine reaction by treatment in powder with hot sulphuric acid. For other characters, see the Table.

Tourmaline may in general be recognized without difficulty by the essentially triangular character of its crystals and crystalline needles, as seen more especially on the transverse fracture. The crystals are generally nine-sided prisms, consisting of three planes of a hemi-hexagonal prism $\frac{V}{2}$, combined with the second hexagonal prism V2, the latter occurring as a bevelment on the vertical edges of the half-form. These prisms, when perfect, are terminated by rhombohedron-planes, with or without a basal plane, or frequently by rhombohedron-planes at one extremity, and by a single large basal plane or dissimilar forms at the other. The rhombohedron-planes belong chiefly to the forms R, $-\frac{1}{2}$R, $-$2R, in which the angle over polar edges equals, respectively, 133° 10' or thereabouts, 155°, and 103°. Black varieties (known as Schorl) and most dark-brown varieties are easily fusible (*see* TABLE XXVI.), but the red, green, blue, clear-brown, and colourless examples are either infusible, or fusible only on the thinnest edges. Some crystals are red internally and green externally, or present different colours at the extremities; and nearly all the clear examples are transparent when viewed across the prism, and opaque longitudinally. All, moreover, exhibit electrical polarity when heated.*

Iolite, known also as Dichroite and Cordierite, is commonly in the form of small, granular, vitreous or resino-vitreous masses, imbedded in granitic and crystalline metamorphic rocks; but is found at some localities in distinct crystals, and occasionally in the form of small rolled pebbles in alluvial deposits. The colour is mostly dark-blue or pale-blue by reflected light, and brownish or yellowish by transmitted light, whence the name Dichroite. Some varieties, however, are colourless, grey, or blueish-brown. The crystals belong to the Rhombic System, but have in general a pseudo-hexagonal aspect: a common combination consisting of the forms V, $\overset{,,}{V}$, P, $\overset{,,}{P}$, and B; with V : V 119° 10'; B : $\overset{,,}{P}$ 150° 49'. Fine splinters melt at the extreme point, but practically the

---

* This may be shewn by suspending a crystal from the ring of the blowpipe-lamp, or other convenient support, by means of a piece of thin silk-thread tied round the centre of the crystal. The latter is heated carefully in a small platinum or porcelain capsule, care being taken not to burn the thread, over a spirit-flame or Bunsen-burner. On the capsule being removed, one end of the prism will be attracted, and the other end repelled, by a glass stirring-rod or stick of sealing-wax rubbed previously for a few seconds on the coat-sleeve.

species may be placed among the infusible silicates. From blue Corundum (Sapphire), and from Sapphirine and blue Spinel, it is readily distinguished by its low sp. gr. (2·6). From blue Tourmaline (Indigolite), also, by lower sp. gr., and by not becoming electric when heated ; and from Quartz, by forming, BB with carb. soda, a slaggy semi-fused mass in place of a clear glass. Many examples of Iolite are partially altered or decomposed, and these give traces of water in the bulb-tube.

Leucite is readily distinguished, as a rule, by its occurence in small rounded grains or crystals of a white, grey, or pale-yellowish tint, in lava. The crystals closely resemble the trapezohedron 2-2 of the Regular System, but have been shewn by Von Rath to be really Tetragonal—at least as regards most examples, if not all. Many crystals contain minute needles and scales of augite, magnetite, &c., scattered through their substance. In powder, leucite is slowly decomposed by hot hydrochloric acid. The solution, rendered pasty by partial evaporation, shews the red K-line in the spectroscope if held on a clean platinum wire for a few seconds in the outer edge of a Bunsen-flame. The K-line is rendered visible also by igniting some of the powder on a loop of platinum wire, and then dipping it into some carbonate of soda or powdered fluor-spar, and again exposing to the flame. The glare from the sodium spectrum may be entirely cut off by the intervention of a piece of deep-blue glass.

Orthoclase ; Albite. These species belong properly to Table XXVI, and their crystallographic and other characters are there described. In general, they form cleavable masses of a white, flesh-red, bright-red, grey, pale-yellowish or clear-green colour ; or occur in crystals of a more or less flattened aspect, often twinned (see Note to TABLE XXVI.). In Orthoclase the principal cleavage planes meet at right angles ; in Albite, at angles of 93° 36′ and 86° 24′, and one of the cleavage planes in the latter species generally shews a delicate striation, best seen under the magnifying glass. Orthoclase, treated in powder with carb. soda (as described under Leucite, above) shews very distinctly the red K-line in the spectroscope.*

---

* This test for the presence of potash in Orthoclase, so far at least as regards the use of carb. soda, was first described by Bunsen. If the mineral, in powder, be fused with fluor-spar the red K-line comes out, I find, still more distinctly ; and many examples, when thus treated, show the Li-line as well. By the intervention of a piece of blue glass the Ca-lines (from the fluor-spar) and the Li-line become obliterated, and only the K-line remains visible. The latter is also brought out in most if not in all cases by simply moistening the test-matter, after ignition, with hydrochloric acid

TABLE XXV.

[Lustre non-metallic (in some cases pseudo-metallic). Slowly or incompletely dissolved BB by phosphor-salt. Infusible, or fusible on thin edges only. Hardness insufficient to scratch ordinary window-glass.]

A.—Occurring in micaceous or foliated masses or crystals, the foliæ elastic or flexible, and easily separable by the finger-nail.

A¹.—FOLIÆ DISTINCTLY ELASTIC.

MUSCOVITE (Potash Mica): Essentially K^2O 9, Al^2O^3 35, SiO^2 46, with small amounts of Fe^2O^3, H^2O, Fluorine, &c. Rhombic or Clino-Rhombic (?), but crystals hexagonal in aspect. Optically biaxial, with large angle of divergence. Structure thin-foliated or scaly, the foliæ easily separable. White, brown, black, green, &c., with metallic-pearly lustre on cleavage-plane; flexible and elastic in thin pieces; H 2·0-3·0; G 2·7-3·1. BB, exfoliates, and melts readily on the edges (if in the form of a thin scale) into a greyish-white enamel.* In acids, insoluble. Fuchsite is a more or less deep-green chromiferous variety, in fine-scaly aggregations. Damourite and Margarodite are hydrated micaceous minerals, apparently derived from Muscovite. See TABLE XXVII. Roscoelite is a Vanadium-mica (in small greenish-brown or green, radiately arranged foliæ of metallic-pearly lustre) from Eldorado Co., California.

PHLOGOPITE (Potassic-Magnesian Mica): K^2O 12·75, MgO 32·55, Al^2O^3 13·95, SiO^2 40·75, with small amounts of H^2O, F, &c. Rhombic (optically biaxial), but essentially hexagonal in aspect; thin-foliated, or scaly; chiefly yellowish-brown, with golden, metallic-pearly lustre on cleavage-face; H 2·5-3·0; G 2·75-2·90; BB, whitens, and melts on thin edges into a greyish-white enamel. In powder decomposed by sulphuric acid, the silica separating in colourless scales. Common in crystalline limestones.†

BIOTITE (Potassic-Ferromagnesian Mica): Closely resembles Phlogopite in composition and general characters, but usually of dark colour—green, black, or brown; optically uniaxial, and of assumed

---

* The Micas (Muscovite, Phlogopite, Biotite and other representatives, Lepidolite excepted), are always placed among the *infusible species*, in works on Determinative Mineralogy. As a rule, however, all melt more or less readily on the edges when tested in the form of a thin scale. In the spectroscope, all shew the red K-line, and many the Li-line also, either *per se*, or when moistened, after ignition, with HCl acid.

† This species is present in great abundance in most of the apatite deposits of Canada.

Hexagonal crystallization. Fusible on edges into a black or dark enamel. Decomposed by sulphuric acid. Commonly found in volcanic and trappean rocks, but many volcanic micas are optically biaxial.

A².—FOLIÆ FLEXIBLE BUT NOT ELASTIC.

† *Yield water by ignition in bulb-tube.*

CHLORITE (Pennine): MgO 13 to 27, FeO 15 to 30, Al^2O^3 19 to 23, SiO^2 25 to 28, H^2O 9 to 12. Hexag. or Hemi-Hex. (crystals mostly tabular), but commonly in foliated and scaly examples of a dark or rich green colour; flexible in thin pieces; H 1·0-1·5; G 2·65-2·95. Fusible on thin edges into a yellowish-grey or dark and often magnetic glass. Decomposed by sulphuric acid. Metachlorite, Prochlorite, Aphrosiderite and Tabergite, are closely related chloritic substances. The latter occurs in coarse, bluish-green, foliated masses.

KÆMMERERITE: A chromiferous chlorite of a red or violet-red colour, or green by reflected, and red by transmitted light. Mostly in hexagonal pyramids and prisms of foliated structure.

RIPIDOLITE (Clinochlore): Clino-Rhombic in crystallization, but identical in general characters and composition with Chlorite proper. Epichlorite, Korundophyllite, Helminthite, are varieties or closely related. Pyrosclerite is a chromiferous variety from Elba. Delessite is an essentially ferruginous chlorite, allied to this or the preceding species, of frequent occurrence in amygdaloidal traps.

PYROPHYLLITE (Foliated Kaolin): Al^2O^3, SiO^2, H^2O, with traces of MgO, &c. Essentially in radio-foliated examples of a clear green or greenish-white colour, and somewhat pearly lustre; flexible in thin pieces; H 1·0; G 2·75-2·95. BB, exfoliates and curls up, but remains unfused, or vitrifies slightly on thinnest edges only. With Co-solution assumes a fine blue colour. Talcosite, from Victoria, is a closely related substance, passing into Kaolin proper.

† † *No water, or traces only, in bulb-tube.*

TALC: Essential composition, MgO 31·7, SiO^2 63·5, H^2O 4·8, but the H^2O is not driven off by moderate ignition, and is thus regarded as basic. Occurs commonly in six-sided tabular crystals and foliated masses of a pearly-white, greenish-white, clear-green, or greenish-grey colour. H 1·0; G 2·67-2·80. BB, exfoliates, becomes opaque-white, and melts on thin edges, but less easily than mica. With Co-solution, becomes pale-red. Insoluble in acids.

B.—Occurring in distinctly schistose or foliated examples, but the component foliæ more or less brittle, not flexible.

B¹.—YIELD WATER BY IGNITION IN BULB-TUBE.*

MARGARITE (Pearl Mica): CaO, Al^2O^3, SiO^2, H^2O, with small amounts of K^2O, Na^2O, Li^2O, MgO, F, &c. Rhombic (?); mostly in six-sided tables and lamellar masses of a pearly-white, pale-green, reddish or greyish colour; the lamellæ more or less brittle. H 3·5-4·0; G 2·95-3·10. BB, melts on the edges, often with slight intumescence. Scarcely attacked by acids. In spectroscope, after ignition and moistening with HCl acid, shews momentary red and green Ca-lines, and, in most examples, red K and Li lines, also. Emeryllite, Euphyllite, Diphanite and Gilbertite, are identical or closely related. Euphyllite, however, is decomposed by sulphuric acid.

ANTIGORITE (A slaty Serpentine): MgO 36 to 37, FeO 6 to 7, SiO^2 41 to 43, H^2O 11·5 to 12·5, with traces of Al^2O^3, &c. In schistose masses of a dark green or greenish-brown colour; H 2·5; G 2·62. Fusible on thin edges. Slowly decomposed by sulphuric acid.

SCHILLER SPAR (Bastite). Probably an altered Bronzite: Contains MgO, FeO, SiO^2, with about 12 per cent. H^2O, and small amounts of K^2O, CaO, Cr^2O^2, Al^2O^3, &c. In schistose or foliated masses of a dark-green colour, with yellowish-brown reflections on the cleavage surfaces. H 3·5-4·0; G 2·6-28; BB, melts on the edges only; becomes brown and sometimes magnetic after ignition. Decomposed by sulphuric acid.

PICROPHYLL: A hydrated magnesian silicate occurring in sub-foliated or coarse-fibrous examples of a greenish-grey colour; H 2·5; G 2·73. Fusible on thin edges. Regarded as an altered Pyroxene.

CHLOROPHYLLITE: Contains MgO, MnO, Al^2O^3, Fe^2O^3, SiO^2, H^2O. In foliated masses or coarse indistinctly formed crystals of a green or brownish colour. H about 3·0; G about 2·7. Fusible on thin edges only. Scarcely attacked by acids. Probably, in part, an altered Iolite.

GROPPITE: Contains K^2O, CaO, MgO, Al^2O^3, Fe^2O^3, SiO^2, H^2O (7 per cent.). In foliated or scaly masses of a rose-red or brownish-

---

* Few, if any, of the minerals belonging to this section can be regarded as true species. As a rule, they consist of altered products of more or less unstable composition, and their determinative characters are commonly ill-defined. This remark applies, with few exceptions, to the representatives of the present Table, generally.

red colour, the foliæ brittle; H 2·5-3·0; G 2·73. BB, whitens, and vitrifies on thin edges.

B².—ANHYDROUS SPECIES: NO WATER, OR TRACES ONLY, EVOLVED IN BULB-TUBE.

BRONZITE (Foliated Enstatite): Contains MgO, FeO, SiO^2. Commonly in schistose or foliated masses of a dark-brown or dark-green colour, with pseudo-metallic bronze-like lustre, and very perfect cleavage in one direction. H 4·0-5·0; G 2·9-3·5. Fusible on thinnest edges only. Not attacked by acids.

ANTHOPHYLLITE: MgO 27·8, FeO 16·7, SiO^2 55·5. Rhombic, but essentially in thin-lamellar and fibrous masses, with tolerably easy cleavage in three directions; yellowish-brown, greenish-grey, bronze-green, with somewhat metallic-pearly lustre. H 5·0; G 3·2. BB, vitrifies only on thinnest edges into a black magnetic enamel: practically, infusible. Very slightly attacked by acids.*

CLINTONITE: Composed essentially of CaO, MgO, Al^2O^3, SiO^2, with traces of H^2O. Chiefly in hexagonal tables of a brown or yellow colour, with metallic-pearly lustre; H 5·0; G 3·0-3·2. Practically infusible. Decomposed by hydrochloric acid. Xanthophyllite (in yellow radiating lamellæ on certain talcose schists), and Brandisite (in dark-green tabular crystals, weathering brownish), are apparently related compounds, but are only partially attacked by hydrochloric acid. In Clintonite and in these related silicates the silica is under 20 or 21 per cent. Ignited and moistened with HCl acid, all shew in the spectroscope red and green Ca-lines in momentary flashes.

C.—Occurring in crystals or in granular, fibrous, compact, or other non-micaceous examples. Streak-powder colourless, pale-green, or lightly-tinted,—not black.

C¹.—YIELDING WATER BY IGNITION IN BULB-TUBE.

† *Form with borax, BB, a deeply coloured glass.*
(*Cu reaction*).

DIOPTASE: CuO 50·44, SiO^2 38·12, H^2O 11·44. In emerald-green crystals—hexagonal prisms with rhombohedral summit-planes—suffi-

---

* In ordinary examples, Bronzite and Anthophyllite can rarely be separately distinguished. The first is regarded as a Rhombic representative of the Pyroxene series, and the latter as a Rhombic Amphibole; but the characteristic pyroxene and amphibole angles (87° 6′ and 124° 30′), or angles approaching these, are rarely determinable. Hypersthene is a very ferruginous and comparatively hard Bronzite, distinctly fusible. *See* TABLE XXVI.

ciently hard (5·0-5·5) to scratch glass slightly: See TABLE XXIV. G 3·3. BB, decrepitates and blackens, but does not fuse. With carb. soda, easily reduced. Gelatinizes in heated hydrochloric acid. A rare species, in crystalline limestone from the Kirghis Steppes of Western Siberia.

CHRYSOCOLLA (including Kupferblau, &c.): Composition somewhat variable, but essentially CuO 45·27, SiO^2 34·21, H^2O 20·52. In amorphous and botryoidal masses, coatings on copper ores, and occasionally in pseudomorphs. Colour, green, greenish-blue, bright-blue; brownish or black from presence of Fe^2O^3, MnO^2, &c.; H 2·0-5·0; G 2·0-2·6. BB, blackens, but does not fuse. On charcoal with carb. soda, reduced to metallic Cu. Decomposed with separation of silica (but as a rule without perfect gelatinization) by hydrochloric acid. Demidowite is a Chrysocolla mixed with copper phosphate. Asperolite a variety with 27 per cent. H^2O. Other varieties are mixed with copper carbonate, opalized silica, &c.

ALLOPHANE (Cupreous varieties): Al^2O^3, SiO^2, H^2O (35 to 36 per cent.), mixed with copper silicate. In amorphous, stalactitic and botryoidal examples, coatings, &c., of a light-blue, green, red, or brownish-yellow colour. H about 3·0; G about 2·0. BB, blackens, and often swells up slightly, but does not fuse. In HCl acid, gelatinizes.

(*Ni reaction: page* 43).

RÆTTISITE: NiO, SiO^2, H^2O (11 per cent.), mixed with Fe^2O^3, copper-phosphate, cobalt-arseniate, &c. Amorphous, incrusting; green of various shades; H 2·0-2·5; G 2·3-2·4.

GENTHITE (Nickel-Gymnite): NiO, MgO, SiO^2, H^2O (19 per cent.). In green and greenish-yellow coatings on some examples of Chromic Iron Ore, and occasionally in soft sub-earthy masses. H 2·0-4·0; G about 2·4. BB, infusible, blackens.

PIMELITE: MgO, NiO, Al^2O^3, SiO^2, H^2O (21 per cent.). In earthy masses, coatings, &c., of an apple-green colour. H 1·0-2·5; G 2·3 (to 2·7?). BB, blackens, and vitrifies on thin edges. Alipite and Chrysoprase-Earth are identical or closely related compounds.

(*Fe reaction*).

ANTHOPHYLLITE: In yellowish-brown, or greenish metallic-pearly examples of lamellar or fibrous structure. Some examples only yield traces of water on ignition. See B^2, above.

HISINGERITE (Thraulite): FeO, Fe^2O^3, SiO^2, H^2O (10 to 20 or 22 per cent.), with small amounts of MgO, Al^2O^3, &c. In earthy and nodular masses of a pitch-black or brownish-black colour, with brownish streak. H 3·0-4·0; G 2·6-3·1. BB, becomes magnetic, and vitrifies on the edges, or in some examples melts into a steel-grey magnetic globule* (*see* TABLE XXVII.). Decomposed by HCl acid with separation of slimy silica.

NONTRONITE: Essential components Fe^2O^3, SiO^2, H^2O (21 to 25 per cent.), but small amounts of Al^2O^3, CaO, &c., are also generally present. In earthy and nodular masses of a yellow, green, greenish-white or brownish colour; H 1·0-1·5; G 2·0-2·4. BB, infusible, or fusible on the edges only, but becomes magnetic. Pinguite and Gramenite are identical or closely related. Chloropal (Unghwarite) is also very similar in general characters and composition, but is somewhat harder, probably from admixture with opalized SiO^2.

(*Cr reaction: see page* 48).

WOLCHONSKOITE: Cr^2O^3, Fe^2O, SiO^2, H^2O (about 20 or 21 per cent.), with small amounts of MgO, MnO, Al^2O^3, &c. In earthy and nodular masses of a grass-green or blackish-green colour; H 1·5-2·5; G 2·2-2·3. BB, practically infusible; gelatinizing in HCl acid.

MILOSCHIN (Serbian): Al^2O^3, Cr^2O^3 (under 4 per cent.), SiO^2, H^2O (about 23 per cent.). In blue or blue-green, earthy and amorphous masses; H 1·0-2·0; G 2·1-2·2; adheres to the tongue. BB, infusible. Partially decomposed by hydrochloric acid.

† † *Form BB with borax an uncoloured or lightly-tinted glass.*

(*The saturated borax-glass becomes opaque-white on cooling or when flamed*).

CERITE: CeO (LaO, DiO) 73·5, SiO^2 20·4, H^2O 6·1. Chiefly in fine-granular masses of a red, brownish, or reddish-grey colour. H 5·0-5·5 (scratches glass feebly); G 4·9-5·0. BB, becomes dull yellow, but remains unfused. Gelatinizes in hydrochloric acid.

THORITE: ThO^2, SiO^2. H^2O. Reg. ? Mostly in small black masses, often fissured, and sometimes with reddish coating; streak, brownish or reddish; H about 4·5; G 4·4-4·7. BB, becomes yellow, but

---

* L. H. Fischer: *Claris der Silicate:* 1864. This work, a Determinative Grouping of the Silicates (containing many original observations), should have been referred to among the list of works on Determinative Mineralogy at page 21.

remains unfused. Gelatinizes in HCl acid. Very rare; commonly regarded as altered Orangite.

ORANGITE: ThO^2, SiO^2, H^2O. Tetragonal? Mostly in small granular or sub-foliated examples of a reddish-yellow or orange-red colour; H 4·5; G 5·2-5·4. Gelatinizes in HCl acid. Very rare; accompanies Thorite in the micaceous zircon-holding syenite of Brevig in Norway.

[NOTE.—Most examples of Cerite, Thorite, and Orangite, when ignited and moistened with hydrochloric acid, shew a momentary Ca-spectrum.]

(*A zinc-sublimate formed on charcoal by fusion with carb. soda and borax*).

CALAMINE: ZnO 67·5, SiO^2 25, H^2O 7·5. Crystallization Rhombic; crystals mostly hemimorphic (with B plane at one extremity only), arranged in drusy or fan-shaped aggregations, and generally flattened from extension of the side vertical or brachy-pinakoid faces $\breve{V}$. The species occurs also very commonly in botryoidal, cavernous, and other examples; colourless, white, yellowish, brown, green, light-blue; H 5·0 (scratches glass feebly); G 3·3-3·5; crystals, pyro-electric. Infusible, BB, or vitrified slightly on thinnest edges, only. only. Gelatinizes with hydrochloric acid. Ignited with Co-solution, becomes green (or partly blue and partly green) on cooling.

(*Slowly attacked BB by borax; the glass not rendered opaque by flaming. With Co-solution, assume a distinct blue colour*).

KAOLIN: Al^2O^3 39·7, SiO^2 46·4, H^2O 13·9. Chiefly in earthy or fine-granular masses made up in part of microscopic scales. White, pale-red, greenish-white; H 1·0 or less; G 2·1-2·3 (or in some varieties slightly higher: 2·3-2·6). Infusible; decomposed by hot sulphuric acid. Cimolite, Anauxite, Pelicanite, Hunterite, &c., are related aluminous compounds, but contain a somewhat higher percentage of silica.

NACRITE or PHOLERITE: A crystalline or sub-foliated Kaolin, in pearly-white scaly masses or six-sided tables, often in fan-shaped groups. Composition and other characters as in Kaolin proper.

AGALMATOLITE (Figure Stone* in part): K^2O, Al^2O^3, SiO^2, H^2O (about 5 per cent.). White, pale-grey, yellowish, pale-red, green,

---

*Although many of the smaller Chinese images are carved out of this stone, a great number (perhaps the greater number) consist of steatite or of serpentine. In these, the substance blackens in the bulb-tube, and assumes a flesh-red colour after ignition with Co-solution.

greenish-white; mostly in fine-granular almost compact masses, but these consist frequently of microscopic scales; H 2·0-3·0; G 2·8-2·9. BB, whitens, and vitrifies on thin edges. Decomposed by sulphuric acid. Shews the red K-line very distinctly in spectroscope, when ignited and moistened with HCl acid.

PINITE: K^2O, MgO, FeO, Fe^2O^3, Al^2O^3, SiO^2, H^2O (4 to 8 per cent.). In six-sided and twelve-sided, more or less opaque crystals, of a greyish-white, grey, brown, greenish or bluish colour; H 2·0-3·5; G 2·5-2·9. BB, vitrifies on thin edges only. In spectroscope shews distinctly the red K-line when ignited and moistened with hydrochloric acid. Apparently an altered Iolite. The following substances, all of which give a K-spectrum, are more or less closely related: *Pyrargillite* from Finland (brown, brownish-red, H^2O 15·5 per cent.); *Fahlunite* (dark-brown, dark-green, greyish, H^2O 8 to 9 per cent.); *Weissite* (grey, brown, H^2O 3 to 5 per cent.); *Iberite* from the vicinity of Toledo (greyish-green, in coarse six-sided prisms, aq. 5 to 6 per cent.); GIESECKITE (greenish-grey, aq. about 6 per cent.); *Liebenerite* (green, greyish, aq. about 5 per cent.). The two latter are regarded as altered nepheline; the others as altered iolite. In all, the hardness is below 4·0, and the sp. gr. below 2·9. *Gigantolite* belongs to the same series, but is readily fusible (*see* TABLE XXVII).

ESMARKITE: MgO, MnO, FeO, Fe^2O^3, Al^2O^3, SiO^2, H^2O (5·5 per cent.). This mineral, like those placed under *Pinite*, above, is also apparently an altered Iolite; but it is placed here, apart, as the representative of a non-potassic series. Occurs mostly in coarse twelve-sided prisms of more or less scaly texture; grey, brown, greenish, &c., in colour; and dull and opaque, or practically so. H 3·0-4·0; G 2·6-2·8; fusible on thin edges only. *Praseolite*, *Aspasiolite*, and *Bonsdorffite*, are identical or closely related substances of a green or greenish-brown colour, occurring mostly in six-sided, eight-sided, or twelve-sided prisms, with dull surface and rounded edges.

HALLOYSITE: Al^2O^3 35, SiO^2 41, H^2O 24. Nodular, earthy: greenish or greyish-white, pale dingy blue; H 1·0-2·5; G 1·9-2·1; feels somewhat greasy, and adheres to the tongue. Infusible. Decomposed by hot sulphuric acid. *Lenzinite* and *Glagerite* are identical or closely related. *Kollyrite* is also very similar in general characters, but contains 40 per cent. H^2O, with 46 Al^2O^3, and only 14 SiO^2.

(*Assume a pale-red colour after ignition with Co-solution, or do not become blue. In the bulb-tube, generally blacken*).

STEATITE (compact or fine-granular Talc): White, greenish, &c., often mottled. More or less soapy-feeling and very sectile. On ignition, yields traces of water only. See C², below.

SERPENTINE: MgO 43·48, SiO² 43·48, H²O 13·04; but part of the MgO very generally replaced by FeO, and small amounts of NiO, Al²O³, and Cr²O³, are occasionally present. In fine-granular or compact masses, or occasionally slaty or fibrous. Sometimes, also, in pseudomorphs after Olivine, Pyroxene, Spinel, and other species. Of various colours, but chiefly some shade of green, greenish- or greyish-yellow, brown, or red, two or more colours in irregular patches being often present in the same specimen; translucent or opaque; H 3·0-4·0; sectile; G 2·5-2·7. BB, whitens, and fuses on thin edges. Deeply-coloured (ferruginous) varieties do not redden distinctly with Co-solution. Decomposed by sulphuric, and also, though less easily, by hydrochloric acid. Picrolite, Picrosmine, Bowenite, Retinalite, Marmolite, Antigorite (*see above*, B¹), Chrysotile (*see below*), and many so-called Soapstones, are varieties.

CHRYSOTILE (Serpentine-Asbestus): Properly, a fibrous asbestiform serpentine, in silky, easily separable fibres, of a yellowish, greenish-white, or oil-green colour. BB, a fine fibre melts at the extreme point. Baltimorite is a bluish, coarsely fibrous variety, often containing Al²O³ and Cr²O³. Metaxite is also a fibrous serpentine.

MEERSCHAUM (Sepiolite): MgO, SiO², H²O (the latter somewhat variable, but usually 11 or 12 per cent.). In fine-granular, more or less compact and very sectile masses of a white, pale-yellow or greyish colour. Sometimes in pseudomorphs after Calcite, &c. H 1·5-2·5; G about 1·0-1·3. BB, hardens, and melts on thin edges. Decomposed by HCl acid, with separation of slimy silica.

DEWEYLITE (Gymnite): MgO 37, SiO² 41, H²O 22. In more or less compact masses of a dingy yellow or yellowish-white colour and somewhat waxy lustre; H 2·0-3·0; G 1·9-2·22. BB, fuses only on the thinnest edges. Decomposed, without gelatinization, by hydrochloric acid. Kerolite is closely related in general characters and composition.

VILLARSITE: MgO, FeO, MnO, SiO², H²O. In pyramidal or thick tabular crystals (apparently rhombic, and probably pseudomorphous after Olivine), arranged generally in compound groups; also in rounded granular masses; green, dingy-yellow, or greyish; H 3·0; G 2·9-3·0. Infusible. Decomposed by acids.

PYRALLOLITE: MgO, CaO, Al^2O^3, SiO^2, H^2O. Commonly in prismatic, coarse-fibrous, or granular masses, rarely in Clino-Rhombic crystals with basal cleavage; green, greenish-white, pale yellowish-grey; H 3·0-4·0; G 2·53-2·73. Fusible on thin edges only. Generally regarded as an altered Pyroxene.

C².—YIELD NO WATER (OR TRACES ONLY) BY IGNITION IN BULB-TUBE.

† *Sectile.*

(*With Co-solution assume a flesh-red colour*).

STEATITE (compact or fine-granular Talc): MgO 31·7, SiO^2 63·5, H^2O 4·8—but the latter is only evolved on intense ignition. Massive: fine-granular or compact; also in pseudomorphs after Scapolite, Orthoclase, Andalusite, Spinel, Pyroxene, and other species; white, grey, greenish, reddish, &c., often mottled; H 1·5-2·5; very sectile; G 2·6-2·8; more or less soapy-feeling. BB, hardens considerably, and fuses on thin edges. Decomposed by hot sulphuric acid.

† † *Not sectile.*

(*Forming zinc-sublimate on charcoal by fusion with carb. soda and borax*).

WILLEMITE: ZnO 73, SiO^2 27. Hemi-Hex. (crystals small, frequently with rounded edges, mostly hexag. prisms terminated by a rhombohedron measuring 128° 30′ over a polar edge*); white, green, brownish, reddish, &c.; H 5·5 (scratches glass feebly); G 3·9-4·2. BB, infusible, or vitrified here and there on surface only.† With Co-solution becomes green, or green and blue. Gelatinizes with hydrochloric acid.

TROOSTITE (Manganesian Willemite): Like Willemite in general composition, but with part of the ZnO replaced by MnO and FeO. Commonly in opaque or semi-opaque yellowish-grey or brown crystals like those of Willemite, but comparatively large. BB, with carb. soda, strong Mn-reaction. Gelatinizes with HCl acid.

---

* This rhombohedron is commonly regarded as the form ⅔ R. In the form R, the angle over a polar edge equals 116°; and in the form -½ R, also often present (especially in the manganese variety Troostite), it equals 143° 24′.

† A small splinter scarcely becomes rounded or changes form, but if examined by the magnifying glass after exposure to the blowpipe, its surface exhibits points of vitrification.

(*Forming with borax, BB, a glass which becomes opaque on flaming. Moistened with sulphuric acid, tinges the flame-point pale-green*).

XENOTIME: YO, CeO, P^2O^5. See under the Phosphates, TABLE XVII., page 164. This rare species is referred to here, as from its general insolubility in acids and its slow solution BB in phosphor-salt, it might escape detection as a phosphate.

(*Slowly attacked, BB, by borax, the bead remaining clear when flamed*).

CHIASTOLITE: Properly a variety of Andalusite, but of lower degree of hardness (5·0-5·5) from incipient alteration. Occurs in slender straw-like prisms, or occasionally in thicker crystals, imbedded chiefly in clay-slate or mica-slate, and presenting on the transverse section a dark cross, or a black lozenge at centre and angles. See TABLE XXIV., page 199.

ANTHOPHYLLITE: In yellowish-brown, greenish-grey, or bronze-green, lamellar or fibrous masses; H 5·0-5·5; G 3·2. BB, vitrifies on thinnest edges only, into a black magnetic enamel. The borax-glass, coloured by iron. See under B^2, above.

D.—Streak-powder, black or greyish-black.

This subdivision includes merely varieties of ANTHRACITE in which the lustre is more or less non-metallic. When pure, Anthracite consists essentially of carbon, but usually contains a small percentage of H, N, and O, besides intermixed mineral matter or so-called "ash." H 2·5-3·25; G 1·2-1·8. BB, in splinters practically unchanged, but in fine powder burns gradually away. In the bulb-tube generally yields a small amount of water. Not attacked by the fluxes. Insoluble in acids and caustic alcalies.

NOTE ON TABLE XXV.

The minerals which belong properly to this Table comprise a series of infusible or difficultly fusible silicates of low or comparatively low degree of hardness, many yielding to the finger-nail, and all being readily scratched by the point of a knife. Whilst some of these silicates are definite species, presenting a fixed composition and well defined physical characters, others are mere mixtures, or more or less unstable products of decomposition. The latter in many cases can only be distinguished from one another by complete chemical analysis; and, as a rule, no two examples of these pseudo-species, unless obtained from absolutely the same spot, will be found to agree exactly in the amount of water or other components. False species of this unstable and indefinite character are easily made by any one capable of performing an ordinary mineral analysis, but their acceptance leads to much confusion, and should therefore be rigorously disallowed. In Tables of the present character, however, products of this kind, already recognized in mineralogical systems and text-books, could not be altogether ignored. By a little latitude, the greater number might be placed under two conventional species : the first including all hydrated magnesian or alumino-magnesian products of the kind in question ; and the second, all the purely or essentially aluminous matters of this kind.

The more common representatives of the Table belong to the following groups :—Micas, Chlorites, Talcs and Steatites, Serpentines, Kaolins, Pinites. Copper Silicates, Zinc Silicates.

The micas are especially characterized by their metallic pearly or general pseudo-metallic lustre, and their ready cleavage into thin, elastic leaves. Those of the present Table include the three species, Muscovite, Phlogopite. and Biotite—the two latter essentially magnesian species. Muscovite, commonly called Potash Mica, although the other species contain an equal or even greater amount of potash, is chiefly distinguished by its want of solubility in sulphuric acid, whilst the other two species, when in fine powder, are decomposed in the boiling acid, with separation of fine scales of silica. Phlogopite is generally of a golden-brown colour ; Biotite, dark-green or black. The optical characters of these micas are also different. Muscovite is biaxial, with angle of divergence 44°-78° ; Phlogopite is also biaxial, but with smaller divergent angle (under 20°, sometimes under 5°) ; and Biotite is (normally) uniaxial. In thin scales, all melt without difficulty on the edges into an opaque-white or greyish enamel ; and when moistened, after ignition, with hydrochloric acid, all shew in the spectroscope the red K-line, with in some cases the Li-line also. Some examples shew one or both of these spectra by simple insertion *per se* in the flame. Red and green Ca-lines sometimes appear from intermixed calcite. Muscovite is commonly present in granites, gneiss and mica slate, as one of the essential components ; Phlogopite is chiefly found in the bands of crystalline limestone associated with many gneissoid rocks ; and Biotite occurs most generally, though not exclusively, in lavas, trachytes, and basalts.

The Chlorites are chiefly distinguished by their dark-green colour and foliated structure ; their flexibility in thin leaves (without the elasticity of the micas) ; their softness ; and the marked amount of water (about 12 p. c.) which they yield by ignition in the bulb-tube. Some chlorites, however, especially chromiferous examples, present a deep-red colour. In thin scales, all fuse more or less readily on the edges into a greyish or black enamel, the latter often magnetic. The original Chlorite has been split up into several species, more or less distinct. The principal comprise Chlorite proper or Pennine (the Ripidolite of Gustav Rose) characterized by its hexagonal or rhombohedral crystallization ; and the clino-rhombic species, Clinochlore or Ripidolite (of von Kobell), for which the old name of Chlorite was retained by Rose. These species closely resemble one another, and in ordinary, uncrystallized examples they can scarcely be distinguished. As a rule, however, Chlorite is a more ferruginous species, and thus generally becomes magnetic after fusion or strong ignition, and its sp. gr. is in some examples as high as 2·9; whilst that of Ripidolite rarely exceeds 2·7. This distinction, however, only applies in special cases, and is practically of little value.

The Talcs and Steatites are exclusively or essentially magnesian silicates, containing 4 or 5 p. c. of apparently basic water, only expelled by intense ignition. Hence, by ordinary ignition in the bulb-tube, these minerals yield, as a rule, merely traces of moisture, and they are thus generally placed among anhydrous species in determinative groupings. The formula may be written (H^2O, 3 MgO), 4 SiO^2. Talc proper is easily recognized by its occurrence in soft, flexible, more or less pearly scales and foliated masses of a white, clear-green or other light colour, combined with its soapy feel, and its property of assuming a flesh-red tint by ignition with cobalt-solution, the latter character serving to distinguish it from pyrophyllite and other foliated minerals of the aluminous Kaolin group. Although very soft and flexible, the foliæ are inelastic. Steatite is a more or less compact Talc, usually white, grey, greenish, reddish, or mottled in colour, and very sectile. It usually gives distinct traces of water on ignition, and, like ordinary talc, it hardens greatly and becomes vitrified on thin edges in the blowpipe flame. Sub-slaty varieties, forming a transition into Talc proper, occasionally occur.

The Serpentine group is closely related to that of the Talcs and Steatites, its included species being essentially hydrated magnesian silicates, comparatively soft and sectile; but (unlike the Talcs) all yield a distinct amount of water on moderate ignition. The group is chiefly represented by Serpentine proper; the asbestiform variety or sub-species of the latter, known as Chrysotile; the foliated or schistose varieties or sub-species, Antigorite, Schiller Spar, &c.; and the related magnesian silicates, Meerschaum, Deweylite or Gymnite, Kerolite, and other similar compounds. Most of these are decomposition products of more or less unstable character. In the Serpentines, the amount of water averages 12 per cent., but in Deweylite and in many Meerschaums it exceeds 20 per cent., and is still higher in Kerolite. Serpentine proper is commonly in beds or masses of fine-granular or occasionally sub-slaty structure, and of dark-green, yellow, brown, red, or other colour, two or more tints or shades of colour frequently occurring in the same specimen. The so-called "Noble Serpentine" is more or less translucent and of rich shades of colour; whilst "Common Serpentine" is opaque or translucent on the edges only, and comparatively dull or muddy in colour. Mixtures of serpentine with calcite or dolomite are known as Ophiolite, Verde Antique, or Serpentine-marble. Serpentine is unknown in true crystals, but frequently occurs in pseudomorphs (essentially pseudomorphs of alteration) derived from Olivine, Pyroxene, Spinel, and other magnesian species.

The Kaolins present a remarkable resemblance in outward characters to many Talcs and Steatites, some representatives of the group (Pyrophyllite, &c.) being made up of soft, flexible, pearly, and foliated masses, whilst others are fine-granular (or microscopically scaly) in structure, and more or less soapy to the touch. But the Kaolins are essentially aluminous, and thus assume a distinct blue colour after ignition with cobalt-solution. The principal representatives of the group comprise Kaolin proper, Nacrite or Pholerite, Pyrophyllite, Agalmatolite, Halloysite, and Kollyrite. These are sufficiently described in the Table. All are essentially decomposition products.

The Pinite group consists of crystallized pseudomorphous products derived from the alteration of Iolite, or apparently in some cases from that of Nepheline or other species. These substances are chiefly in six-sided or twelve-sided prisms, often more or less ill formed, with dull lustre, and dingy-white, pale-grey, greyish-green, dull-blue, reddish, or dark-brown colour. The hardness is under 4·0 (usually 2·5-3·5), and the sp. gr. about 2·6 or 2·8. They may be grouped conveniently under three series, typified respectively by Pinite, Esmarkite and Gigantolite. The minerals referrible to Pinite and Esmarkite are fusible on the edges only; those referred to Gigantolite melt before the blowpipe more or less readily. These latter, therefore, come under notice in Table XXVII. In the Pinite series, a certain amount of potash is always present (although that alkali has not been found in the supposed parent-stock, Iolite), and the included forms (Pinite, Weissite, Fahlunite, Pyrargillite, Iberite, &c.) shew very distinctly the red K-line in the spectroscope, after being ignited and then moistened with hydrochloric acid, or by fusion with carbonate of soda or fluor-spar. The yellow Na-line, and the green and red Ca-lines from the fluor-spar, may be entirely cut off by the intervention of a piece of deep-blue glass. The representatives of the Esmarkite series, on the other hand (including Esmarkite, Bonsdorffite, Praseolite, Aspasiolite, &c.), do not contain potash.

The group of Copper Silicates inclu : the rare Dioptase and the comparatively common Chrysocolla, the latter including in the Table both the green and blue varieties. The characters of these are sufficiently given in the text. The amorphous Chrysocolla, as a rule, will alone come under the student's observation.

The Zinc Silicates, which include the anhydrous Willemite, with its manganese-holding variety, Troostite, and the hydrous species Calamine, are also described in sufficient detail in the Table. They find a place also in Table XXIV., as in most examples they are sufficiently hard to scratch glass slightly. They do not readily yield a zinc sublimate on charcoal, unless fused in powder with a mixture of carb. soda and borax, or treated according to the method recommended at page 39. With cobalt-solution they assume partly a green and partly a blue colour, the latter, more especially, after strong ignition.

TABLE XXVI.

[Lustre non-metallic (in some cases pseudo-metallic). Slowly or incompletely dissolved, BB, by phosphor-salt. More or less readily fusible. Yielding no water (or merely traces) on ignition].

A.—Fusible into a black or very dark bead, magnetic or non-magnetic.

A¹.—OCCURRING IN SCALY, MICACEOUS, OR ASBESTIFORM EXAMPLES.

† *Scaly or micaceous. Readily decompos* ∴ *by hydrochloric acid.*

LEPIDOMELANE: K^2O 9·20, FeO 12·43, Al^2O^3 11·60, Fe^2O^3 27·66, SiO^2 37·40, with traces of H^2O, &c. In hexagonal tables and scaly masses of a black colour with greenish streak, the scales somewhat brittle; H 2·5-3·0; G 3·0-3·2. BB forms a black magnetic glass or enamel.

ASTROPHYLLITE: K^2O, Na^2O, CaO, MgO, MnO, FeO, Fe^2O^3, Al^2O^3, SiO^2, with 7·66 per cent. TiO^2 and a little H^2O, according to Pisani's analysis. In six-sided tables and micaceous prisms of a bronze-yellow colour and metallic-pearly lustre. Foliæ slightly elastic. BB easily fusible with some bubbling into a black, more or less magnetic bead. The HCl solution, slightly diluted and boiled with a piece of tin, assumes an amethystine colour.

*** In the spectroscope both Lepidomelane and Astrophyllite, when moistened, after ignition, with HCl acid, shew the red K-line.

† † *Readily decomposed by sulphuric acid. Structure micaceous.*

BIOTITE (Potassic Ferro-magnesian Mica): Mostly in dark-green or black micaceous examples, with flexible foliæ. Fusible on the edges only: *See* TABLE XXV.

† † † *Fibrous. Insoluble in acids.*

BYSSOLITE (Ferruginous, asbestiform Amphibole): In fibrous masses of a green or greenish-brown colour. BB, fuses into a black and often magnetic bead.

A².—OCCURRING IN CRYSTALLIZED, LAMELLAR, GRANULAR, OR OTHER NON-MICACEOUS EXAMPLES.

† *Easily decomposed, with gelatinization, by hydrochloric acid.*
(*Fusion-bead magnetic.*)

FAYALITE: FeO 70·6, SiO^2 29·4, but part of the Fe in some examples replaced by Mn; intermixed FeS or FeS^2 also frequently

---

* The silicates of this Division form also in most cases a black glass by fusion with carb. soda.

present. In black or greenish-black masses, commonly magnetic from intermixed pyrrhotine or magnetite; H 6·0-6·5; G 4·0-4·2. BB, easily fusible into a black magnetic bead.

HYALOSIDERITE (Ferruginous Chrysolite): MgO, FeO, SiO^3. In small prismatic crystals of the Rhombic System, yellowish-brown in colour; H 6·0-6·5; G 3·4-3·5; BB, fusible only in fine splinters into a black more or less magnetic slag.

ILVAITE or LIEVRITE: CaO 13·7, FeO 35·2, Fe^2O^3 19·6, SiO^2 29·3 (with 2·2 basic water?). Rhombic; crystals essentially prismatic, with V:V 112° 38', and V2:V2 106° 15', the V planes in most crystals longitudinally striated; also in coarsely fibrous, columnar, and granular masses; black, brownish-black, with dark streak; H 5·5-6·0; G 3·8-4·1. Easily fusible into a black magnetic bead. Moistened with HCl acid, shews red and green Ca-lines in spectroscope very distinctly.

ORTHITE or ALLANITE (Cerine): CaO, CeO, LaO, FeO, Fe^2O^3, Al^2O^3, SiO^2, with, in some examples, YO, MgO, H^2O, &c. Clino-Rhombic: crystals in general transversely elongated, but sometimes tabular; occurs also in columnar and fine granular examples, mostly of a pitch-black colour and somewhat sub-metallic aspect; but also brown or dull greyish-yellow, and then more or less resinous in lustre; H 5·5-6·0; G 2·8-3·8 or 4·0. BB melts easily, with bubbling, into a dark, generally magnetic, bead. Bodenite, Bagrationite, Erdmannite, and Muromontite, are probably varieties.

ALLOCHROITE (Ferro-calcareous Garnet): CaO, FeO, SiO^2. Chiefly in rhombic dodecahedrons of a dark-red, dark-green, or brown colour. Easily fusible. Decomposed, with gelatinization, by hydrochloric acid in some examples, only. See under Garnet, below, page 230.

SIDEROMELANE: CaO, Fe^2O^3, Al^2O^3, SiO^2, with small amounts of MgO, MnO, K^2O, NaO. In black amorphous masses resembling black Obsidian; H 6·0; G 2·55-2·60. Easily fusible into a black magnetic slag. Practically identical with Tachylite, but distinguished by its larger amount of iron, and by dissolving somewhat less readily in hydrochloric acid.

(*Fusion-globule not magnetic. No sulphur-reaction.*)

TACHYLITE: CaO, FeO, Al^2O^3, SiO^2, with, in general, small amounts of K^2O and Na^2O, MnO, MgO, and sometimes TiO^2. In black or brownish-black amorphous masses of vitreous lustre, much resembling some Obsidians. H 6·0-6·5; G 2·51-2·60. BB, easily fusible with bubbling into a black (non-magnetic) glass or enamel. In spectroscope, shews Ca-lines, and in many examples the red K-line also. An essentially volcanic or trappean product.

TEPHROITE: MnO 70·3, SiO^2 29·7. In granular, cleavable masses of a reddish-grey or dull reddish-brown colour, weathering brownish-black: the cleavage rectangular. H 5·5-6·0; G 4·0-4·12. Easily fusible to a black slag. With carb. soda gives strong manganese-reaction. Knebelite is probably identical, although said to be infusible.* Tephroite differs essentially from the more common manganese silicate Rhodonite, by its ready gelatinization in HCl acid. Rhodonite being practically insoluble.

(*Fusion-product not magnetic. Strong sulphur-reaction.*†)

HELVINE: BeO, MnO, FeO, SiO^2 with Mn, Fe, S. Reg.; crystals chiefly tetrahedral; occurs also, though rarely, in botryoidal masses: H 5·5-6·5; G 3·2-3·4; yellow, brownish, yellowish-green. BB, in O.F. a dark bead, dull yellow in R.F. In hydrochloric acid, evolves odour of sulph. hydrogen, and is decomposed with gelatinization.

DANALITE: A flesh-red or yellowish-grey Helvine, with MnO largely replaced by ZnO. Crystallizes in regular octahedrons, sometimes with truncated edges, and occurs also in small, disseminated grains. Blowpipe and acid reactions like those of Helvine proper, but a zinc-sublimate formed (with carb. soda and borax) on charcoal.

† † *Decomposed imperfectly by hydrochloric acid, but completely by sulphuric acid.*

SPHENE (Titanite): CaO (partly replaced by FeO and MnO) 28·57. TiO^2 40·82, SiO^2 30·61. Clino-Rh.; crystals mostly small, with more or less ortho-rhombic aspect, often tabular and frequently twinned: see Note at end of Table; brown, grey, yellow, green, &c.; occurs also in cleavable and fine-granular masses; dark-brown, light-brown, grey, yellow, green, &c.; H 5·0-5·5; G 3·4-3·6; lustre vitreo-resinous. BB, commonly becomes yellow and melts with bubbling to a dark enamel. The sulphuric acid solution (or the aqueous solution obtained by fusing the finely ground mineral with bisulphate of potash)

---

* Judging from its stated characters and composition, its infusibility is most improbable. I have tried without success to procure a specimen for comparison.

† See page 61, Experiment 1. The carb. soda should be used somewhat in excess. These minerals give also a strong Mn-reaction.

assumes a violet colour if boiled with a few drops of hydrochloric acid and a piece of tin*. In spectroscope, shews red and green Ca-lines if moistened with HCl acid after strong ignition.

KEILHAUITE (Yttro-titanite): CaO, YO, Al^2O^3, Fe^2O^3, TiO^2, SiO^2. Commonly in dark reddish-brown twin-crystals resembling those of Sphene, but often of comparatively la/ ;e size; H 6·0-7·0; G 3·5-3·72. BB, like Sphene.

SCHORLAMITE (Ferro-titanite): CaO ?·38, Fe^2O^3 20·11, TiO^2 21·34, SiO^2 26·09, with small amounts of M_cO, FeO, and alkalies. Reg.; crystals rare, commonly the Rhombic Dodecahedron, or that form with the trapezohedron 2-2, hence much resembling garnet crystals. Occurs mostly in small granular masses of a pitch-black colour; H 7·0-7·5; G 3·78-3·86. BB, fuses on the edges, or entirely, into a black slag or bead; other reactions like those given under Sphene.

† † † *Partially or slightly attacked in normal condition by hydrochloric acid, but readily decomposed by that acid after fusion.*†

(*During fusion, impart a red colour to the flame.*)

FERRUGINOUS LEPIDOLITE: In brown, grey, or greyish-red scaly aggregations; H 2·5; G 2·9-3·0. BB, fusible with great bubbling into a dark magnetic bead. See Lepidolite proper, under B^2., page 234.

(*During fusion, impart a green colour to the point of the flame.*)

AXINITE: CaO 20·2, MnO 2·6, FeO 2·8, Fe^2O^3 6·8, Al^2O^3 16·3, B^2O^3 5·61, SiO^2 43·5, with small amounts of MgO, K^2O, and basic H^2O. Anorthic; crystals essentially flat or very thin rhomboidal prisms, replaced only on single edges and angles; brown, violet-brown, green, pearl-grey, amethystine, different tints often shewing in different directions; H 6·5-7·0; G 3·27-3·33. BB, easily fusible, with green coloration of the flame-point, to a black bead, which generally becomes green and translucent in the inner flame.

(*No green or red coloration of flame during fusion. Never in fibrous, acicular, or prismatic examples.*)

GARNET: *Dark sub-species* (Almandine, Aplome, Andradite, Pyrope, Melanite, Spessartine, &c.): Average composition, RO 33 to 43, R^2O^5

---

* In fine powder, Sphene is also sufficiently decomposed by hydrochloric acid to give this characteristic reaction when the solution is boiled with a piece of metallic tin.

† The fused bead or slag must be crushed under paper on the anvil, or in a small steel mortar, and then ground to a fine powder.

21 to 32, SiO^2 35 to 40 ($RO = CaO$, MgO, FeO, MnO ; $R^2O^3 = Al^2O^3$, Fe^2O^3). Reg.; principal forms, the rhombic dodecahedron and the trapezohedron 2-2 (*see* Note at end of Table). Frequently in rounded grains and indistinct crystals; red, brown, black, dark-green, &c.; H 6·5-7·5; G 3·6-4·3 (in dark varieties). BB, fusible more or less readily into a dark and generally magnetic bead. The Bohemian garnet, Pyrope, which occurs chiefly in small grains of a deep-red colour, contains a small amount of chromium (Cr ?), and becomes black and opaque on gentle ignition, but recovers its red colour and translucency on cooling. As shewn by Dr. L. H. Fischer, it is only decomposed to a slight extent, after fusion, by hydrochloric acid.

(*Essentially in fibrous, acicular, or prismatic examples.*)

EPIDOTE (Pistacite, Thallite, Bucklandite, Piedmontite, Withamite, &c.): CaO 36 to 40, Al^2O^3 18 to 30, Fe^2O^3 7 to 20 or Mn^2O^3 10 to 25, SiO^2 36 to 40, with traces of MgO, &c., and about 2 per cent. basic water. Clino-Rh.; crystals in general elongated parallel to the ortho-axis, with cleavage planes meeting at angle of 115° 24': *see* Note at end of Table; occurs also in acicular, fibrous, and other examples; green of various shades, greenish-yellow, black. (In manganese varieties, blackish-red or dull cherry-red.) H 6·0-7·0; G 3·3-3·5. BB, swells up, and forms a dark cauliflower-like slag, or in some cases a black glass, generally magnetic. In phosphor-salt, somewhat easily decomposed, differing remarkably in this respect from examples of Pyroxene and Amphibole of similar aspect,

† † † † *Very slightly attacked by hydrochloric acid, both before and after fusion.*

(*In triangular or nine-sided prisms; or in acicular, columnar, or fibrous examples, triangular on cross-fracture.*)

SCHORL; BLACK or DARK-BROWN TOURMALINE: Approximate composition: MgO 7 or 8, FeO 5 to 10, Al^2O^3 30, B^2O^3 9 or 10, SiO^2 38, with small amounts of K^2O, Na^2O, Li^2O, KaO, MnO, F, and basic water. Hemi-Hexagonal (*see* Note at end of Table); also very commonly in columnar and fibrous masses, the component fibres shewing under the magnifying glass a triangular cross section; Black, dark-brown, with vitreous external lustre; H 7·0-7·5; G 3·03-3·20; pyro-electric. BB, melts more or less easily to a black slag or glass, which often attracts the magnet. The fused bead reduced to fine powder is decomposed by strong sulphuric acid. Alcohol added to the solution, and ignited, burns with the green flame characteristic of B^2O^3. The crushed bead made into a paste with sulphuric acid, imparts this colour to the blowpipe-flame. A drop of glycerine intensifies the reaction : *see* page 28.

(*Essentially in lamellar or foliated masses with strongly pronounced cleavage in one direction.*)

HYPERSTHENE (Ferruginous Bronzite): MgO, FeO, SiO^2. Rhombic, but crystals of quite exceptional occurrence ; essentially in bronze-brown, green, or greenish-black, lamellar masses, with metallic-pearly lustre on cleavage plane ; H 5·0-6·0 ; G 3·3-3·4. BB, fusible more or less easily into a black magnetic bead or slag. *See* under Bronzite in Table XXV.

(*In lamellar or fibrous masses or distinct crystals, with cleavage-angle and principal prism-angle near* 87°.)

AUGITE (DARK PYROXENE): Average composition, MgO 12 to 18, CaO 18 to 20, FeO 10 to 13, Al^2O^3 4 to 8, SiO^2 47 to 50, with small amounts of MnO, &c. Clino-Rh. ; the more common crystals are eight-sided prisms, composed of the forms V, $\bar{V}$, and $\acute{V}$, with two inclined summit-planes, or large basal plane*. Often twinned parallel to $\bar{V}$ (*see* Note at end of Table). V : V 87° 6′; $\bar{V}$ on $\acute{V}$ 90°; angle over summit-planes 120° 48′. Commonly, in cleavable, fibrous, or granular masses. Black, greenish-black, dark-green, dark-brown ; H 5·0-6·0 ; G 3·0-3·4. BB, fusible more or less easily into a black, generally magnetic bead. Hedenbergite is a non-magnesian augite, consisting of CaO 22·18, FeO 29·43, SiO^2 48·39: black, blackish-green, in cleavable masses. Coccolite is a dark-green augite, occuring in granular masses or small crystals with rounded edges and angles. Breislakite is an acicular variety from Italian lavas. Fassaite (Pyrgom), and some Sahlites also belong to the present sub-species.

ACMITE: Na^2O 13·88, FeO 6·45, Fe^2O^3 28·64†, SiO^2 51·03, with small amounts of K^2O, MnO, TiO^2, &c. Clino-Rh.; crystals long and thin ; striated longitudinally, and, as regards the typical examples, imbedded in quartz ; V on V 87° 15′ ; H 6·0-6·5 ; G 3·4-3·53. Easily

---

* This plane is regarded by most German crystallographers, and by many others, as a front-polar or hemi-orthodome. *See* the note on the crystallization of Pyroxene at the end of the present Table.

† Some mineralogists make all the iron Fe^2O^3, but FeO is certainly present in Acmite as well.

fusible into a black magnetic bead. Ægirine is identical or closely related.

JEFFERSONITE: CaO, MgO, MnO, ZnO, FeO, SiO^2, with small amounts of Al^{2}O^3, &c. Clino-Rh., but occurring only in granular examples with cleavage-angle of about 87° 30'. Dark-green, brown. greenish-black; H 4·5; G 3·3-3·5. BB, fusible into a black bead. With carb. soda and borax on charcoal, gives a zinc sublimate and strong manganese reaction. Hitherto, only met with at Sparta, New Jersey.*

BABINGTONITE: CaO 19·32, MnO 7·91, FeO 10·26, Fe^{2}O^3 11·00. SiO^2 51·22, with traces of MgO, &c. Anorthic (crystals mostly short. eight-sided prisms, with two summit-planes). Occurs also in radiating groups. Black, greenish-black; H 5·5-6·0; G 3·3-3·4. Easily fusible into a black magnetic bead. Generally associated with Albite or Orthoclase. Distinguished from black augite only by its crystallization.

RHODONITE (Silicate of Manganese): MnO 54·2, SiO^2 45·8, but part of the MnO commonly replaced by CaO, FeO, or MgO. Anorthic, but crystals of exceedingly rare occurrence; commonly in cleavable masses, with cleavage-angle of 87° 38'; rose-red, greyish-red, weathering dark-brown; H 5·0-5·5; G 3·5-3·65. BB, fusible into a dark-red or amethystine glass which becomes black and opaque in the outer flame. With carb. soda, strong Mn reaction. Bustamite. in radiated-fibrous examples of pink or pale greenish-grey colour, is a calcareous variety; Fowlerite, in coarse crystals and cleavable masses of a reddish-brown or dull-red colour, has the MnO largely replaced by FeO, CaO, and ZnO.

(*In lamellar or fibrous masses or in distinct crystals with cleavage-angle and principal prism-angle near* 124°).

HORNBLENDE; DARK OR STRONGLY-COLOURED AMPHIBOLE (Includes Common Hornblende, Basaltic Hornblende, Pargasite, and most examples of Actynolite): Average composition, CaO 9 to 12. MgO 10 to 20, FeO 8 to 20, Fe^{2}O^3 5 to 6, Al^{2}O^3 5 to 15, SiO^2 40 to 44; but in non-aluminous or slightly aluminous varieties, the SiO^2 generally exceeds 50 per cent. Small amounts of Na^{2}O, K^{2}O, and Fluorine are also usually present. Clino-Rhombic; crystals mostly six-sided prisms, composed of the forms V and V̌, terminated generally

---

* As the composition of Jeffersonite does not appear to be at all constant, the mineral may perhaps be nothing more than a mixture of Pyroxene and Franklinite.

by three comparatively flat rhombiform faces ($=$ B and P), also sometimes consisting of the prism V alone, terminated by two triangular planes $\acute{P}$. The front prism-angle V on V equals $124° 30'$: V on V' $= 117° 45'$; P on $P = 148° 30'$; P on $B = 145° 35'$; $\acute{P}$ on $\acute{P}$ (over summit) $= 148° 16'$. Occurs also very abundantly in lamellar, fibrous and granular masses; colour, dark-green, black, dark-brown; H 5·0-6·0; G 3·0-3·4. BB, fusible more or less easily into a black, usually magnetic bead. Cummingtonite is a brown, fibrous variety, containing very little lime. Arvedsonite is a closely-related species or variety containing 10·60 per cent. Na^2O. Mostly in black cleavable masses, with greenish streak; H 6·0 : G 3·33-3·60. Very easily fusible, with much bubbling, into a black, magnetic bead*. See also Glaucophane, under B^3, below.

(*In amorphous, obsidian-like masses.*)

WICHTISITE (Wichtyne): Na^2O, CaO, MgO, FeO, Fe^2O^3, Al^2O^3, SiO^2. In black, more or less dull, amorphous masses, with well-marked conchoidal fracture; H 6·0-6·5; G 3·0-3·1. Fusible, with bubbling, into a black opaque bead.

(*In deep-red grains and rounded crystals.*)

PYROPE (Bohemian Garnet): *See* under Garnet, above.

(*In flat tabular crystals or granular masses. Sp. gr. over 3·5.*)

CERINE : Black, brownish-black ; scarcely attacked by hydrochloric acid. *See* under Orthite, above, page 228.

B.—Fusible into a colourless or lightly-tinted bead or glass.

B¹.—IMPART A DISTINCT RED OR GREEN COLOUR TO THE BLOWPIPE-FLAME.

† *BB, flame coloured red.*

(*Soft ; scaly or foliated.*)

LEPIDOLITE (Lithionite, Lithia Mica): K^2O 4 to 11, Na^2O 1 to 3, Li^2O 1·5-5 ; MnO 2 to 5, Al^2O^3 14 to 29, Fe^2O^3 0 to 28, SiO^2 40 to 52, with from 4 to 8 per cent. Fluorine. Essentially in scaly aggregations or micaceous masses of a rose-red, pale-red, pearl-grey, or greyish-white colour ; H 2·0-4·0 (commonly 2·5); G 2·8-3·0. BB, very easily fusible with great bubbling into a colourless blebby glass

---

* Very thin splinters fuse without the aid of the blowpipe, as first pointed out by Dr. L. H. Fischer: *Clavis der Silicate*, p. 11.

(or, as regards ferruginous examples, into a dark metallic bead), with crimson coloration of the flame. In the spectroscope, the red Li-line and yellow Na-line come out very prominently, the red K-line subordinately*. After fusion, completely decomposed by hydrochloric acid.

CRYOPHYLLITE : K^2O, Li^2O, MgO, MnO, FeO, Fe^2O^3, Al^2O^3, SiO^2 (53·46) with 2 to 3 Fluorine. Essentially in dark-green, six-sided, micaceous prisms and scaly masses; G 2·9. BB, colours the flame red, and fuses with great bubbling.

(*Hard. Not micaceous in structure.*)

PETALITE : Li^2O (with small amount of Na^2O) 4·42, Al^2O^3 17·80, SiO^2 77·96. Essentially in lamellar masses (Clino-Rh.) with cleavage-angles of 117°, 141° 23′ and 101° 30′, but the two latter often indistinct ; colour, pale-red, reddish-white, or nearly colourless ; H 6·0-6·5 ; G 2·4-2·6. BB, colours flame pale-red, and melts to a colourless glass. In the spectroscope, especially if the test-matter be moistened with hydrochloric acid, the red Li-line comes out very distinctly. Insoluble in acids. Kastor is a variety in coarse Clino-rhombic crystals from Elba : V on V 86° 20′.

SPODUMENE (Triphane): Li^2O 6·73, Al^2O^3 29·21, SiO^2 64·06 ; but part of the Li^2O commonly replaced by small amounts of Na^2O and K^2O and traces of CaO. Clino-Rhombic, with V : V 87°, but crystals comparatively rare. Commonly in cleavable masses with cleavage-angles of 87° = V : V, and 133° 30′ = V : $\overline{V}$. Pale-green, greenish-white, or greenish-grey ; H 6·0-7·0 ; G 3·12-3·20. BB, colours flame distinctly red, and melts easily, with much expansion and bubbling, into a colourless glass. Insoluble in acids. In spectroscope, shews red Li-line and yellow Na-line distinctly.

† † *Flame coloured green.*

(*Very easily fusible.*)

AXINITE : Essentially in groups of thin sharp-edged crystals, brown, green, brownish-violet, pearl-grey, or amethystine in colour. BB, melts in the outer flame into a black glass, and with carb. soda gives manganese reaction. *See* above, page 230.

---

* The K-line is scarcely visible unless the Na and Li lines be cut off by the intervention of a piece of deep-blue glass.

DANBURITE : $CaO\ 22·75$, $B^2O^3\ 28·45$, $SiO^2\ 48·80$. Anorthic; but mostly in lamellar massess with cleavage-angles of 110°, 126° and 93°, the two latter more or less indistinct. Yellowish-white, pale-yellow; H 7·0; G 2·95-2·96. BB, easily fusible, with green coloration of the flame. The powder moistened after ignition with hydrochloric acid, shews in the spectroscope green B-lines with transitory flashes of the red Ca-line.

(*Fusible with difficulty or on the edges only.*)

HYALOPHANE (Barytic Feldspar): $K^2O\ 7·82$, $Na^2O\ 2·14$, $BaO\ 15·05$, $Al^2O^3\ 21·12$, $SiO^2\ 52·67$, with traces of CaO, MgO, &c., but the composition, more especially as regards the amount of baryta, appears to be somewhat variable. Clino-Rhombic; crystals practically identical with those of Orthoclase; cleavage very perfect parallel with basal plane; white, pale-reddish; H 6·0-6·5; G 2·80. BB, fusible on edges only, unless in thin splinters. Distinguished from the feldspars, generally, by the green colour imparted to the point of the flame. In acids scarcely attacked.

B2.—YIELD STRONG REACTION OF SULPHUR OR CHLORINE.*

† *Give sulphur reaction, BB, with carb. soda.*

HELVINE; DANALITE : Essentially in small tetrahedrons or octahedrons, or in small grains, of a yellow, brownish, yellowish-green, or reddish-grey colour. H 5·5-6·5; G 3·2-3·4. Gelatinize and evolve odour of sulph. hydrogen in hydrochloric acid. BB, in outer flame give a black or dark fusion-product. *See* under A2, page 229.

. HAUYNE: $K^2O\ 4·96$, $Na^2O\ 11·79$, $CaO\ 10·60$, $Al^2O^3\ 27·64$, $SiO^2\ 34·06$, $SO^3\ 11·25$. Reg.; chief crystal form, the rhombic dodecahedron; occurs also in small grains. Essentially blue or bluish-green, rarely colourless (Berzeline); H 5·0-5·5; G 2·4-2·5. BB, decrepitates, and melts slowly into a pale-blue or colourless glass. Gelatinizes in hydrochloric acid.

NOSINE (Nosean): NaO, Al^2O^3, SiO^2, SO^3. Closely resembles Hauyne in crystallization, and in its blowpipe and acid reactions, but commonly ash-grey, greyish-blue, or greenish-white in colour, and with larger percentage of soda (24·89).

---
* See page 61, Experiments 1 and 3. In testing for sulphur, the reagent, carb. soda, should be used somewhat in excess.

LAPIS-LAZULI : NaO, CaO, SiO², SO³, &c. Essentially in granular masses of a rich blue colour, frequently intermixed with calcite, grains of iron pyrites, and other substances. When crystallized, in rhombic dodecahedrons. H 5·5 ; G 2·38-2·45 ; BB, melts easily to a colourless glass. Gelatinizes in hydrochloric acid, most examples evolving sulph. hydrogen during decomposition.

MICROSOMMITE : Gives feeble S-reaction, but strong reaction of chlorine : *see* below.

† † *Give Cl-reaction with cupreous phosphor-salt bead.*

SODALITE : Na²O, Al²O³, SiO², NaCl. Reg. ; chiefly crystallized in rhombic dodecahedrons, or in combinations of that form and the cube ; occurs also in granular examples ; mostly colourless or greenish-white, less commonly blue or bluish-green. H 5·5 ; G 2·13-2·30. BB, a colourless glass. In hydrochloric acid, gelatinizes.

MICROSOMMITE : K²O, Na²O, CaO, Al²O³, SiO², NaCl, with small percentage of SO³ in most examples. Hexagonal ; chiefly in minute six-sided prisms on certain Vesuvian lavas ; H 6·0 ; G 2·6. BB, according to Sacchi, difficulty fusible. Gelatinizes in hydrochloric acid. The spectroscope should shew Na, K, and Ca lines, but the writer has not been able to procure a specimen for examination.

EUDIALYTE : Na²O, CaO, FeO, ZrO², SiO², with small amounts of CaO, MnO, &c., and about 2 per cent. NaCl. Hemi-Hexagonal ; crystals, acute rhombohedrons with extended basal plane ; R : R 73° 30', B : R 112° 18' and 67° 42'. Dark purplish-red, brownish-red. H 5·0-5·5 ; G 2·8-3·0. Melts easily to a greyish-green glass or enamel. Gelatinizes in hydrochloric acid. Eucolite from Norway is closely related. Both are rare species.

B.—NO DISTINCT (RED OR GREEN) FLAME-COLORATION. NO REACTION OF SULPHUR OR CHLORINE.

† *Decomposed with gelatinization by hydrochloric acid.*

(*BB, with carb. soda on charcoal, a distinct sublimate*).

EULYTINE (Bismuth Blende) : Bi²O³ 83·75, SiO² 16·25, but generally intermixed with Fe²O³, Mn²O³, P²O⁵, Fl, &c. Reg. : crystals essentially tetrahedral, very small, in drusy aggregations ; occurs also in botryoidal masses ; H 4·5-5·0 ; G about 6·1. Fusible into a dull brownish bead. With carb. soda forms on charcoal a deep-yellow sublimate. Gelatinizes in hydrochloric acid.

WILLEMITE: ZnO, SiO^2. White, brownish, &c. Fuses on edges or surface only. With carb. soda and borax on charcoal gives a zinc-sublimate. With Co-solution, coloured blue or bluish-green. See TABLE XXV.

(BB, with carb. soda no sublimate. Colour, black).

GADOLINITE: Essentially in small, vitreo-resinous masses of a black colour and greenish-grey streak. BB, generally swells up, but vitrifies on edges only. See TABLE XXIV.

TSCHEWKINITE: CaO, MnO, FeO, CeO, LnO, DO, TiO^2, SiO^2, with traces of K^2O, Na^2O, &c. In more or less compact masses; velvet-black, with brownish streak; H 5·0-5·5; G 4·5-4·8. BB. swells up into a porous mass, and then melts slowly into a dull yellowish enamel. Gelatinizes in hydrochloric acid. The diluted solution boiled with a piece of metallic tin assumes a violet colour. A very rare species.

(Colourless or lightly-tinted species. Fusible on thin edges, only).

GEHLENITE: Essentially in greenish-grey, or pale-brownish, square prisms of small size. Ca-lines in spectroscope readily brought out by moistening the ignited test-substance with hydrochloric acid. See TABLE XXIV.

MONTICELLITE (Batrachite): Essentially in small crystals of the Rhombic System. V:V 98° 8', V_2^1 : V_2^1 133°, P̈ : P̈ over summit 82° nearly, P : P over front edge 141° 50'; over side edge 82°. Colourless, pale-green, pale-brownish. Other characters as in Gehlenite. See TABLE XXIV.

(In platinum forceps, more or less readily fusible. In spectroscope, after ignition and moistening with hydrochloric acid, shew distinct red and green Ca-lines).

WOLLASTONITE (Table Spar): CaO 48·28, SiO^2 51·72. Clino-Rhombic. but crystals comparatively rare; commonly in lamellar and fibrous masses, with cleavage angles of 95° 30' and 84° 30' (= B on V̄); colourless, pale-reddish or yellowish-white, &c.; H 4·5-5·0; G 2·75-2·92; in the forceps, thin splinters fuse more or less readily. Decomposed, with gelatinization by hydrochloric acid.

HUMBOLDTILITE (Melilito): Na^2O, CaO (31 or 32), MgO, Fe^2O^5, Al^2O^3, SiO^2. Tetrag.; crystals mostly tabular, with large basal plane; occurs also in fibrous and columnar examples; yellowish-white, pale-yellow, brownish, &c.; H 5·0-5·5; G 2·9-2·95. BB.

fusible with slight bubbling into a colourless or yellowish glass. Gelatinizes in hydrochloric acid.

SARCOLITE: K^2O 1·20, Na^2O 3·30, CaO 32·36, Al^2O^3 21·54, SiO^2 40·51 (Rommelsberg). Tetrag.; crystals, mostly small square prisms with replaced angles (= $\overline{V}$, B, P); also sometimes with hemihedral polar planes; pale-red, reddish-white; H 5.5-6·0; G 2·55-2·95; fusible into a white blebby glass or enamel. In hydrochloric acid, gelatinizes.

DAVYNE; CANCRINITE: *See* under Nepheline, below.

(*No Ca-lines brought out in spectroscope by moistening with hydrochloric acid* *).

NEPHELINE (Elæolite): K^2O 4·5 to 6·5, Na^2O 15.5 to 17, Al^2O^3 34·5 to 35·5, SiO^2 41 to 45. Hexag.; crystals mostly small hexagonal prisms with replaced basal. edges; occurs also in lamellar masses; colourless, white, pale-brownish, with vitreous lustre (Nepheline proper); and greyish-blue, bluish-green, or red, with vitreoresinous lustre (Elæolite); H 5·5-6·0; G 2·55-2·65. Fusible, with more or less bubbling, into a blebby glass. Gelatinizes in hydrochloric acid. Most examples shew the red K-line distinctly in the spectroscope if moistened with HCl acid after fusion or ignition. Davyne and Cancrinite are partly altered varieties, containing intermixed CaO, CO^2, and a small percentage of H^2O.

† † *Decomposed by hydrochloric acid, but without gelatinization.* †

(*The hydrochloric acid solution boiled with tin assumes a blue or violet colour*).

SPHENE (Titanite): CaO, TiO^2, SiO^2. In Clino-Rhombic crystals and cleavable masses of a brown, yellow, yellowish-grey or green colour; H 5·0-5·5; G 3·4-3·6. BB, melts generally into a black or dark enamel, but in some cases the fusion product is dull-yellow. *See* under B', above, page 229.

GUARINITE: CaO, TiO^2, SiO^2. Rhombic, but hitherto only recognized in apparently square tables. Sulphur-yellow. Fusible into a yellow glass.

---

* Unless intermixed calcite be present, as in many examples of the Davyne and Cancrinite varieties.

† In some cases, the decomposition, although sufficiently marked, is more or less incomplete. If decomposition ensue at all, the supernatant liquid, diluted slightly and filtered from the undissolved residuum, will yield a distinct precipitate with ammonia, or with oxalate of ammonia added subsequently.

WÖHLERITE: Na^2O, CaO, FeO, ZrO^2, Nb^2O^5, SiO^2. Rhombic or Clino-Rhombic, but crystals mostly indistinct; commonly in small angular grains, or in sub-columnar masses and indistinct tabular forms. Yellow of various shades, yellowish-brown; H 5·0-6·0; G 3·41. BB melts easily into a yellowish bead. Hitherto only found in the Zircon-syenite of Norway.

(*Giving BB with fused phosphor-salt in open glass tube a strong Fluorine-reaction*).

LEUCOPHANE: CaO, BeO, SiO^2, NaF. Essentially in cleavable lamellar masses of a pale yellow or greenish-grey colour. H 3·5-4·0; G 2·9-3. Strongly phosphorescent, and very easily fusible. Slowly decomposed by hydrochloric acid. *See* under the Fluorides, in Table XX, page 178. A rare species.

MELINOPHANE (Meliphanite): CaO, BeO, SiO^2, NaF. Occasionally in Tetragonal crystals, but commonly in lamellar masses and disseminated grains of a yellow colour; H 5·0; G 3.02. BB, easily fusible (but is said not to phosphoresce?). Very rare, and still imperfectly known.

(*Fusible on charcoal into a glassy bead*).

PREHNITE[*]: CaO 27·14, Al^2O^3 24·87, SiO^2 43·63, H^2O 4·36. Rhombic; crystals tabular or short prismatic, generally aggregated in groups; occurs also abundantly in fibrous-botryoidal masses, and sometimes in pseudomorphs after calcite, analcime, &c.; H 6·0-7·0; G 2·8-3·0; generally greenish-white, also colourless and light-green. Fuses very easily and with much bubbling. In the bulb-tube, gives off a small amount of water, but only at a comparatively high temperature. After fusion or strong ignition, decomposed with gelatinization by hydrochloric acid, and then shews in spectroscope momentary red and green Ca-lines.

WERNERITE (Scapolite, Paranthine, Meionite, &c.): Contains CaO, Al^2O^3, SiO^2, in somewhat variable proportions, with small amounts of K^2O, Na^2O, and H^2O. Tetragonal; crystals, commonly, eight-sided prisms composed of the two square prisms V and V̄, with terminal polar planes, P, P̄, &c. (*See* note at end of Table). P : P over middle edge 63° 42′, over polar edge 136° 11′; cleavage parallel with V, less distinct parallel with V̄; crystals often large, and fre-

---

[*] Belongs properly to Table XXVII., but is referred to also here, as the small amount of water which it contains might in certain cases escape detection.

quently more or less weathered; occurs also in columnar, sub-fibrous, granular, and other masses; colourless, white, greenish-white, green, pale-reddish, greyish, &c.; H 5·0-6·0; G 2·6-2·8. BB, easily fusible with more or less bubbling. In the spectroscope, after ignition and moistening with hydrochloric acid, shews red and green Ca lines, in most cases, very distinctly. Meionite (often classed as a distinct species) and Mizzonite are varieties from Monte Somma. Nuttalite, Dipyre, Couseranite, Passauite, are varieties from other localities. Wilsonite, in pale purplish-red, cleavable and sub-fibrous masses, is probably an altered Wernerite containing intermixed $CaOCO^2$.

GROSSULAR, and most other light-coloured GARNETS:* CaO, Al^2O^3, SiO^2, &c. In crystals of the Regular System, chiefly the rhombic dodecahedron or the trapezohedron 2-2, and in small rounded grains; H 6·5-7·5; G 3·15-3·8 (in grossular, proper, usually about 3·4 or 3·5); light-green (grossular proper), red, yellow, brown, &c., rarely colourless. BB, more or less readily fusible into a lightly-tinted or uncoloured glass.

(Fusible in the forceps, but not fusible into a bead on charcoal).

ANORTHITE (Lime-Feldspar in part, Indianite, Christianite): CaO 20·10, Al^2O^3 36·82, SiO^2 43·08. Anorthic; crystals often large, with B and V́ planes predominating; frequently twinned parallel to one or the other of these forms, to which the cleavage planes are also parallel; cleavage-angles, 85° 50′ and 94° 10′; Right V on Left V 120° 30′. Occurs also in lamellar and granular masses; H 6·0; G 2·66-2·80; colourless, white, pale-reddish, with pearly lustre on cleavage planes and vitreous lustre on other planes. BB, fusible into a clear glass. Completely decomposed by hydrochloric acid, but without gelatinization. In the spectroscope, the Ca lines come out distinctly after ignition and moistening with acid.

LABRADORITE (Lime Feldspar, Lime-soda Feldspar, Labrador Feldspar): NaO, CaO, Al^2O^3, SiO^2. Anorthic; but commonly in cleavable masses, with cleavage-angles of 86° 40′ and 93° 20′; mostly light or dark grey, with play of green, blue, violet, red, or orange, in certain directions, but sometimes white, and without or with very

---

* The deep-red and most dark garnets fuse into a black and generally magnetic bead, and are thus placed in section A of the present Table. Many light garnets, again, are partially decomposed by hydrochloric acid, whilst others are scarcely attacked by that reagent. These latter are referred to, consequently, under the next sub-section † † †.

feeble play of colour. H 6·0; G 2·6-2·8. Fusible into a clear glass. Slowly and only partially decomposed by hydrochloric acid. Spectroscope reaction as in Anorthite.

† † † *Scarcely attacked by hydrochloric acid.*
(*Micaceous species: flexible in thin leaves. Fusible on edges or in thin scales, only*.)

MUSCOVITE (Potash Mica); PHLOGOPITE (Potassic Magnesian Mica): In thin leaves, flexible and elastic; lustre more or less metallic-pearly. Phlogopite is decomposed by sulphuric acid; Muscovite, not. Both fuse in general on the edges into a grey enamel. Biotite is also decomposed by sulphuric acid, but melts on the edges, as a rule, into a black ferruginous glass or slag. *See* TABLE XXV., A^1.

TALC: MgO, SiO^2. White, greenish, &c.; very soft. Flexible in thin pieces, but not elastic; H 1·0. More or less soapy to the touch. Reddens by ignition with Co-solution. *See* TABLE XXV., A^2.

(*Foliated species, with marked cleavage in one direction, but not flexible in thin leaves*).

MARGARITE (Pearl Mica): White, reddish, greenish, &c., with pearly lustre; H 3·5-4·0. Fusible on the edges, often with more or less intumescence, into a greyish enamel. *See* TABLE XXV.

DIALLAGE: MgO, CaO, SiO^2, with, commonly, small amounts of FeO, MnO, Al^2O^3, and H^2O. In foliated or sub-foliated masses or indistinct tabular crystals of a greyish-green or greenish-brown colour and metallic-pearly lustre; H about 4·0; G 3·2-3·4. Fusible more or less easily into a greyish enamel. An aberrant, schistose variety of Pyroxene.

(*Very sectile: readily cut by the knife. Fusible on edges only*).

STEATITE (Soapstone in part): MgO, SiO^2. In white, grey, greenish, reddish, or mottled masses of more or less compact structure, or occasionally sub-slaty. Sometimes, also, in pseudomorphs after Scapolite, Spinel) and other species; H 1·5-2·5. BB, hardens greatly, but only fuses on thin edges. Reddens by ignition with Co-solution. Generally gives off traces of water in the bulb-tube. Decomposed by sulphuric acid. A compact or non-foliated variety of Talc. *See* TABLE XXV., C^2.

---

* Most of these species, when ignited in the bulb-tube, give off traces of water.

(*Asbestiform: in soft, fibrous masses*).

ASBESTUS (Amianthus): Essential components, CaO, MgO, SiO^2. In white, grey, brownish, greenish-white, or green masses of fibrous structure, more or less soft and silky. Readily fusible into a colourless or pale greenish glass. A fibrous variety of Amphibole or Pyroxene. Passes into fibrous serpentine, but distinguished properly from the latter by not being decomposed by sulphuric acid, and by yielding merely traces of water in the bulb-tube. Also by its greater fusibility.

(*Sp. gr. 2·9 or higher. In most cases, distinctly over 3·0*).

DIOPSIDE, and other light-coloured PYROXENES (Malacolite, Alalite, Sahlite in part): Average composition, MgO 18, CaO 26, SiO^2 56; but in some cases 5 or 6 per cent. Al^2O^3, and only 50 or 51 per cent. SiO^2 are present. Clino-Rhombic; crystals, as in Augite (*see above*), commonly eight-sided prisms made up of the forms V, $\bar{V}$, and $\acute{V}$, and terminated by several polar forms or by a large basal plane.* V:V 87° 6'; V:$\bar{V}$ 133° 33'; V:$\acute{V}$ 136° 27'; B:$\bar{V}$ 105° 30'. Occurs also abundantly in lamellar and other conditions, with cleavage angles of about 87° and 93°. Usually greenish-white or some light shade of green, passing into deeper green; H 5·0-6·0; G 3·0-3·4. BB, in thin splinters, fuses more or less readily into a colourless or lightly-tinted glass.

TREMOLITE, and other light-coloured AMPHIBOLES (Grammatite, Actinolite in part, Nephrite in part, Smaragdite): Average composition, CaO 13·5, MgO 28·5, SiO^2 58; but in some varieties a small amount of Al^2O^3 is present, with corresponding decrease of SiO^2. Many examples also contain 1 or 2 per cent. of fluorine. Clino-Rhombic; crystals commonly oblique-rhombic prisms composed of the four planes V, with two depressed triangular planes or side-polars $\acute{P}$ at each extremity; or sometimes six-sided, from presence of $\acute{V}$; the basal form B also often present. V:V 124° 30'; V:$\acute{V}$ 117° 45'; $\acute{P}$:$\acute{P}$ 148° 16'. Occurs likewise very abundantly in fibrous and lamellar masses, with cleavage-angle of 124° 30'; H 5·0-6·0; G 2·9-3·2; colourless, but more generally greenish-white or some pale

---

* This plane is regarded by many crystallographers as a front-polar or hemi-orthodome. *See* the note on Pyroxene at the end of the present Table.

shade of green, passing into grass-green and other deeper shades. BB, in thin splinters, more or less easily fusible.

GLAUCOPHANE: Na^2O 7·33, CaO 2·20, MgO 13·07, FeO 5·78, Al^2O^3 12·03, Fe^2O^3 2·17, SiO^2 57·81. Clino-Rhombic, with V on V (as in Amphibole) 124° 30'-125°; crystals, mostly long flat prisms, vertically striated, and passing into fibrous masses; H 5·5-6·5; G 3·1-3·2; dark greyish-blue, bluish-black; BB, easily fusible into a greenish glass. A rare species, hitherto only obtained from the Island of Syra.

ZOIZITE: CaO 24, Al^2O^3 30, SiO^2 41, with small amounts of MgO and Fe^2O^3, and about 2 per cent. of basic water, the latter not revealed by ordinary ignition in the bulb-tube. Rhombic or Clino-Rhombic (but crystals more or less indistinctly formed; commonly in bladed or sub-columnar examples, longitudinally striated. White, pale-grey, pale-greenish, yellowish, or red; H 6·0; G 3·1-3·4. BB, swells up, emits a few bubbles, and melts, if in thin splinters, into a colourless glass. After fusion or strong ignition, is decomposed with gelatinization by hydrochloric acid, and then shews in the spectroscope momentary red and green Ca-lines. Thulite is a rose-red variety, containing a small percentage of Mn^2O^3. Unionite is a white variety.

TOURMALINE: Black varieties (SCHORL) and some Brown varieties: MgO, FeO, MnO, Al^2O^3, B^2O^3, SiO^2. Hemi-Hexagonal, mostly in three-sided or nine-sided prisms, or in fibrous and columnar masses of a jet-black or brown colour; H 7·0-7·5; G 3·0-3·2. Fuses generally into a black or dark slag, but sometimes into a dull-yellowish or more or less uncoloured glass or enamel. The fused mass crushed to powder and moistened with sulphuric acid imparts a distinct green coloration to the flame-border. See above, under A^2 ††††.

VESUVIAN (Idocrase, Egerane): Average composition, CaO 30 to 34; MnO, FeO, MgO, 5 to 8; Fe^2O^3, Al^2O^3, 18 to 20, SiO^2 37 to 39, with small amount of alkalies and basic H^2N. Tetragonal; crystals, commonly, square prisms (or 8-sided prisms composed of the two square prisms V, V̄) terminated by the pyramid P and a well-developed base, B: the latter form very rarely absent. B on P 142° 45' to 142° 57'. Occurs also in columnar and granular masses; H 6·5; G 3·33-3·45; dark-brown, yellowish-brown, brownish-red, yellow, green of various shades, rarely blue. BB, melts, usually with slight bubbling, into a lightly-tinted glass. This, when crushed, dissolves with gelatinization in hydrochloric acid, and then shews momentary red and green Ca-lines in the spectroscope. Cyprine is a blue variety containing a small percentage of CuO. Wiluite, Egerane, Xanthite, Loboite, Frugardite, Heteromerite, are other varieties. Colophonite, in yellow or brown grains and rounded masses, commonly referred to Garnet, is also as regards most examples a Vesuvian. Practically, however, Vesuvian and Garnet can scarcely be distinguished from each other, except by their crystallization.

GARNET: Light-coloured varieties: CaO, Al^2O^3, SiO^2, &c. Essentially in rhombic dodecahedrons or trapezohedrons, or in rounded grains, of a red, yellow, brown, or green colour; H 7·0-7·5; G 3·2-3·8. More or less readily fusible into a colourless or lightly tinted glass. See also under †† and A^2 of this Table.

(*Sp. gr. under* 2·8, *in most cases about* 2·6. *Fusible, unless in fine splinters, on the edges only*).

ORTHOCLASE (Common or Potash Feldspar): K^2O 16·9, Al^2O^3 18·4, SiO^2 64·7, but, very generally, small portions of Na^2O, &c., are also present. Clino-Rhombic; crystals frequently flattened parallel with the side-vertical planes, and often extended in that direction; twins very common: see Note at end of TABLE. Prism-angle 118° 47'. Occurs also abundantly in cleavable, lamellar masses, the cleavage planes (= B, V) meeting at right-angles. H 6·0; G 2·53-2·58; colourless, white, flesh-red, bright-red, light-green, pale-yellowish, light-grey; somewhat pearly on cleavage-planes: iridescent in some varieties, and occasionally opalescent. BB, fusible on the edges only, unless in the form of a thin pointed splinter, in which case the extremity is quickly rounded into a clear glass. Ignited, and then fused with carb. soda or fluor spar, or simply moistened with hydrochloric acid after ignition, shews in spectroscope the red K-line very distinctly. All other lines (derived from the soda or fluor spar) may be entirely obliterated by the intervention of a piece of blue glass. Adularia, Sanidine or Ryacolite (often called glassy feldspar), Pegmatolite, &c., are varieties. Loxoclase is also a variety, but resembles Oligoclase in composition. Perthite is a dark red-brown iridescent mixture of Orthoclase and Albite, the iridescence derived from minute scales of Iron Glance.

MICROCLINE :* A potassic feldspar closely allied to Orthoclase, but apparently anorthic (triclinic) in crystallization. The cleavage angle only differs, however, from a right angle by 15 or 16 minutes; and the prism-angle (118° 31') and other angles scarcely differ from corresponding angles in Orthoclase. Most of the green feldspars (commonly called Amazon Stone) are supposed to be referrible to Microcline; but the species (?) can only be distinguished from Orthoclase by minute optical investigation.

HYALOPHANE : A barytic feldspar, almost identical with Orthoclase in crystallization. G 2·8; white or flesh-red. BB, tinges the flame-point pale-green. See under †† of this section.

ALBITE (Soda Feldspar) : Na^2O 11·82, Al^2O^3 19·56, SiO^2 68·62, but 1 or 2 per cent. of the Na^2O commonly replaced by K^2O. Anorthic; but crystals generally clino-rhombic in aspect, and much like those of Orthoclase : see Note at close of Table. Prism-angle 120° 47'; cleavage-angles 86° 24' and 93° 36'. Crystals commonly in twinned or compound forms, rarely simple. Occurs also abundantly in lamellar masses, with cleavage as above; colourless, white, light-red, light-green, yellowish, brownish, &c.; H 6·0 (or 6·0-6·5); G 2·59-2·64. BB, like Orthoclase, but colours the flame more or less strongly yellow : the two species, however, can only be distinguished by their crystallization, or by accurate chemical analysis, although if the red K-line be distinctly obtained in the spectroscope, the substance, as a rule, may be safely regarded as Orthoclase (or Microcline). See under Orthoclase, above. Pericline is a white opaque or feebly-translucent variety in crystals elongated more or less in the direction of the right-and-left axis, frequently twinned, and strongly striated on the side-vertical faces V́. Peristerite is a white, slightly iridescent variety. Olafite, Cleavelandite, and Zygadite, are other varieties.

OLIGOCLASE (Soda-lime Feldspar) : Na^2O (slightly replaced by K^2O), CaO, Al^2O^3, SiO^2. Anorthic; crystals much like those of Albite, with prism-angle 120° 42' to 120° 53', and cleavage-angles of about 86° 10' (or 86° 30'), and 93° 50' (or 93° 30'). Principal cleavage-plane (B), delicately striated; twin-crystals, very frequent. Occurs also in lamellar and fine-granular masses. H 6·0; G 2·6-2·66; white, pale-red, greenish-grey, &c., with somewhat waxy lustre;

---

* Of Des-Cloizeaux, not Breithaupt. The Microcline of the latter is the iridescent Orthoclase from the zircon-syenite of Norway.

occasionally iridescent. BB, fuses, in thin splinters, into a colourless glass. Apart from its more ready fusibility, this species can scarcely be distinguished from Albite, except by actual analysis.

(*Sp. gr. under* 2·5. *Compact structure. Very easily fusible*).

OBSIDIAN : K^2O, Na^2O, Al^2O^3, SiO^2, with small amounts of CaO, Fe^2O^3, &c. In amorphous masses, breaking with conchoidal fracture into glassy sharp-edged fragments. H 6·0-7·0 ; G 2·2-2·4. Black, brown, grey, greenish, &c., sometimes striped or zoned in different shades; translucent to opaque. Easily fusible with bubbling into a white glass or enamel. Pitchstone is a less vitreous, coarser variety. Pearlstone is a closely related substance, made up essentially of small pearly concretions, or containing these in a vitreous obsidian-like paste. All are volcanic products: rather rocks than minerals proper.

NOTE ON TABLE XXVI.

This Table consists entirely of silicates, distinguished from other compounds of that class by being distinctly fusible, and by yielding no water (or merely traces) when ignited in the bulb-tube. All give the characteristic reaction of silicates by fusion with phosphor-salt—a silica-skeleton separating, whilst the bases dissolve in the flux. In some cases, a portion of the silica is dissolved also, but this precipitates on cooling, and the bead becomes more or less opalescent or clouded. The more commonly occurring minerals of the Table comprise representatives of the following series: Micas, Boro-Silicates, Garnets, Epidotes, Iron Chrysolites, Pyroxenes and Amphiboles, Scapolites, Feldspars.

The Mica Group, as regards the present Table, is chiefly represented by Lepidolite—the ordinary micas, Muscovite, Phlogopite, and Biotite, being as a rule fusible only when in very thin scales, and often on the edges only. Hence, these latter species are described in Table XXV., and in the Note to that Table. Lepidolite is easily recognized (in ordinary examples) by its delicate red or reddish-grey colour, and its occurrence in aggregations of soft, pearly scales. Also by its intumescence and ready fusion in the blowpipe-flame, or even in the flame of the Bunsen burner, and by the crimson coloration which it imparts to this. In the spectroscope, the crimson Li-line and yellow Na-line come out at once with great brilliancy, but the red K-line is generally overpowered by the intensity of the lithium spectrum, unless this be cut off by the intervention of a blue glass between the spectroscope and the flame.

The Boro-silicates of this Table include the dark, fusible Tourmalines, represented essentially by Schorl, and the anorthic species, Axinite. These, however, have no very close relations as minerals, beyond the presence in both of boracic acid, an exceptional component. The silica percentage is comparatively low, averaging 38 or 39 in Tourmaline, and about 44 in Axinite. The boracic acid apparently replaces alumina.

Schorl may generally be distinguished by its jet-black colour and triangular cross fracture. The crystals are sometimes simple three-sided prisms; but these are bevelled, in general, on their vertical edges—a combination of $\frac{\overline{V}}{2}$ and V2 being thus formed—and they are usually terminated by the planes of a rhombohedron (R) with polar angle, i.e., angle over a polar edge, of about 133° 30'. Frequently also the planes of a second rhombohedron ($-2R$) with polar angle of about 103° or 103° 20', alternate with the latter; and crystals often shew dissimilar forms at their extremities: see the Note to TABLE XXIV.

Axinite is readily distinguished by its flattened, sharp-edged, anorthic crystals (brown, violet, pinkish-grey, in colour, or sometimes green from intermixed chlorite), and by the green coloration which it communicates to the blowpipe-flame during fusion. The crystals are essentially oblique rhomboidal prisms with only the diagonally-opposite edges and angles replaced. The prism-angle equals 135° 31'; B on one prism-plane, 134° 45'; and on the other prism-face, 115° 38'.* The two prism planes are vertically striated, i.e., parallel with their combination edges, whilst the B plane is striated transversely.

The Garnet group is represented in this Table by the different varieties or sub-species of Garnet (the infusible chrome-garnet Uwarowite [TABLE XXIV.] excepted), and by the related species Vesuvian.

The specific name of Garnet includes a great number of related silicates of regular crystallization and common formula—the latter, empirically, 3 RO, R^2O^3, 3 SiO^2. The RO represents CaO, MgO, MnO, FeO; and the R^2O^3 equals Al^2O^3, Fe^2O^3, &c. The varieties which result from the preponderance of one or the other of these isomorphous bases necessarily present different colours, and, within certain limits, different degrees of specific gravity.† The colour thus varies, as a rule, from light tints of red, yellow, and green, through deep-red and olive-green into brown and black; and, occasionally, colourless examples are met with. The more common garnets are dark-red or red-brown, and nearly or quite opaque. The average sp. gr. is about 3·5 for the lighter coloured varieties, and 3·9 or 4·0 for the dark garnets, the limits lying between 3·15 and 4·25 or 4·3. The crystallization is comparatively uniform, consisting essentially of the rhombic dodecahedron or of the trapezohedron 2-2, or of the two combined. In the trapezohedron, the angle over a long or axial edge equals 131° 49'. In combination, the trapezohedron replaces the edges of the dodecahedron, and thus presents a cruciform four-planed point-

---

* By most German crystallographers B is made the face of a tetarto-pyramid, P. The angles given above are those of Von Rath, but they fluctuate within 30 or 40 minutes in crystals from different localities.

† This latter character, however, does not depend absolutely on composition, as regards minerals generally. A striking instance is afforded by ordinary Iron Pyrites and Copper Pyrites. The former, consisting of Fe 46·67, S 53·33, has an average sp. gr. of 5·0; whilst the latter, with less sulphur (34·9), and with the heavier metal copper forming part of the base (Cu 34·6, Fe 30·5), shews a maximum density of only 4·3.

ment at each pole of the crystal. Occasionally also, the edges of the rhombic dodecahedron are bevelled by the planes of the adamantoid 3-$\frac{3}{2}$ or 4-$\frac{4}{3}$.

Vesuvian or Idocrase closely resembles Garnet in general composition, and until recently the two were thought to present the same atomic constitution. This is probably not the case, although the formula of Vesuvian is still doubtful. But the two minerals apart from crystallization are evidently nearly allied. The more common crystals of Vesuvian are composed of the two square prisms V and $\bar{V}$, striated longitudinally, and terminated by a square pyramid, P, more or less deeply truncated at the apex by the basal form B. Frequently the vertical edges of V are bevelled by the planes of an octagonal prism V2 or V3; and the polar edges of the pyramid are replaced by a front-polar or front-pyramid $\bar{P}$. Angular measurements are slightly variable, but average as follows: P : P over polar edge 129° 29', over middle edge 74° 14'; B : P 142° 53'; $\bar{P}$: $\bar{P}$ over polar edge 141° 1', over middle edge 56° 8'; B : $\bar{P}$ 151° 56'. For other characters, see the Table.

The Epidote Group is represented in the Table by Epidote, Zoizite, and Allanite or Orthite. The latter in most examples is decomposed with gelatinization by hydrochloric acid, and is black and almost sub-metallic in aspect. Commonly in columnar and fine-granular masses; more rarely in clino-rhombic crystals, with V : V 70° 48' and 109° 12'; V : $\bar{V}$ 125° 24'; and B : $\bar{V}$ 115°. This latter is also the cleavage-angle, but the cleavage is very indistinct. Zoizite and Epidote are not decomposed by hydrochloric acid until after fusion, when they also gelatinize. Zoizite is light-coloured, mostly grey or greyish-white, and chiefly in columnar masses. Its crystallization, long considered identical with that of Epidote, is now regarded as Rhombic, but crystals are rare and more or less indistinctly formed. Epidote is usually distinctly coloured; the tints ranging from light yellowish-green to dark green, brown, and black. Many examples are fibrous and acicular, and closely resemble examples of pyroxene and amphibole, and also schorl. From these, however, Epidote is readily distinguished by its peculiar reaction under the blowpipe. In place of forming a single bead or fused globule, it swells up into a cauliflower-like mass, the separate portions of which become rounded, but cannot with ordinary blowing be brought into a bead, properly so called. Crystals are of frequent occurrence. They are clino-rhombic, and practically identical with those of Orthite, but are not easily made out by the unpractised eye. In their conventional position, they form transversely elongated prisms, the extension being in the direction of the ortho-diagonal or right-and-left axis, with usually two (or several) inclined planes at the side. The horizontally extended planes usually comprise the basal plane B, and the front-vertical $\bar{V}$, with interfacial angle (which is also the cleavage angle) of 115° 24'. In the same zone with these planes, several intermediate planes (the faces of front or ortho-polars) also frequently occur; and in most cases the planes of this zone are striated parallel with their combination-edges. The more common forms of the zone are B (the chief cleavage-plane), V (the second cleavage-plane), and P; with consecutive interfacial angles of 115° 24' as stated above, 128° 18', and 116° 18'. The two predominating planes at the lateral ends of the crystal are sometimes the prism-planes V, with angles of 110° on adjacent faces, and 70° in front or over V̆, and 125° on V̄. In other crystals, these end planes are those of the hemi-pyramid P, and they meet at an angle of 109° 35'. Both V and P are also sometimes present together, meeting at angles of 150° 57' and 117° 40'. Twin combinations, with twin-face parallel to V̆, are of frequent occurrence.

The so-called Iron Chrysolites are represented by Fayalite, Hyalosiderite, and Lievrite or Ilvaite, the latter, only, of general occurrence. This species, by its black colour and general aspect somewhat resembles Orthite. Like Orthite also, it melts readily into a black magnetic glass, and is decomposed with separation of gelatinous silica by hydrochloric acid. The crystallization however is Rhombic, and the crystals are elongated vertically. In most cases they are eight-sided prisms, composed of the two rhombic prisms V and V̆2, terminated by the four planes of a rhombic pyramid P, the front polar edges of which are replaced by a plane of the form P̄. The chief angles are as follows : V : V 112° 38' ; V̆2 : V̆2 106° 15' ; P : P, over front edge or over P̄, 117° 30' ; over side edge 139° 30' ; over middle edge 77° 12' ; P̄ : P̄, over summit, 112° 49'. The prisms, in general, shew strong vertical striæ, indicating additional prismatic forms, V½, &c.; and crystals thus affected often become more or less cylindrical, and pass into columnar masses.

The Pyroxene series comprises a group of species and sub-species (essentially bisilicates of RO, typically MgO, FeO, CaO) in which the crystallization is either Clino-Rhombic or Rhombic, with the chief prism-angle and cleavage-angle approximating to 87° (or its supplement 93°). The Rhombic species comprise Enstatite, with Bronzite and Hypersthene. The typical Clino-Rhombic forms, in which, as in the Rhombic group, alumina is either absent or only subordinately present, include Pyroxene proper, with Acmite and other rarer species (Jeffersonite, &c.); and also the manganese species, Rhodonite, and the more or less aberrant Wollastonite, the latter a purely calcareous species differing essentially from the ordinary pyroxenes by being readily decomposed, with separation of gelatinous silica, in hydrochloric acid. The lithia-holding and aluminous Spodumene or Triphane is also commonly referred to the Pyroxene group from its cleavage-angle and lately determined crystallization ; but its composition (Li^2O 4·5 to 6·5, Al^2O^3 25·3 to 29, SiO^2 63 to 66) and its general aspect, are more feldspathic than augitic. The prism-angle (and corresponding cleavage-angle) V : V, scarcely differs from the principal cleavage-angle in Albite. Its distinctive characters, and those of the other minerals of the group, are given sufficiently in the Table, but some additional remarks on the commonly occurring species Pyroxene are here appended. This species is commonly subdivided into Non-aluminous and Aluminous Pyroxene. The non-aluminous pyroxenes (apart from the ferruginous sub-species Hedenbergite) are chiefly of a light colour, and the aluminous varieties, mostly (though not exclusively) deep-green or black, and more or less ferruginous ; but even in these, the alumina is always under 10, and generally under 7, per cent. The old name of *Diopside* may serve conveniently to include all the light-coloured non-aluminous pyroxenes (Malacolite, Alalite, &c.), and that of *Augite* to denote the dark and generally aluminous varieties. In both diopside and augite the crystals are prismatic and essentially eight-sided, or (as regards these prismatic planes) made up of the four planes of the rhombic prism V, truncated on its obtuse vertical edges by the two planes of the Front-Vertical $\bar{V}$, and on its acute edges by the Side-Vertical or Clino-Vertical $\acute{V}$. The prism-angle in front equals 87° 6′; $\bar{V}$ on $\acute{V}$ of course equals 90°; and V on $\bar{V}$, 133° 33′. But apart from these vertical planes, Pyroxene crystals present three more or less distinct types. In one, common to both light and dark varieties, the crystals are simply 8-sided prisms terminated by the basal plane,* with B on $\acute{V}$ equal to 105° 30′. These crystals are sometimes flattened parallel to $\bar{V}$ (the ortho-pinakoid); but in general they are remarkably symmetrical, and as the pinakoids or Front and Side Verticals, $\bar{V}$ and $\acute{V}$, which meet at right angles, frequently preponderate, this type of crystal looks remarkably like a square prism with truncated vertical edges. In the second type, especially characteristic of augite, proper, the crystals are almost invariably flattened parallel with $\bar{V}$, and are surmounted by the two planes of the clinodome or side-polar $\acute{P}$, meeting over the summit at an angle of 120° 48′.† In this type, the base is also occasionally, but only subordinately, present, together with other slightly developed polar forms; and its crystals are often twinned parallel with $\bar{V}$. The crystals are then terminated by four planes, and shew re-entering angles at one extremity. In the third type, the crystals are largely terminated by the planes of a hemi-pyramid, with angle over front polar edge of 93° 48′, a second hemi-pyramid, with front angle of 131° 30′, often appearing at the lower extremity. Other combinations occur, but are comparatively rare.

The Amphiboles form a parallel series with the Pyroxenes, and like the latter are essentially bisilicates of CaO and MgO, with part of these bases replaced in dark varieties by FeO; and with Al^2O^3 (5 to 15 per cent.) frequently replacing a portion of the silica, the latter in non-aluminous amphiboles varying from about 55 to 59 per cent., and in aluminous varieties from 39 to 49 per cent. Small amounts of fluorine and alcalies are also commonly present, especially in the darker amphiboles; and magnesia always exceeds lime in the base, whereas in the pyroxenes the lime predominates. Corresponding varieties shew in amphibole a slightly lower sp. gr. than in pyroxene; but, practically,

---

* By German and many other crystallographers this is not regarded as the base, but as an orthodome or front-polar $\bar{P}$. By making it the base, however, the two sloping planes by which the common augite crystals are always terminated, become clinodomes or side-polar planes $\acute{P}$, in place of being the planes of a hemi-pyramid P; and in that manner, as pointed out by Von Rath, the correspondence between pyroxene and amphibole crystals is rendered much more apparent. This view has been always held by French crystallographers, and is recommended by its greater simplicity. It was departed from, apparently, in the first instance by German crystallographers in order to obtain an imaginary *Grundform* or triaxial pyramid.

† See the preceding foot note. These terminal planes are regarded by most German crystallographers as the planes of a hemi-pyramid, P,

the two species can only be distinguished by their crystallization and cleavage-angles. In Amphibole proper, two leading varieties or sub-species may be recognized. *Tremolite* or *Grammatite*, including all the white, grey, and pale-green amphiboles; and *Hornblende*, including the deep-green, dark-brown and black kinds. These are connected by the variety known as *Actynolite*, which presents a more or less bright-green colour, and usually occurs in fibrous masses and long prismatic crystals and aggregations. These forms are also generally presented by Tremolite; whilst Hornblende is usually in dark-green lamellar or granular masses, or in thick crystals (commonly known as Basaltic Hornblende) of a dark-brown or black colour. The System, as in Pyroxene, is Clino-Rhombic, and viewed generally, amphibole crystals present three leading types. The first and simplest type is comparatively rare. It consists of an eight-sided prism composed of the vertical forms V, $\overline{V}$, and $\acute{V}$, terminated by a large basal plane; and it thus represents the simple Pyroxene type described under that species. V : V = 124° 30′; V : $\acute{V}$ 117° 15′; B : $\overline{V}$ 104° 50′. A second and much commoner type consists of a six-sided prism composed of the rhombic prism V (with angle as above) truncated on its acute vertical edges by the form $\acute{V}$, and terminated by two nearly flat side-polar planes $\acute{P}$, meeting at an angle of 148° 16′. Sometimes, also, the base, in the form of a narrow plane, replaces the common edge of these terminal planes; and occasionally the prism is eight-sided from the presence of $\overline{V}$; but this front-vertical form, so characteristic of Pyroxene crystals, is comparatively rare in Amphibole. The third type, exhibited especially by the so-called Basaltic Hornblende, consists of a six-sided prism, composed of V and $\acute{V}$ (with planes of practically equal width), surmounted by three rhombiform planes, consisting of two planes of a hemi-pyramid P, and the basal plane B—these three terminal planes being also in general of equal or nearly equal size. A marked pseudo-hexagonal aspect is thus imparted to the crystal. P : P 148° 30′, P : B 145° 35′. Other polar forms are sometimes subordinately present; and the crystals of this type are frequently twinned parallel to the position of the front-vertical $\overline{V}$. In these twins there is no re-entering angle, but the four planes of the hemi-pyramid P are brought together at one end of the crystal, and the two B planes at the other. The interfacial angle of the Basal planes, thus brought together, equals 150° 20′.

The Scapolites are essentially lime-alumina silicates of Tetragonal crystallization. They have been separated into various species or sub-species, but all may fairly be referred to a single representative, Scapolite proper or Wernerite. In this species, the crystals consist commonly of combinations of the two square prisms V and $\acute{V}$, forming an eight-sided prism, terminated by the four planes of the pyramid P, or by those of P and $\bar{P}$, the common summit of these being frequently truncated by the basal form B. The angles fluctuate somewhat in different varieties, but average as follows: B : P 148° 9′; P : P over polar edge 136° 11′; P : V 121° 51′; B : P 156° 14′; $\bar{P}$: $\bar{P}$ over polar edge 147°. In addition to these forms, many crystals shew an octagonal prism V2 (slightly developed), and an octagonal pyramid 3P3, but these are usually in a hemihedral condition. Many crystals, again, are much distorted from inequalities in the size of corresponding planes. Apart from its crystallization, Wernerite is distinguished from light-coloured Pyroxenes and Amphiboles by its lower specific gravity (2·6 to 2·8, in place of 2·9 to 3·4), and by its partial decomposition in hydrochloric acid. From the Feldspars it differs essentially by its want of sharply-defined, smooth and lustrous cleavage-planes, and by its ready fusion. The more typical feldspars, moreover, Orthoclase and Albite, are not attacked by hydrochloric acid.

The Feldspars are essentially aluminous silicates of potash, soda, or lime, characterized by the general absence of iron oxides and magnesia, by their light coloration, their non-fibrous, cleavable structure, the latter an especially salient character, and by their clino-rhombic or triclinic (anorthic) crystallization. As a rule, they are difficultly fusible, and the lime species only are decomposed by acid. In the more typical or alcaline feldspars, the amount of silica exceeds 60 per cent. It is now very generally thought that three species only of feldspar should be admitted, viz.: the potassic species Orthoclase, the soda species Albite, and the lime species Anorthite, the other so-called species being regarded as isomorphous mixtures or combinations of these. This view is probably correct, but in the present state of our knowledge it seems necessary to recognize (as in the Table) the following compounds as constituting distinct feldspathic types: The potash feldspars *Orthoclase* and *Microcline*; the baryto-potassic feldspar *Hyalophane*; the soda feldspar *Albite*; the soda-lime feldspar *Oligoclase* (including Andesine); the lime-soda feldspar *Labradorite*; and the lime feldspar *Anorthite*. The more distinctive characters of these are given fully in the Table; but some additional remarks on the crystallization of the two more important species Orthoclase and Albite are here appended.

Orthoclase crystals fall under three comparatively distinct types. The crystals of the first or simplest type are short rhombic-prisms terminated by two sloping planes. The latter are frequently of nearly similar size and shape, but consist of the base, B, and a hemi-orthodome or ortho-polar $\bar{P}$, of course in alternate positions. V : V 118° 47'; B : $\bar{P}$ 129° 43'; B : V 112° 13'; $\bar{P}$: V 110° 41'. $\bar{P}$ is often transversely striated, and is sometimes much larger than B, in which case its planes resemble the V planes in shape, and the crystal has much the aspect of a truncated rhombohedron. Occasionally, the side-vertical $\acute{V}$ is also present. This type frequently occurs in twin forms, with twin-face,* a face of B. It might be termed the Adularia or St. Gothard type. Its crystals are in general more or less translucent, and are always in druses or attached to the sides of clefts and cavities of the rocks in which they occur. In the second or Baveno type, the crystals are usually six-sided prisms, composed of four V planes and the two planes of the side or clino-vertical $\acute{V}$, terminated by the basal plane and a second ortho-dome or ortho-polar $2\bar{P}$.

---

* Throughout these notes, the term "twin-face" always denotes the face or plane of junction of the united crystals.

These crystals, as a rule, are greatly elongated in the direction of the clino-diagonal, and thus the two B planes and the two $\acute{V}$ planes become drawn out backwards and upwards, so as to mask the true symmetry of the crystal to an unpractised eye. V : V and B : V, as above; V : $2\bar{P}$ 134° 20'; B : $2\bar{P}$ 99° 38'; B : $\acute{V}$ 90°. The cleavage is parallel to the latter planes. Very frequently the edges between B and $\acute{V}$ are replaced by the side-polar or clino-dome $2\acute{P}$, the planes of the latter inclining on B and $\acute{V}$ at angles respectively of 135° 4' and 134° 56'. Occasionally also, the vertical edges between V and $\acute{V}$ are replaced by the planes of the prism V3. Crystals of this type occur very commonly in twins, with the twin-face a plane of the side-polar or clino-dome $2\acute{P}$. In these crystals, consequently, two long B planes, and two long $\acute{V}$ planes, come together, and the crystals are rectangular in aspect. In other twins—with marked re-entering angle—the basal plane is the twin-face or plane of junction. These crystals are sometimes translucent, but are commonly opaque, and are often rough or dull on their external surfaces. Crystals of the third or Carlsbad type possess the same forms as those of the preceding type, but present a very different aspect from the predominance of the side-vertical planes $\acute{V}$, and the apparent flattening of the crystals parallel with these. The elongation moreover is essentially vertical. Simple crystals are much less common than interpenetrating twins, with twin-face parallel with $\acute{V}$. These crystals are always imbedded, and they are commonly quite opaque and more or less rough and dull. Very often they are partially altered into Kaolin, and sometimes into impure Calcite, without change of form; and in Cornwall, tin-stone pseudomorphs have assumed their shape. A fourth type is presented, according to Gustav Rose, by the orthoclase twins from the syenite of southern Norway, in which the form $\acute{V}$ fails, and the crystals are united parallel to the ortho-vertical $\bar{V}$.

In Albite, simple crystals are of rare occurrence. Crystals which appear to be simple, are in most cases really compound, as shewn by the striation of the basal plane. One of the more common combinations consists of a six-sided prism composed of the three forms V, (V), and $\acute{V}$, terminated by three other forms, the base B, a front polar $(\grave{P})$, and a tetarto-pyramid (P) : each of these six forms, of course, consisting of a pair of opposite planes only. When the crystal is in position, B appears at the top in front, and $(\grave{P})$ and (P) at the back; these positions being necessarily reversed as regards the bottom of the crystal. V : (V) 120° 47'; B : $(\grave{P})$ 52° 17' and 127° 43'; B : $\acute{V}$ 86° 24' and 93° 36' (= the cleavage angles); B : V 110° 50'; B : (V) 114° 42'. The side-vertical planes V commonly preponderate and impart a flattened appearance to most crystals. In the more common twins, two B planes, two $\acute{V}$ planes, and two $(\grave{P})$ planes come together. The re-entering angle between B and B equals 172° 48', and these planes are delicately or strongly striated. Double or multiple twins of this character, with two B planes and two (P̀) planes alternating at both extremities of the crystal, are not uncommon.

In the variety of Albite known as Pericline the crystals are more or less elongated in a transverse or right-and-left direction, but the interfacial angles are practically identical with those given above. The forms B and (P̀) predominate, and the short, side-vertical planes V́ are strongly striated; but the striæ arise, here, from an oscillation between the latter form and another vertical prism V̀3, the planes of which occasionally replace the combination edges of V̀ and V, or V̀ and (V). In the twinned Periclines, the plane of junction is parallel to the base.

TABLE XXVII.

[Lustre non-metallic. Slowly attacked or only in part dissolved, BB, by phosphor-salt. Fusible. Yielding water on ignition].

A.—Fusion-product, magnetic.*

A¹.—DECOMPOSED WITH GELATINIZATION BY HYDROCHLORIC ACID.

† *In masses of essentially leafy or scaly structure, or in crystals with marked basal cleavage. Hardness less than that of calcite.*

CRONSTEDITE: MgO, MnO, FeO, Fe^2O^3, SiO^2, H^2O (10 to 12 per cent.). Hemi-Hex.; crystals very small, often acicular, mostly very acute rhombohedrons and scalenohedrons with basal plane; cleavage parallel to the latter; in thin leaves somewhat flexible; also in radiated-fibrous examples. H 2·5; G 3·3-3·3; black; streak, dark-green. Fusible with intumescence into a black magnetic bead. Sideroschizolite is identical or closely related. In both, the crystal-planes shew a strong tendency to curvature, and in Cronstedite the R planes are longitudinally striated.

VOIGTITE (Altered Biotite?): CaO, MgO, FeO, Fe^2O^3, Al^2O^3, SiO^2, H^2O (9 per cent.). In green or dark-brown scaly and foliated examples, resembling an ordinary dark-coloured mica. Fusible into a black, more or less magnetic bead.

THURINGITE: MgO, MnO, FeO, Fe^2O^3, Al^2O^3, SiO^2, H^2O (10 to 12 per cent.). In dark-green, scaly-granular and micaceous masses, with greyish-green streak and pearly lustre. H 2·0-2·5; G 3·1-3·2. Fusible into a black, magnetic bead. Owenite is identical or closely related.

METACHLORITE: FeO, Al^2O^3, SiO^2, H^2O. In dark-green, radiated, leafy masses, resembling ordinary Chlorite, but differing by its larger percentage of FeO, and by gelatinizing in hydrochloric acid.

† † *Occurring in earthy or uncrystalline masses.*

CHAMOISITE (Chamosite): FeO, Al^2O^3, SiO^2, H^2O, often mixed with calcite, &c. In dark-green or greenish-black, fine-granular,

---

* The minerals of this subdivision are for the greater part of more or less indefinite composition. Very few can be ranked, properly, as distinct species. In most cases, therefore, only their essential components are stated in the Table. As a rule, lime is not present normally in these minerals, but many, after prolonged ignition, shew a calcium spectrum from the presence of intermixed calcite.

oolitic, or earthy masses; H 2·0-3·0; G 3·0-3·4. Easily fusible into a magnetic bead. Gelatinizes in hydrochloric acid.

LILLITE: FeO, Fe²O³, SiO², H²O (about 11 per cent.). In blackish-green, earthy rounded masses; H 2·9; G 3·04. BB, fuses with difficulty into a dark magnetic slag. Gelatinizes in hydrochloric acid.

PALAGONITE: CaO, MgO, Al²O³, Fe²O³, SiO², H²O. In granular masses of a yellow or dark-brown colour and vitreo-resinous lustre; streak, dull yellow; H 3·0-5·0; G 2·4-2·6. Easily fusible with intumescence into a more or less magnetic bead. Rapidly decomposed by hydrochloric acid, with separation of gelatinous silica, as regards most examples.

† † † *In distinct crystals or in fibrous and columnar masses which scratch glass readily.*

ILVAITE; ORTHITE: See TABLE XXVI. Some examples, only, evolve traces of water on ignition.

A².—DECOMPOSED BY HYDROCHLORIC ACID, WITH SEPARATION OF SCALY OR GRANULAR SILICA.

† *In leafy or scaly masses, or in tabular or prismatic crystals with marked basal cleavage.*

CHLORITE (Aphrosiderite and other essentially ferruginous varieties): MgO, FeO, Fe²O³, Al²O³, SiO², H²O (about 9 to 12 per cent.). In tabular (Hexagonal) crystals, and in foliated and fine-scaly masses, of a dark or bright-green colour; H 1-1·5; G 2·75-2·95. BB, melts as a rule on the edges and surface, only, into a dark magnetic slag. Strigovite is closely similar in general characters; but its sp. gr. is slightly lower, 2·59, and its water percentage equals 14·80 according to Websky's analysis. Delessite is another dark-green chloritic mineral, occurring in scaly and fine-fibrous masses and coatings in amygdaloidal traps.

ASTROPHYLLITE (Titaniferous Mica): In golden or bronze-yellow foliated masses, often radiately grouped, and in tabular Clino-Rhombic crystals; H 3·5. Most examples yield only traces of water on moderate ignition. See TABLE XXVI., page 228.

PYROSMALITE: Essential components—MnO, FeO, SiO², H²O (about 8 per cent.), Cl. Hexagonal: crystals mostly six-sided prisms or tables with strongly-marked basal cleavage; occurs also in granular masses; brown, dark-green, with metallic-pearly lustre on cleavage plane; H 4·0-4·5; G 3·0-3·2. In bulb-tube yields water, and on stronger ignition, yellow drops of ferrous chloride. BB, fuses easily into a steel-grey or black magnetic globule.

† † *In granular, fibrous, or earthy masses.*

PALAGONITE: In granular, vitreo-resinous masses, of a yellow or brown colour with dull-yellow streak. Commonly gelatinizes in hydrochloric acid, but some examples are decomposed without gelatinization. See above, page 257.

DELESSITE: In dark-green scaly and short-fibrous masses and coatings in amygdaloidal trap. See above, under Chlorite.

ANTHOSIDERITE: Fe^2O^3, SiO^2, H^2O (about 3·6 per cent.). In tough, fibrous masses of ochre-yellow or brown colour, associated with magnetic iron ore. H 6·5; G about 3·0. BB fuses with difficulty to a grey magnetic slag.

XYLOTILE (Mountain Wood, &c.): MgO, Fe^2O^3, SiO^2, H^2O (about 10 per cent.). In light-brown or dark-brown fibrous or ligniform masses; H 1·5-2·5; G 1·5-2·6, commonly about 2·2. Some examples melt, BB, quite easily, others with difficulty, to a more or less magnetic bead. Mountain Cork is a related substance; also Xylite; but, in all, the composition is indefinite. Some varieties do not give BB, a magnetic product. Others are scarcely attacked by hydrochloric acid.

HISINGERITE (Thraulite): Essential components—FeO, Fe^2O^3, SiO^2, H^2O (19 to 22 per cent.), with small amounts of MgO, Al^2O^3, &c. In rounded masses with rough surface and compact structure, conchoidal in fracture, and pitch-black colour, with brown or greenish streak; H 3·0-4·0; brittle; G 2·6-3·1; BB melts difficultly (in some cases on the edges only) into a grey or dark magnetic slag.

MELANOLITE: Na^2O, FeO, Al^2O^3, Fe^2O^3, SiO^2, H^2O (about 10 per cent.). In black, sub-fibrous coatings of waxy lustre and somewhat greasy feel; H 1·5-2·0; G 2·7-2·9. Easily fusible into a black magnetic globule.

SELADONITE (Green Earth): K^2O, MgO, FeO, Al^2O^3, SiO^2, H^2O, mixed with $CaOCO^2$, &c. In earthy or compact masses and coatings in amygdaloidal traps, and also frequently in pseudomorphs after augite. Green of various shades; somewhat shining in the streak; H 1·0-2·0;

G 2·8-2·9. BB, melts into a black magnetic bead. In hydrochloric acid loses its colour, and is slowly decomposed with separation of fine-granular silica. Glauconite or Green-Sand, in disseminated particles and grains in cretaceous and other strata, is of generally similar character. Both substances, when ignited in the Bunsen-flame, shew the red K-line, in the spectroscope, very distinctly.

A³.—INSOLUBLE IN HYDROCHLORIC ACID, OR SCARCELY ATTACKED BY THAT REAGENT.

† *In masses or crystals of leafy or scaly structure with strongly-marked cleavage in one direction.*

(*Fusible in thin pieces*).

FERRUGINOUS MICAS (BIOTITE, &c.): Yield traces of water in some examples, only; as a rule, fuse merely on the edges. *See* TABLE XXV.

(*More or less brittle. Hardness insufficient to scratch glass*).

STILPNOMELANE: Essential components—MgO, FeO, Al²O³, SiO², H²O (about 9 per cent.). In dark-green or greenish-black radio-foliated masses or small scaly particles. H 3·0-3·5; G 2·8-3·4. Fusible (in some cases readily, in others slowly) into a magnetic slag or globule. Scarcely attacked by acids.

(*Hardness sufficient to scratch glass slightly*).

CHLORITOID: Average composition—MgO 3·0, FeO 27·0, Al²O³ 39·0, SiO² 26, H²O 7·0. In dark or blackish-green, foliated and scaly-granular masses, the foliæ more or less curved and brittle. H 5·5; G 3·5-3·6. BB, slowly fusible (often on the edges only) into a black magnetic slag. Slightly attacked by hydrochloric, but readily decomposed by sulphuric acid. Sismondine (blackish-green), and Masonite (dark greenish-grey) are apparently identical. Ottrelite (greenish-grey to greenish-black, in small six-sided tables with rounded angles, in certain clay slates) is also closely related. It gives, BB, with carb. soda a strong manganese-reaction.

† † *In fibrous masses.*
(*Easily fusible*).

KROKYDOLITE (Crocidolite): Na²O, MgO, FeO, SiO², H²O (2·5 to 5 per cent.). In deep-blue or lavender-blue fibrous masses, the fibres tough and flexible. H 3·0-4·0; G 3·2-3·3. Easily fusible into a black magnetic globule.

KIRWANITE ; CaO, FeO, Al^2O^3, SiO^2, H^2O (about 4 per cent.). In opaque dark-green nodular masses of radiated-fibrous structure. H 2·0 ; G 2·9. BB blackens, and melts.

(*Fusible on edges only.*)

XYLITE : CaO, MgO, Fe^2O^3, SiO^2, H^2O (4·7 per cent.), with small (accidental ?) amount of CuO. In opaque nut-brown, fibrous or ligniform masses. H 3·0 ; G 2·93. Fusible on the edges only. Distinguished from Xylotile or Mountain Wood, proper, by its resistance to acids. It differs also from the latter mineral by containing (according to Hermann's analysis) a certain amount of lime.

† † † *In more or less earthy or compact masses.*

SORDAWALITE : MgO, FeO (or Fe^2O^3), Al^2O^3, SiO^2, H^2O, with intermixed ferrous phosphate, &c. In black or dark-green coatings and earthy masses, weathering brown. H 4·0-4·5 (?); G 2·6; fusible into a black magnetic globule. Partially decomposed by hydrochloric acid. Hitherto, from Finland only.

CHLOROPHÆITE : MgO, FeO, SiO^2, H^2O (about 42 per cent.). In green or brownish-green amygdaloidal masses in trappean rocks. Weathers brown and black. H 1·0-2·0 ; G about 2·0. BB, forms a black magnetic slag or globule. Distinguished from Delessite, Lillite, Chamoisite, &c., by its resistance to hydrochloric acid, and by the large amount of water which it yields on ignition. Nigrescite, a green amygdaloidal mineral, blackening on exposure, is identical or closely related.

B.—Fusion-product, non-magnetic.

B¹.—FUSIBLE ON THE EDGES OR IN FINE SCALES OR SPLINTERS ONLY; BUT EXFOLIATING AND CURLING UP, IN SOME CASES, ON IGNITION.

† *Micaceous or scaly minerals.*

(*Water under 5·5 per cent. In bulb-tube little more than traces evolved.*)

MUSCOVITE (Ordinary or Potash Mica): Elastic in thin leaves. Not decomposed by sulphuric acid. *See* TABLE XXV., page 213.

DAMOURITE : K^2O 11·20, Al^2O^3 37·85, SiO^2 45·22, H^2O 5·25. In yellowish-white pearly scales and foliated masses, associated (as regards known localities) with Staurolite and Cyanite, or with Corundum. H 1·5-2·5 ; G 2·8. BB exfoliates, and melts on edges. Decomposed, with separation of silica scales, by sulphuric acid.

Margarodite and Sericite are closely allied micaceous substances, apparently altered Muscovite, with variable amounts of water. All shew the red K-line in the spectroscope very distinctly.

PARAGONITE (Hydrous Soda-Mica): Na^2O, K^2O, Al^2O^3, SiO^2, H^2O (2·5 to 4·5 per cent.). In scaly or schistose masses of a yellowish-white, pale-grey, or light-green colour, and pearly lustre. H 2·0-3·0; G 2·79. Fusible on the edges into a white enamel; decomposed by sulphuric acid. Pregrattite, distinguished by marked exfoliation BB, is closely related.

OELLACHERITE (Hydrous Barium-Mica): K^2O, Na^2O, SrO, BaO, CaO, MgO, Al^2O, SiO^2, H^2O (about 4 or 4·5 per cent.). In white or pale-green scaly masses of pearly lustre. H 1·5-3·0 (?); G 2·8-2·9. Fusible into a white enamel. Should be readily distinguished by its spectroscopic reactions, but the author has not been able to procure a specimen for examination.

PHLOGOPITE (Potassic-Magnesian Mica): In golden-brown, micaceous crystals and masses. Decomposed by sulphuric acid. *See* TABLE XXV., page 213.

COOKEITE: A hydrous mica, giving marked lithium reaction BB, or in spectroscope. Forms red or reddish-grey scaly aggregations. Probably altered Lepidolite.

RUBELLANE: Na^2O, K^2O, MgO, Fe^2O^3, Al^2O^3, SiO^2, H^2O. In red or brownish-red hexagonal tables with pearly lustre on cleavage plane. H about 2·5; somewhat brittle. BB, melts (in some cases on edges only) into a dark ferruginous glass. Regarded as an altered Mica. Occurs in certain trachytes and other volcanic rocks. Helvetane (copper-red, yellow, green) is closely related.

MARGARITE (Pearl Mica): In white or light-coloured scaly and foliated masses with strong pearly lustre. Fusible on edges, only, but in some cases with slight bubbling. Moistened with hydrochloric acid, shews momentary red and green Ca-lines in spectroscope. *See* TABLE XXV., page 215.

TALC: MgO, SiO^2, with small amount of basic water. In white, light-green or other foliated or scaly examples, with pearly lustre. H 1·0; very sectile, flexible, and soapy to the touch. BB, exfoliates, but melts on thin edges only. Evolves merely traces of water in the bulb-tube. With Co-solution becomes flesh-red. *See* TABLE XXV., page 214.

(*Water*, 5·6 to 14 per cent.: *evolved in marked quantity in bulb-tube.*)

PYROPHYLLITE : Al^2O^3, SiO^2, H^2O, with traces of MgO, &c. In light-green or greenish-white radio-foliated and scaly masses. H 1·0. BB, exfoliates and curls up, but remains practically unfused. Becomes blue by ignition with Co-solution. Belongs properly to Table XXV.: *see* page 214. *See* also Nacrite or Pholerite, page 219.

VERMICULITE : MgO, FeO, Al^2O^3, SiO^2, H^2O, with traces of CaO, K^2O, &c. In scaly and coarsely-foliated examples and six-sided micaceous tables of a yellowish-brown, yellow or green colour. H 1·0-1·5 ; G 2·2-2·4 ; slightly flexible in thin leaves. BB, expands and curls up greatly, and melts subsequently to a white or greyish enamel. According to Prof. Cooke, should form three species : Jefferisite, Culsageeite, Hallite.

CHLORITE (Pennine); and RIPIDOLITE or CLINOCHLORE : In green, scaly or foliated masses and micaceous crystals. As a rule, fusible on the edges only, in many cases into a black, slightly magnetic enamel. Belong properly to Table XXV.: *see* page 213.

† † *Minerals of compact, fibrous, or other non-micaceous structure.*
More or less distinctly sectile.*

(*Assume a blue colour after ignition with Co-solution.*)

AGALMATOLITE : Massive, fine-granular, or compact in structure ; white, greyish, greenish, &c. The substance of many Chinese " Figure-stones." Fusible on thin edges, only. *See* page 219, TABLE XXV.

PINITE, FAHLUNITE, PYRARGILLITE ; WEISSITE ; IBERITE ; ESMARKITE ; BONSDORFFITE : In more or less dull and opaque crystals—essentially six-sided, eight-sided or twelve-sided prisms—of a greyish-white, grey, brown, green or dull-bluish colour. Fusible on the edges only. *See* page 220, TABLE XXV.

KILLINITE : K^2O, FeO, Al^2O^3, SiO^2, H^2O (about 9 or 10 per cent., or less in some cases). Chiefly in greenish-grey or brownish-yellow columnar or broad-prismatic aggregations, translucent in thin pieces. H 3·0-4·0 ; G about 2·7. BB, expands somewhat, and melts slowly (in some cases on the edges and surface only) into a white or greyish enamel. Decomposed, in powder, by sulphuric acid.

SCHRÆTTERITE (Hydrargillite (?) mixed with a lime or other sili-

---

* The minerals of this section are fusible, as a rule, upon the edges only. They belong properly, therefore, to TABLE XXV. *See* pages 219-222.

cate, traces of copper sulphate, &c. Yields on ignition from 36 to 41 per cent. water). In earthy and botryoidal masses, coatings, &c., of a green, pale-yellow, grey, or brownish colour, with more or less conchoidal fracture. H 3·0-4·0 ; G about 2. BB, whitens, and fuses slowly (often on edges only) into a white or light-grey enamel. Decomposed (with gelatinization, according to Fischer) by hydrochloric acid.

PYKNOTROPE : K^2O, MgO, Al^2O^3, SiO^2, H^2O (about 7 or 8 per cent.). In greyish-white, pale-greenish, or brownish-red, coarse-granular masses, with cleavage in two directions at right-angles. H 2·0-3·0 ; G 2·6-2·72. BB, fusible only in thin splinters or on the edges. Associated with serpentine.

(*Assume a flesh-red colour by ignition with Co-solution, or do not, otherwise, become blue*).*

STEATITE (Compact or Granular Talc): In masses and pseudomorphous crystals of a white, grey, greenish or other colour, often mottled. Very sectile ; yields very little water in bulb-tube, but blackens more or less. BB, hardens, and fuses on thin edges. *See* page 222.

SERPENTINE : Forms compact, fine-granular, or other masses, of a green, red-brown, yellowish-grey, or variegated colour. In bulb-tube, yields about 12 or 13 per cent. water. Fusible on thin edges only. *See* page 221, TABLE XXV.

CRYSOLITE (Fibrous Serpentine) : In silky, parallel-fibrous masses of a yellowish-white or green colour, the fibres easily separable. Melts at the point of a fine fibre into a white or greyish enamel. *See* page 221, TABLE XXV.

MEERSCHAUM (Sepiolite) : In fine-granular or compact masses of a white or pale yellowish colour, adherent to the tongue. BB, hardens, but fuses on thin edges only. *See* page 221, TABLE XXV.

† *Not sectile. Hardness sufficient to scratch glass.*

POLLUX : In translucent, camphor-like masses and small crystals (combinations of cube and trapezohedron 2·2). Yields traces only of water in bulb-tube, and fuses only, BB, on thinnest edges. *See* page 203, TABLE XXIV.

---

* In the bulb-tube, all blacken on evolving water.

B². —FUSIBLE WITHOUT MARKED BUBBLING OR PREVIOUS INTUMESCENCE.

† *Insoluble in hydrochloric acid.*

DIALLAGE (Schistose and more or less altered Pyroxene): In foliated or sub-foliated masses of a greyish-green or greenish-brown colour and metallic pearly lustre. Yields often merely traces of water: in no case more than 3 or 4 per cent. *See* page 242, TABLE XXVI.

† † *Decomposed, by hydrochloric acid, with production of chlorine fumes.*
(*BB, with carb. soda, strong Mn-reaction.*)

KLIPSTEINITE: MgO, MnO, Mn^2O^3, Fe^2O^3, SiO^2, H^2O (9 per cent.). In amorphous masses of a brown or brownish-grey colour, with reddish-brown streak; H 5·0; G 3·5. Fusible into a dark slag.

† † † *Decomposed, with or without gelatinization, by hydrochloric acid.*
(*BB, with borax, a chrome-green glass.*)

PYROSCLERITE: MgO, FeO, Al^2O^3, Cr^2O^3 (1·43 per cent.), SiO^2, H^2O (11 per cent.), von Kobell. In cleavable masses, indicating Rhombic crystallization; in thin pieces somewhat flexible; H 3·0; G 2·7-2·8; green of various shades, with pearly lustre on cleavage-planes. Fusible quietly, or with slight bubbling only, into a greenish-grey enamel. Hitherto, from Elba only.

(*In spectroscope, marked Ba-reaction when moistened with hydrochloric acid.*)

HARMOTOME: K^2O 3·3, BaO 20, Al^2O^3 15·7, SiO^2 46, H^2O 15. Rhombic (?): commonly in groups of small, cruciform crystals, with calcite, &c., in trap amygdaloids. Generally colourless, otherwise white, grey, reddish, brown, &c.; H 4·5; G 2·4-2·5. Fuses quietly, with pale-green coloration of the flame-border. Decomposed by hydrochloric acid, with separation of fine-granular silica. *See* Note at end of Table.

EDINGTONITE: BaO 26·84, Al^2O^3 22·63, SiO^2 36·98, H^2O 12·46, Heddle. Tetragonal; crystals, mostly, small square prisms with hemihedral polar planes; greyish-white, pale-red; H 4·0-4·5; G 2·7. Gelatinizes in hydrochloric acid. Hitherto, from Scotland only, accompanying harmotome, analcime, calcite, &c.

(*In spectroscope, marked Ca-reaction when moistened, after ignition, with hydrochloric acid.**)

PECTOLITE : Na^2O, CaO, Al^2O^3, SiO^2, H^2O. Clino-Rh. ; but commonly in cleavable fibrous or sub-fibrous masses, with cleavage angle of 95° 23' ($= B : \acute{V}$). Colourless, or greyish or pale greenish-white, often opaque and more or less earthy from alteration. H (in unweathered examples) 5·0 ; G 2·74-2·88. Fuses quietly. Yields as a rule only 2 or 3 per cent. water on ignition. Decomposed without gelatinization by hydrochloric acid, but gelatinizes after fusion.

CHALILITE : Na^2O, CaO, MgO, Al^2O^3, Fe^2O^3, SiO^2, H^2O (about 16 per cent.). Reddish-brown, massive ; H 4·5 ; G 2·25. An imperfectly-known mineral, hitherto from Antrim only.

ANALCIME ; NATROLITE : Normally lime free, but some examples of exceptional occurrence shew momentary Ca-lines in spectroscope, see below.

(*In spectroscope, no Ca-lines, but strong Na-reaction.*)

ANALCIME : Na^2O 14·0, Al^2O^3 23·3, SiO^2 54·5, H^2O 8·2 ; but a small percentage of CaO present in some varieties and K^2O in others. Regular ; crystals either small cubes with angles replaced by the planes of the trapezohedron 2-2, or the latter form alone. Colourless, white, light-grey, flesh-red ; H 5·5 ; G 2·1-1·3. Fusible without intumescence into a more or less clear glass. Decomposed by hydrochloric acid with separation of slimy silica. *See* Note at end of Table. Cuboite is a green or greenish-grey variety. Eudnophite is regarded as a Rhombic Analcime. Cluthalite is a somewhat decomposed variety.

NATROLITE (Mesotype in part) : Na^2O 16·30, Al^2O^3 26·96, SiO^2 47·29, H^2O 9·45, but traces of CaO, &c., occasionally present. Rhombic ; crystals very small, often acicular ; essentially Rhombic (almost rectangular) prisms, terminated by the planes of a rhombic octahedron. V : V 91° ; P : P, over polar edges, 143° 20' and 142° 40'. Occurs also, and more commonly, in radio-fibrous masses, often with crystalline-botryoidal surface. Colourless, white, yellow, light-brown,

---

* If a zeolitic mineral do not shew these spectroscopic reactions very distinctly when simply moistened by hydrochloric acid, a portion in fine powder should be dissolved in the acid in a small porcelain capsule with attached handle (like that figured on page 20) over the spirit-lamp or Bunsen-flame. A drop of the solution may then be taken up by a platinum wire (bent at the extremity into a small loop or ear) and held within the edge of the flame, care being taken to test the wire previously for negative results. By this treatment, distinct although more or less transitory spectra are always obtained when lime, baryta, potash &c., are present in the mineral.

red; two or more tints frequently present in concentric zones in the same example. H 5·0-5·5; A 2·17-2·27. Very easily fusible in the simple candle or Bunsen-flame, without intumescence, into a colourless glass. Decomposed, with gelatinization, by hydrochloric acid, Radiolite or Bergemanite, Lehuntite, Galactite, Brevicite, Fargite, are varieties. Mesolite (Antrimolite, Harringtonite) is a closely related zeolitic mineral, but contains both lime and soda, and is thus intermediate between Natrolite and Scolecite. It occurs essentially in radio-fibrous masses and acicular crystals. Yields 12 to 14 per cent. water; gelatinizes in hydrochloric acid, and fuses quietly or with very slight intumescence.

B³.—FUSIBLE WITH MUCH BUBBLING OR WITH PREVIOUS INTUMESCENCE.

† *Undissolved or scarcely attacked by hydrochloric acid.*

(*In yellow, fibrous examples. BB, strong Mn-reaction.*)

CARPHOLITE : MnO, FeO, Fe^2O^3, Al^2O^3, SiO^2, H^2O (10 to 11 per cent.), with small amounts of MgO, F, &c. Acicular, or in radio-fibrous aggregates of a straw-yellow or greenish-yellow colour and silky lustre; H 4·5-5·0; G 2·9-3·0. The water evolved by strong ignition deposits spots of silica on the sides of the bulb-tube, and attacks the glass. BB, intumesces and forms a dull-brownish bead.

(*In opaque, prismatic crystals.*)

GIGANTOLITE : Na^2O 1·2, K^2O 2·7, MgO 3·8, MnO 0·9, Al^2O^3 25·0, Fe^2O^3 15·6, SiO^2 46·3, H^2O, 6·0. Rhombic; crystals (probably pseudomorphous after Iolite), thick, twelve-sided prisms, more or less dull; green, greenish-grey; H 3·5; G 2·8-2·9. Fusible with bubbling into a greenish slag. When ignited and moistened with hydrochloric acid, shews red K-line distinctly in spectroscope.

(*In pale-red cleavable masses.*)

WILSONITE : K^2O, CaO, MnO, FeO, Al^2O^3, SiO^2, H^2O. In rose-red or pale purplish-red cleavable masses; slightly fibrous and pearly in the cleavage directions, lustreless and more deeply-coloured transversely; cleavage rectangular; H 3·0-3·5 on cleavage surfaces, otherwise 5·0-5·5; G 2·75-2·8. BB, expands or increases in volume, and fuses with slight bubbling into a very blebby glass or white enamel.

Moistened with hydrochloric acid, shews Ca-lines in flashes, and red K-line persistently.*

† † *Decomposed by hydrochloric acid, with separation of granular or slimy silica.*
(*Hardness* 6·0 *or* 7·0. *Scratch glass strongly.*)

PREHNITE: CaO 27·14, Al²O³ 24·87, SiO² 43·63, H²O 4·36. Rhombic; crystals tabular or short-prismatic, in aggregated groups (*see* Note at end of Table). Occurs also, and more commonly, in radio-fibrous masses with botryoidal and crystalline surface; greenish-white passing into distinct shades of green; H 6·0-7·0; G 2·8-3·0. Fusible with continued bubbling. In spectroscope, when moistened with hydrochloric acid, especially after fusion, shews red and green Ca-lines in flashes. The fused bead gelatinizes in the acid, but in its normal state Phrenite is more or less slowly and incompletely decomposed, with separation of fine granular silica. Koupholite is a thin tabular variety, which blackens on ignition from the presence of intermixed dust or organic matter. Chloristrolite from Isle Royale, Lake Superior, in small nodular masses of green colour and radio-fibrous structure, is also a variety or related substance, intermixed with grains of magnetic iron ore, &c.

FAUJASITE: K²O 4·36, Na²O 4·84, CaO 4·36, Al²O³ 16·00, SiO² 46·77, H²O 28·03. Regular; crystals, small octahedrons (or according to Knop, very flat-planed trapezohedrons), sometimes twinned; white or brownish; H 6·0; G 1·9-1·95. BB, intumesces, and fuses readily. In the bulb-tube yields a large amount of water.

(*H* 4·5 *to* 5·0. *No essential precipitate formed in the diluted solution on addition of ammonia.*)

APOPHYLLITE (Ichthyopthalmite): CaO 24·72, SiO² 52·97, H²O 15·90, KF 6·40. Tetragonal; crystals commonly square prisms with truncated angles, or acute square-based octahedrons, mostly with basal plane (*see* Note at end of Table); colourless, pale-red, brownish, &c., with pearly lustre on basal plane, the latter also frequently iridescent; H 4·5-5·0; G 2·3-2·4. BB, exfoliates and melts with

---

* Whilst this mineral has much the composition of a Pinite, its general aspect and physical characters are very different, and have caused it to be regarded as an altered Scapolite. The presence of potash is the chief objection to the latter view. Were it not for its sub-fibrous structure, as seen on the cleavage surface more especially, it might be considered an altered Orthoclase.

bubbling to a white glass. Gives fluorine reaction with fused phosphor-salt in open tube (page 26). Moistened with hydrochloric acid, shews Ca-lines in flashes, and persistent red K-line. Albin is an opaque-white, slightly weathered variety. Oxhaverite, Tesselite, are also varieties.

OKENITE : CaO 26·42, SiO^2 56·60, H^{2}O 16·98. Rhombic in crystn., but chiefly in fibrous masses, more or less tough ; colourless, pale-bluish or yellowish-white; H 5·0; G 2·28-2·36. Fusible with bubbling into a white glass or enamel. In spectroscope, no red K-line.

PECTOLITE : See under B^2, above.

(*H 5·0 to 5·5. A marked precipitate [insol. in acids] formed in the diluted solution by sulphuric acid.*)

BREWSTERITE : BaO, SrO, Al^{2}O^3, SiO^2, H^{2}O (13·6 per cent.), with traces of CaO, &c. Clino-Rh.; crystals, small, vertically-striated prisms, terminated by the two planes of a very flat side-polar or clino-dome ; V : V 136°; $\acute{P}$ on $\acute{P}$ over summit 172°. Yellowish-white, pale-brown ; H 5·0-5·5 ; G 2·2-2·45. Fusible with intumescence and bubbling. Moistened with hydrochloric acid, shews in spectroscope transitory Ba and Sr lines (*see* page 56), but in some examples the reaction is not very strongly marked.

(*H 4·5 or less. Crystalline and clearable. A copious precipitate in diluted solution thrown down on addition of ammonia.*)

CHABASITE : Average composition : K^{2}O 1·98, CaO 9·43, Al^{2}O^3 17·26, SiO^2 50·50, H^{2}O 20·83. Hemi-Hexagonal ; crystals, commonly small rhombohedrons, often twinned, the twin-axis corresponding with the vertical axis ; R : R 94°-95°, commonly 94° 46' (*see* Note at end of Table); colourless, white, pale-red, &c.; lustre, vitreous ; H 4·0-4·5 ; G 2·6-2·2. BB intumesces and fuses into a very blebby glass or white enamel. Decomposed by hydrochloric acid, with separation of slimy silica. In spectroscope, the solution, or a splinter moistened with the acid, shews red and green Ca-lines in flashes, with feeble and very transitory display of the red K-line. Acadialite is a reddish Chabasite from Nova Scotia. Phacolite is a variety in inter-penetrating very obtuse twelve-sided pyramids (with other accompanying forms), often lenticular from distortion. Haydenite and Seebachite are also varieties. Levyne and Herschellite are closely related compounds, occurring mostly in hexagonal or pseudo-hexagonal tabular crystals with large basal plane. Gmelinite is also very similar, but gelatinizes in hydrochloric acid : *see* below, page 271.

STILBITE (Desmine of German systems): CaO 9, Al^2O^3 16, SiO^2 58, H^2O 17, with, occasionally, traces of Na^2O and K^2O. Rhombic : crystals small and commonly in groups, consisting usually of a rectangular prism (= $\bar{V}$, $\ddot{V}$, the $\bar{V}$ planes vertically striated), terminated by a rhombic octahedron P, the latter measuring 119° 16' and 114° over polar edges, and occasionally having its apex truncated by a small basal plane. Cleavage very perfect parallel to the side vertical or brachypinakoid $\ddot{V}$, the cleavage-lustre strongly pearly. Occurs also abundantly in radio-fibrous and leafy aggregations. Colourless, white, red, brown, &c.; H 3·5-4·0 ; G 2·1-2·2. BB, intumesces, and fuses into a very blebby glass. Decomposed by hydrochloric acid, with deposition of slimy silica. Epistilbite agrees in composition and general characters, but its crystals are small rhombic prisms terminated by the front and side polars $\bar{P}$ and $\ddot{P}$, the latter predominating. V : V 135° 10'; $\bar{P}$: $\bar{P}$ over summit 109° 46', $\ddot{P}$: $\ddot{P}$ 147° 40'. Colourless or bluish-white. In hydrochloric acid, decomposed with separation of fine granular silica.

HEULANDITE (Stilbite of most German systems): CaO (with small amount of Na^2O and K^2O) 9·34, Al^2O^3 16·83, SiO^2 59·06, H^2O 14·77. Clino-Rhombic; crystals mostly tabular parallel to the side or clino-vertical plane; commonly made up of the front and side verticals $\bar{V}$ and $\acute{V}$ (the latter predominating) with a front polar $\bar{P}$, and narrow Base. When lying consequently with $\acute{V}$ upwards, the crystals present a pseudo-hexagonal aspect. $\bar{P}$: $\bar{V}$ 129° 40'; B : $\acute{V}$ 116° 20' and 63° 40'. Cleavage very perfect parallel to $\acute{V}$, the planes, as in Stilbite, strongly pearly. Colour, hardness, and other characters, physical and chemical, like those of Stilbite. Euzeolite, Lincolnite, Beaumontite (?) are varieties.

(*In amorphous examples without distinct cleavage.*)

CHONIKRITE: CaO, MgO, Al^2O^3, SiO^2, H^2O (9 per cent.). In snow-white or pale-yellowish, disseminated masses; H 2·5-3, more or less sectile ; G 2·9. BB, fusible with bubbling into a greyish-white glass or enamel. Decomposed by hydrochloric acid, with separation of granular silica. Hitherto, from Elba only. Related to Pyrosclerite, page 265 above.

† † † *Decomposed, with perfect gelatinization, by hydrochloric acid.*

(*BB, sulphur-reaction with carb. soda.*)

ITTNERITE : K^2O, Na^2O, CaO, Al^2O^3, SiO^2, H^2O (9·8 per cent.) In small, granular masses, with dodecahedral cleavage, of a grey or blue-grey colour. H 5·0-5·5 : G 2·3-2·4. Fusible with strong bubbling into a blebby semi-opaque glass or enamel. Yields gypsum to boiling water, as recognized by the precipitates formed in the solution by oxalate of ammonia and chloride of barium, respectively (Fischer). Decomposed by hydrochloric acid, with emission of sulphuretted hydrogen and separation of gelatinous silica. An altered Hauyne or Nosean, *see* page 236.

(*BB, flame-border coloured distinctly green.*)

DATOLITE : CaO 35·0, B^2O^3 21·9, SiO^2 37·5, H^2O 5·6. Clino-Rhombic (or Ortho-Rhombic ?); occurs commonly in groups of small vitreous crystals, rich in planes (*see* Note at end of Table), or in coarsely granular masses. Greenish-white, colourless, green, reddish-white. H 5·0-5·5 ; G 2·8-3·0. Fuses very easily, with much bubbling, and green coloration of the flame, to a colourless or very lightly-tinted glass. Gelatinizes in hydrochloric acid. In spectroscope, shews *per se* two vivid green lines with one pale-green and a faint blue line, from presence of B^2O^3. When moistened with hydrochloric acid, a test-fragment shews also red and green Ca-lines in flashes; but the presence of lime is best shewn by a drop of the solution, taken up in a double-loop of clean platinum wire and held against the edge of the Bunsen-flame. Humboldtite is a variety in small crystals, associated with lamellar Apophyllite, from the Tyrol.

BOTRYOLITE : Contains 10·64 per cent. water, and occurs in fibro-botryoidal examples of a greenish, pale-grey, or reddish colour; otherwise like Datolite.

(*Moistened with hydrochloric acid, shew distinct red K-line in spectroscope.*)

PHILLIPSITE (Lime-Harmotome, Christianite): Average composition, K^2O 7, CaO 6, Al^2O^3 21·5, SiO^2 48·5, H^2O 17. Rhombic (?); commonly in cruciform crystals resembling those of Harmotome (*see*

Note at end of Table). Colourless, white, reddish-white, pale-grey, &c.; H 4·5-5·0; G 2·15-2·2. Fusible with intumescence and bubbling. Gelatinizes in hydrochloric acid. The moistened test, or the solution, shews K and Ca-lines in spectroscope, the latter in flashes only.

GISMONDINE: K^2O 2·85, CaO 13·12, Al^2O^3 27·33, SiO^2 35·88, H^2O 21·10. Tetragonal (or Rhombic?); crystals small, and often imperfectly formed or sub-spherical, consisting commonly of a simple pyramid or octahedron (with angle of 118° 30' over polar edge, and 92° 30' over middle edge), or of this form combined with the prism $\overline{V}$; greyish or reddish-white; H 5·0-6·0; G 2·27. Fusible with intumescence. See Zeagonite, below.

ZEAGONITE: K^2O 11·09, CaO 5·31, Al^2O^3 23·34, SiO^2 43·95, H^2O 15·31. Rhombic; crystals mostly, rectangular prisms (composed of $\overline{V}$ and $\overset{\shortmid\shortmid}{V}$) with angles replaced by a rhombic octahedron P, measuring 121° 44' and 120° 37' over polar edges, and 89° 13' over middle edges, the planes often rounded and the crystals in sub-spherical groups. Colourless, white, pale-bluish; H 5·0-6·5 or 7, the latter at the points and edges. Fusible with intumescence. Probably identical with Gismondine, both being Rhombic, with pseudo-tetragonal aspect. In spectroscope, the red K-line comes out very distinctly.

GMELINITE; THOMSONITE: Shew sometimes in spectroscope a feeble or indistinct K-line: *see* below.

HYDROTACHYLITE: In vitreous, amorphous masses: *see* below, page 273.

(*No distinct K-line brought out in spectroscope*).

GMELINITE: Average composition—Na^2O (with small amount of K^2O) 5, CaO 5, Al^2O^3 20, SiO^2 48, H^2O 21. Hexagonal or Hemi-Hex.; crystals, commonly, very short six-sided prisms (horizontally striated), combined with a six-sided pyramid measuring 142° 33' over polar edges, and 79° 54' over middle edges; but the planes of the latter often alternate in size, and hence the pyramid is regarded as consisting of two complementary rhombohedrons, with R : R = 112° 26'. Colourless, greenish-white, yellowish-white, pale-red. H 4·5; G 2·0-2·1. Fusible with intumescence. Is closely allied to Chabasite, but is distinguished by the presence of Na^2O, and by its perfect gelatinization in hydrochloric acid. Ledererite is a variety.

THOMSONITE (Comptonite): Na^2O (with small amount of K^2O, 4·4, CaO 13·3, Al^2O^3 30·6, SiO^2 38·7, H^2O 13. Rhombic in crystallization, but crystals usually small or acicular (*see* note at end of Table), essentially eight-sided prisms composed of the forms $\bar{V}$, V, and $\check{V}$, with V planes vertically striated (V : V 90° 40'). Occurs chiefly in fibrous and fibro-spherical masses. Colourless, white, reddish-white, brown. H 5·0-5·5 ; G 2·35-2·4. Fusible with intumescence. Gelatinizes in hydrochloric acid. In spectroscope the solution or moistened fragment shews red and green Ca-lines. Ozarkite, according to Dana, is a massive Thomsonite. Faroelite, Scoulerite, Chalilite, are also varieties ; the latter red-brown, and partially altered.

SCOLECITE (Mesotype in part): CaO 14·26, Al^{2}O^3 26·13, SiO^2 45·85, H^{2}O 13·76. Clino-Rhombic ; crystals mostly rhombic prisms (with V : V 91° 35') with low pyramidal terminations (P and – P'), hence much resembling an ortho-rhombic combination. In general, however, crysts. very small or acicular. Occurs commonly in fibrous and radio-spherical examples. Colourless, white, reddish-white, &c. H 5·0-5·5 ; G 2·2-2·4. Fusible with intumescence, the more typical examples curling up greatly. Acid and spectroscope reactions like those of Thomsonite. Poonahlite is a variety. Mesolite is also closely related, but contains both soda and lime, and fuses more or less quietly. *See* under Natrolite, page 266.

L$\overset{u}{A}$MONTITE : CaO 12, Al^{2}O^3 22, SiO^2 50, H^{2}O 16 ; but the latter usually less, from the ready efflorescence of the mineral. Clino-Rhombic ; crystals essentially rhombic prisms, with V : V (in front) = 86° 16', terminated by a very oblique front-polar or hemi-ortho-dome[*] inclined on the V planes at angle of 113° 30'. Cleavage very perfect parallel to V. Occurs also very commonly in columnar, fibrous, and sub-earthy masses. White, yellowish or reddish-white, pale-red, pale-grey. H 3·5-4·0 normally, but often less from partial disintegration. G 2·25-2·36. Fusible with intumescence into a white enamel or very blebby glass. Gelatinizes in hydrochloric acid. A drop of the solution on loop of platinum wire, or a moistened fragment of the mineral, shews in spectroscope red and green Ca-lines. Leonhardite, Caporcianite, and Ædelforsite,[†] are identical or closely related.

---

[*] The Basal plane of French crystallographers.

[†] This is the so-called "Red Zeolite of Ædelfors." Its hardness is usually stated in text-books to equal 6·0, an error arising from a confusion of names—the degree of hardness in question applying to an older ".Ædelforsite," since shewn to be an impure Wollastonite containing intermixed quartz.

(*In vitreous, amorphous masses*).

HYDROTACHYLITE: K^2O, Na^2O, CaO, MgO, FeO, Fe^2O^3, Al^2O^3, TiO^2, SiO^2, H^2O (12·90 per cent.), according to Peterson and Senfter. Forms nodular and other masses of uncrystalline structure in basalt. Dark-green or black. H 3·5; G 2·13. Fusible with more or less bubbling. Decomposed, with gelatinization, by hydrochloric acid. See Tachylite, page 228.

NOTE ON TABLE XXVII.

Many of the minerals placed (to avoid risk of error in their determination) in the present Table, belong properly—on account of their difficult fusibility or slight percentage of water—to preceding Tables, and are described more fully in these latter. The various Micas, Talc and Steatite, Agalmatolite, the Pinites, &c., are examples. See more especially the Note to TABLE XXV.

The minerals which belong essentially to the present Table consist for the greater part of zeolites—hydrated silicates of very characteristic occurrence in trappean or basaltic rocks. With these, in a Determinative grouping, the boro-silicate Datolite may be conveniently placed, as it resembles many zeolites in general characters, and is also frequently present in amygdaloidal traps. The zeolites, as the name implies, either swell up or intumesce on the first application of the blowpipe-flame, or otherwise melt very easily, and generally with bubbling. All, when reduced to powder, are readily decomposed by boiling hydrochloric acid, the silica separating in many cases in a gelatinous form. The presence of CaO, BaO, or K^2O. is easily ascertained by the pocket-spectroscope, if a drop of the solution be taken up in a small loop of platinum wire and held within the edge of a Bunsen-flame. As a rule, when lime and potash are present together, the red and green Ca-lines come out first, and then, as these fade away, the red K-line comes into view.

In the present Note, only the more common of these minerals are referred to, the crystallographic and other characters of the less important species being given in sufficient detail in the Table. The commonly occurring species, as regards their blowpipe reactions, fall into three series, as follows:

§ 1. Fusible quietly: (*a*) soda-species: Analcime, Natrolite; (*b*) barytic species: Harmotome.

§ 2. Fusible with much bubbling, but without (or without marked) intumescence* on first application of the flame: Datolite ; Prehnite.

§ 3. Curling up or intumescing on first application of the flame: (*a*) lime-potash species: Apophyllite, Phillipsite ; (*b*) gelatinizing lime-species: Thom-

---

* By "intumescence" is meant, here, not a mere expansion of the substance, but a throwing out of excrescences or curling up after the manner of borax. Minerals which intumesce in this manner on the first application of the flame, fuse afterwards in general without bubbling, and, as a rule, somewhat slowly.

sonite, Scolecite, Laumontite; (c) non-gelatinizing lime-species: Chabasite, Stilbite, Heulandite.

The leading characters of these species are given in the Table, but necessarily in brief form only; a few additional references to their crystallization are therefore appended.

Analcime, in most examples, is at once recognized by its crystals, as these are generally well-formed and easily made out. They belong to the Regular System, and consist either of the trapezohedron 2 - 2 (measuring 131° 48′ 36″ over long or axial edges, and 146° 26′ 33″ over intermediate edges), or of a combination of this form with the cube, the latter commonly predominating and thus having each angle replaced by three triangular planes (with inclination of cube-face on abutting 2 - 2 face measuring 144° 44′). The cleavage is cubical, but very indistinct. In the spectroscope, as a rule, no other line than a strong Na-line is observable if the test-matter be carefully freed from accompanying calcite.

Crystallized Natrolite was formerly and is still often known as Mesotype, the term Natrolite having been originally limited to the yellowish-brown, concentric-fibrous variety, then regarded as distinct. The crystals belong to the Rhombic System, but are frequently acicular, or are only partially formed (as polar planes) at the extremities of the fibres of which ordinary examples are so commonly composed. When distinctly formed, they consist of a nearly rectangular prism with front angle ($=\check{V}:\check{V}$) of about 91°, terminated by the planes of a somewhat low pyramid or octahedron measuring 143° 20′ and 142° 40′ over polar edges, and 53° 20′ over middle edge. P on V, consequently, measures 116° 40′. The prism-planes in most examples are striated vertically (sometimes very coarsely), and occasionally either the front or side edges are replaced by $\overset{..}{V}$ or $\overset{..}{V}$. In the spectroscope, pure examples as a rule shew only a strong Na-line, but transitory flashes of red and green Ca-lines sometimes appear.

Harmotome, a barytic zeolite, is in general readily recognized by its small, symmetrically formed cruciform crystals, although, occasionally, re-entering angles in these are more or less inconspicuous or are indicated only by striæ. The crystallization is apparently Rhombic, but the crystals have to some extent a Tetragonal aspect. They consist commonly of a rectangular prism (composed of the forms $\bar{V}$ and $\overset{..}{V}$), terminated by the planes of an octahedron or pyramid, P, or occasionally by those of a side-polar or brachydome $\overset{..}{P}$. In some crystals, the polar planes are simply striated; in others, the V planes shew a lozenge-shaped striation.[*] Two (or four) of these crystals form interpenetrating twins, with vertical axis in common. P : P, over polar edges, 120° 1′ and 120° 42′; $\overset{..}{P}:\overset{..}{P}$ 110° 20′. Cleavage, $\overset{..}{V}$ distinct, $\bar{V}$ somewhat less apparent. A drop of the hydrochloric acid solution, taken up in a loop of

---

[*] Some crystallographers (after Des Cloizeaux) make the System Clino-Rhombic, and regard this front-vertical form as the basal form. On that view, most of the crystals will be elongated in the direction of the clino-axis.

platinum wire, shews the green Ba-lines in the spectroscope very distinctly. The diluted solution gives also a marked precipitate with a drop of sulphuric acid.

Datolite—a hydrated boro-silicate of lime—is described fully, as regards its more distinctive characters, apart from crystallization, in the Table. Its crystals belong to the Clino-Rhombic system, but many (the Arendal crystals, especially) are strikingly Ortho-Rhombic in aspect. These latter are chiefly in the form of rhombic or six-sided tabular crystals, composed of the forms V and V̈, with broadly-extended basal plane, and commonly with a front-polar or orthodome ($-2\dot{P}$) and other polar planes subordinately developed. In many crystals these polar planes appear equally at corresponding extremities, with but little if any difference in their angle values, and thus impart an Ortho-Rhombic character to the crystal. In crystals from other localities, however, and in some of the Arendal crystals, they are developed only at one extremity. In the Andreasberg and most other crystals, the basal plane is also well-developed as a rule, but the prism-planes (V, V½, and V̄) and certain polar planes (especially $-2\bar{P}$, $-P$, and the side-polars or brachydomes $2\dot{P}$ and $4\dot{P}$) are also well formed, and the crystals are thus more short-prismatic than tabular. In some crystals, again, the basal form is entirely absent. The principal angles are as follows: V : V 76° 38′; V½ : V½ 115° 22′; V : V¼ 160° 38′; B : V̄ 90° 6′ (and 89° 54′); B : $-2\dot{P}$ 135° 4′. The marked green coloration (from the presence of B^2O^3) which datolite imparts to the flame of the blowpipe or Bunsen burner serves at once to distinguish it from other minerals of similar aspect.

Prehnite is distinguished from other Zeolites by its high degree of hardness (= 6 to 7), and its small percentage of water. It occurs most commonly in botryoidal masses with crystalline surface and radio-fibrous structure, the colour varying from pale greenish-white to deep apple-green. Distinct crystals are comparatively rare. They belong to the Rhombic System, and present four types: (1), The symmetrically tabular type—in which the crystals are thin rhombic tables composed of the forms V and B; or six-sided tables composed of V V̈ and B; or eight-sided tables made up of V, V̄, V̈, and B, the basal form in each case greatly preponderating. (2), The tabular type with brachydiagonal elongation—in which the thin crystals contain the forms V, V̈ and B, and are greatly extended along the two latter, thus passing at times into fibrous aggregations with the two front planes of V at the free end of the fibres. (3), The short-prismatic type with development of side or brachy-forms—the crystals of this type being composed essentially of the forms V and B, with V̈ and $3\dot{P}$ at the sides, the planes of the rhombic prism V preponderating; and (4), The short-prismatic type, with front or macro-forms—the crystals presenting the forms V and B, as preponderating forms, with the front-vertical V̄, and the front-polar or macrodome $\tfrac{3}{4}\dot{P}$ subordinately developed, in addition occasionally to the planes of the rhombic pyramid P,

forming a narrow border to the basal plane. V : V 99° 56′ ; B : 3P̈ 106° 30′ ;
B : ⅔P̈ 134° 52′. The vertical faces are frequently convex, whilst the basal
plane is more or less concave, and from the aggregation of these curved crystals, parallel to B, globular or spheroidal examples commonly arise. For
other characteristics, see the Table.

* Apophyllite is distinguished chemically by its fluorine reaction, by the
absence of alumina, and by the persistent K-line which it exhibits in the spectroscope when moistened with hydrochloric acid. Its Tetragonal crystals are
in general distinctly formed, and are thus easily recognized. They present
three more or less distinct types : (1), A prismatic type—in which the crystals
are simple square prisms (V̄, B), with angles replaced by the triaxial pyramid
P ; (2), A tabular type—in which the crystals present a large base, with V
and P depressed to little more than a narrow border around it ; and (3), A
pyramidal type—in which the pyramid P essentially predominates, although
combined with the front-vertical form or pinakoid, V̄, and occasionally with
the octagonal prism V2 (which appears as a bevelment on the vertical edges of
V̄). The basal plane, with its peculiar iridescent-pearly lustre, is also frequently present in this type, but it is always of small size, and the general
aspect of the crystals is essentially pyramidal. P : P over polar edge 104° to
104° 20′, over middle edge 120° to 121°. B : P about 119° 30′. The cleavage
is basal and very perfect, the points of the pyramid consequently are commonly broken off. Twin crystals, so common in many Zeolites, are in this
species all but unknown.

Phillipsite is also a potassic species, but differs from Apophyllite by containing alumina, as well as by the absence of fluorine, and essentially by its
crystallization. It differs also by its complete gelatinization in hydrochloric
acid. Its crystals are practically identical with those of Harmotome (see
above), and thus consist essentially of a rectangular prism (V̄, V̈) terminated
by the polar forms P, P̋ ; two (or four) crystals being united in cruciform
twins. In some crystals, the vertical planes look like those of a simple prism,
but the compound nature of the crystal is revealed by the re-entering angles
at the summit. In general, however, the cruciform character of the crystals
is sufficiently distinct. The planes of the forms P, P̋, and V̈, are transversely
striated.

Thomsonite occurs chiefly in fibrous and acicular forms, but is also found in
small, distinct crystals. These belong to the Rhombic System, and present
two types or varieties : (1), The Thomsonite type, proper, in which the crystals are short, large-based, vertically-striated rhombic prisms, V, replaced on
the acute edges by the side or brachy-vertical V̈, and on the obtuse edges and
angles by the front-vertical V, and front-polar or macrodome mP̌ ; and (2),
The Comptomite type, in which the crystals form short eight-sided prisms
(composed of the forms V, V̄, V̈) with the two planes of an exceedingly flat
brachydome or side-polar $^{1}/_{m}$P̈ entirely occupying the position of the base. The

prism V is nearly square, its front-angle measuring 90° 40'. The flat brachy-dome planes meet (according to Des Cloiseaux) at an angle of 177° 23'.

Crystals of Scolecite very closely resemble those of Natrolite or Mesotype, as they consist of nearly square prisms terminated at each extremity by four pyramidal planes. But whilst Natrolite crystals are clearly Ortho-Rhombic, Scolecite crystals are regarded as Clino-Rhombic, the pyramidal planes at the top and bottom of the crystal, respectively, differing slightly in their interfacial angles. These angles, nevertheless, closely correspond to those of Natrolite. V : V = 91° 35' (in Natrolite 91°); P : P, over polar edge in front, 144° 20' (in Natrolite 143° 20'); -P : -P 144° 40'. Occasionally the prism is six-sided, its acute edges being replaced by the side-vertical V́. Scolecite differs, however, essentially from Natrolite in being a lime-species in place of a soda-species, and by its remarkable blowpipe comportment: as, whilst Natrolite fuses quietly, Scolecite expands and curls up or throws out excrescences on the first application of the flame, at least in all typical examples. Some examples are said to fuse without intumescence, but these are probably soda-holding varieties, or Mesolite. All essentially calcareous zeolites exfoliate or intumesce before the blowpipe, or otherwise fuse with continued bubbling. Purely alcaline zeolites, on the other hand fuse quietly.

Laumontite when in crystals is easily recognized, but when in fibrous masses it is distinguished with difficulty from other calcareous zeolites. A somewhat salient character is its great tendency to fall into a white, earthy powder from efflorescence. The crystals are Clino-Rhombic, and they consist most commonly of a simple rhombic prism terminated obliquely by a single plane. The latter is the basal plane of most French crystallographers, but is commonly made the plane of a hemi-orthodome or front-polar - P. The prism-angle V : V, in front, equals 86° 16'; V : -P = 113° 30'. Very frequently the opposite angle of the prism is replaced by the corresponding hemi-orthodome P̀, the latter inclining to a face of the prism at an angle of 104° 20'. Often, also, other polar planes (P, &c.) are subordinately present, and the vertical edges of the prism are sometimes slightly truncated by V̄ and V́. Spectroscopic and other characters are given in the Table.

Chabasite is easily distinguished from other zeolites by its rhombohedral crystallization. The crystals, although small, are in general distinctly formed. They consist essentially of cuboidal rhombohedrons, with R : R measuring over polar edges 94° to 95°, usually 94° 46', whence the old French name of zéolite cubique by which the species was at one time known. In many examples, this rhombohedron occurs in the simple state, but very often its polar edges are replaced by an obtuse rhombohedron $-\frac{1}{2}$R, and its middle angles by the acute form -2R, measuring respectively over their own polar edges, 125° 13', and 72° 53'. R on $-\frac{1}{2}$R = 136° 23'; R on -2R = 119° 42'. The planes of the chief rhombohedron, R, are sometimes striated parallel to the polar edges, the striæ meeting in the line of the longer diagonal of each plane. These striæ indicate a very obtuse scalenohedron, occasionally present in Chabasite crystals. An obtuse twelve-sided pyramid $\frac{3}{2}$P2 (with angle of 145° over polar edges) is the predominating form in the Bohemian variety known as Phacolite. This variety occurs in interpenetrating twins; and twin-forms, with the vertical axis in common, are of frequent occurrence in crystals of Chabasite generally. The solution in hydrochloric acid, in which the silica separates in a slimy or at times in almost a gelatinous condition, shews in the spectroscope a vivid calcium spectrum, and as this fades out a transitory red K-line generally comes into view.

Stilbite and Heulandite may in general be distinguished easily from other zeolites by their almost constant occurrence in bladed or narrow-foliated examples, with very perfect cleavage in one direction and strong pearly lustre on the cleavage surface. The latter is parallel to a side-vertical, $\overset{\prime\prime}{V}$, or (in Heulandite) $\overset{\prime}{V}$. The hardness, also, is lower than in most other zeolites, viz.: 3·5-4·0. The free ends of the foliæ generally shew crystalline facets. The colour is commonly either white, red, or light-brown. In Stilbite, the crystal-system is Rhombic, and the more common crystals consist of a rectangular prism ($\overline{V}$, $\overset{\prime\prime}{V}$, usually flattened parallel to $\overset{\prime\prime}{V}$, the cleavage plane), with the planes of a rhombic octahedron, P, at each extremity. Occasionally, the vertical edges of the rectangular prism are slightly replaced by the rhombic prism V, and the point of the octahedron is truncated by the basal form B. The prism-angle, V : V, equals 94° 16′ ; P : P over front polar edge, 119° 16′ ; over side polar edge, 114° ; over middle edge, 96°.

In Heulandite, the system is Clino-Rhombic. The more commonly-occurring crystals are made up of the front-vertical form $\overline{V}$, the side or clino-vertical $\overset{\prime}{V}$, the front-polar or hemi-orthodome $\overline{P}$, and the basal form B. The side-vertical $\overset{\prime}{V}$ (the cleavage plane) generally predominates, the crystals being usually much flattened in that direction ; but occasionally, crystals are elongated transversely, i.e., in the direction of the ortho-diagonal or right-and-left axis, in which case the frontal forms $\overline{V}$ and $\overline{P}$ preponderate. The hemi-pyramids 2P and ⅜P, and the clinodome or side-polar 2P′, also occasionally occur as subordinate forms. $\overline{P} : \overline{V}$ equals 129° 40′ ; B : V, 116° 20′ ; 2P : 2P, in front, 136° 4′ ; ⅜P : ⅜P, 146° 52′ ; 2P′ : 2P′, over summit, 98° 44′.

Although both Stilbite and Heulandite are essentially lime species, they usually contain small amounts of soda and potash. When a drop of the hydrochloric-acid solution (taken up in a loop of clean platinum wire) is examined by the spectroscope, the red K-line, therefore, almost always appears for an instant, as the vivid red and green Ca-lines fade out of view.

INDEX

TO THE MINERALS IN PART II.

Abichite, 144.
Abrazite (v. Gismondine).
Acadialite, 268.
Acanthite, 107.
Acanticone (v. Epidote).
Acmite, 232.
Actinolite, 233.
Adamantine Spar, 196, 207.
Adamite, 145.
Adularia Feldspar, 245, 253.
Ædelforsite, 272.
Ægirine, 233.
Æschynite, 126.
Agalmatolite, 219.
Agaric Mineral (v. Calcite).
Agate, 208.
Aikinite, 107.
Alabandine, 108, 124, 152.
Alalite, 243.
Albertite, 132.
Albin, 268.
Albite, 246, 254.
Alexandrite (v. Chrysoberyl).
Algodonite, 101.
Alipite, 217.
Alisonite, 107.
Allanite, 228.
Allemontite, 101.
Allochroite, 228.
Alloclase, 103.
Allophane, 217.
Almandine, 230.
Alstonite, 137.
Altaite, 113.
Alum, 154.
Alumstone, 155.
Aluminite, 155, 162.
Alunite, 155.
Alunogene, 154.
Amalgam, 115.
Amazon-stone, 246.
Amber, 132, 133.
Amblygonite, 164.
Amethyst, 202, 208.
Amianthus, 243.
Ammonia-alum, 154.
Amphibole, 233, 243, 251.

Amphigene (Leucite), 203.
Analcime, 265, 274.
Anatase, 127, 129, 196, 206.
Anauxite, 219.
Andalusite, 199, 210.
Andesine, 253.
Andradite, 230.
Anglarite (v. Vivianite).
Anglesite, 151, 160.
Anhydrite, 152, 161.
Ankerite, 136.
Annabergite, 145.
Anorthite, 241.
Anthophyllite, 216.
Anthosiderite, 257.
Anthracite, 128, 223.
Anthraconite (v. Calcite).
Antigorite, 215.
Antimony, 113.
Antimony Blende, 149.
Antimony Glance, 110, 112.
Antimonial Silver (v. Dyscrasite).
Antimonial Nickel Glance, 110.
Antimonial Nickel Ore, 114.
Antimony Ochre, 149.
Antimonite, 149.
Antrimolite, 266.
Apatelite, 158.
Apatite, 163, 169.
Aphanese, 144.
Aphrodite (Meerschaum?)
Aphrosiderite, 257.
Aphthalose, 153.
Apthonite (Tetrahedrite?)
Apjohnite, 158.
Aplome, 230.
Apophyllite, 267, 276.
Aquamarine (Beryl), 200.
Aræoxene, 144.
Aragonite, 138, 141.
Arcanite, 153.
Arfvedsonite, 234.
Argentite, 107, 109.
Arkansite, 127.
Arksutite, 178.
Arquerite, 115.

Arragonite, 138, 141.
Arsenic, 101.
Arsenical Iron, 101.
Arsenical Pyrites, 103, 104.
Arsenious Acid, } 143.
Arsenolite,
Arseniosiderite, 146.
Asbestus, 243.
Asbolan, 189.
Asmanite, 201.
Asparagus stone v. Apatite).
Aspasiolite, 220.
Asperolite, 217.
Asphalt, 132.
Aspidolite (Magnesia Mica).
Astrakanite, 154.
Astrophyllite, 227.
Atacamite, 176, 177.
Atelesite, 145.
Atelite, 176.
Atheriastite.
Atlasite, 176.
Auerbachite, 198.
Augelite, 168.
Augite, 232.
Auricbalcite, 136.
Auripigment (v. Orpiment).
Automolite, 197.
Autunite, 166.
Avanturine (= Quartz with interspersed scales of mica, iron-glance, &c.).
Axinite, 230, 248.
Azurite, 135, 142.

Babingtonite, 233.
Bagrationite, 228.
Baikalite (Amphibole).
Baltimorite, 221.
Bamlite (var. Sillimanite), 200.
Barnhardtite, 105.
Barrandite (Strengite ?), 165.
Barytine, 152, 160.
Barium Mica, 261.
Baryto-calcite, 137, 153.
Baryto-celestine, 152.
Bastite, 215.
Batrachite, 204.
Baudisserite, 139.
Beaumontite, 269.
Bechilite, 172.
Beraunite, 165.
Bergemannite, 266.
Berlinite, 168.
Berthierite, 110.
Beryl, 200, 209.
Berzelite, 146.
Berzeline, 106, 236.

Beudantite, 144.
Beyrichite, 105.
Bieberite, 157.
Bindheimite, 150.
Binnite, 103, 104.
Biotite, 213.
Bismuth, 115.
Bismuthine, } 107, 109,
Bismuth Glance,
Bismuth Ochre, 188.
Bismutite, 136.
Bitter Salt (Epsomite), 154.
Bitter Spar, 138.
Bitumen, 132.
Bituminous Coal, 132, 134.
Black Band, 141.
Black Oxide of Copper, 116.
Black Jack, 109.
Blende, 108, 109, 151, 159.
Blœdite, 154.
Bloodstone (v. Quartz).
Blue carb. copper, 135.
Blue Vitriol, 156.
Bodenite, 228.
Bog Iron Ore, 193.
Bog Manganese Ore (Wad), 188.
Bolognese Spar (Barytine), 152.
Boltonite, 204.
Bombiccite, 133.
Bonsdorffite, 220.
Boracite, 171, 173.
Borax, 171, 173.
Bornite, 105, 108.
Borocalcite, 172.
Boronatrocalcite, 172.
Boracic Acid, 171.
Botryogene, 157.
Botryolite, 270.
Boulangerite, 111.
Bournonite, 111, 112.
Bowenite, 221.
Bragite, 127.
Brandisite, 216.
Braunite, 125.
Breislakite, 232.
Breithauptite, 114.
Breunnerite (Mesitine), 136.
Brevicite, 266.
Brewsterite, 268.
Brittle Silver Ore, 110.
Brochantite, 158.
Bromargyrite, 175.
Bromlite, 137.
Brongniardite, 111.
Brongniartine (v. Glauberite).
Bronzite, 216.
Brookite, 127, 199.
Brown Coal, 132.

INDEX. 281

Brown Iron Ore, 125, 128, 187, 192.
Brucite, 190, 194.
Brushite, 168.
Bucholzite, 200.
Bucklandite, 231.
Bunsenite, 190.
Buratite, 136.
Buntkupfererz, 105.
Bustamite, 233.
Byssolite, 227.
Bytownite (var. Anorthite).

Cabrerite, 145.
Cacholong (var. Opal).
Cacoxene, 165.
Cairngorm, 208.
Calaite, 167, 170.
Calamine, 205, 219.
Calamite (Tremolite), 243.
Calaverite, 113.
Calcedony, 208.
Calcite, } 137, 140.
Calc Spar, }
Caledonite, 135, 151.
Calomel, 176.
Canaanite (var. Pyroxene).
Cancrinite, 239.
Cantonite (Covelline ?).
Caporcianite, 272.
Carbonado, 196.
Carminite, 144.
Carnallite, 174.
Carnelian, 208.
Carpholite, 266.
Cassiterite, 127, 195, 206.
Castor, 235.
Cat's-Eye, 208.
Celestine, 152, 160.
Cerargyrite, 175, 177.
Cerine, 228.
Cerite, 205, 218.
Cerussite, 135, 142.
Cervantite, 149.
Ceylanite, 197.
Chabasite, 268, 277.
Chalcanthite, 156.
Chalcedony, 208.
Chalkosine, 106.
Chalilite, 265.
Chalcanthite, 156.
Chalcophanite, 126.
Chalcophyllite, 144.
Chalcopyrite, 105, 108.
Chalcolite, 166.
Chalcotrichite (= Acicular Cuprite).
Chalcosiderite, 166.

Chalcosine, } 106, 109.
Chalkosine, }
Chalcostibite, 111.
Chalibite, 265.
Chalybite (Siderite), 136.
Chatamite, 101.
Chamosite, 256.
Chessylite, 135.
Chesterlite (Orthoclase).
Chiastolite, 199, 223.
Childrenite, 165.
Chile Saltpetre, 181.
Chiolite, 178.
Chladnite (Meteoric Enstatite).
Chloanthite, 101.
Chlor-Apatite, 163.
Chlorastrolite, 267.
Chlorides (v. TABLE XIX.).
Chlorite, 314, 224, 257.
Chloritoid, 259.
Chlorocalcite, 174.
Chloromelane (Cronstedite), 256.
Chloropal, 218.
Chlorophæite, 260.
Chlorophane Fluor Spar).
Chlorophyllite, 215.
Chlorotile, 144.
Chondro-arsenite, 146.
Chondrodite, 204, 211.
Chonikrite, 269.
Christophite, 151.
Chromic Iron Ore, 118, 124, 128.
Chromite, 118, 124, 128, 186, 192.
Chrome Garnet, 198.
Chrome Mica, 213.
Chrysoberyl, 197, 207.
Chrysocolla, 217.
Chrysolite, 204, 210.
Chrysoprase, 208.
Chrysotile, 221.
Churchite, 168.
Cimolite, 219.
Cinnabar, 121, 122, 130, 131.
Cinnamon-stone (Garnet).
Clarite, 103.
Claudetite, 148.
Clausthalite, 106.
Clay Ironstone, 136, 141.
Cleavelandite, 246.
Clinoclase, 144, 147.
Clinochlore, 214.
Clino-Humite, 204.
Clintonite, 216.
Cluthalite, 265.
Coals, 132, 134.
Cobalt Bloom, 145.

INDEX.

Cobalt Spar, 137.
Cobalt Vitriol, 157.
Cobaltine, 103, 104.
Coccinite, 176.
Coccolite, 232.
Collyrite, 220.
Colophonite, 245.
Columbite, 126.
Comptonite, 271, 276.
Cookeite, 261.
Copiapite, 158.
Copper, 116.
Copper Binnite, 103.
Copper Glance, 106, 109.
Copper Mica, 144.
Copper Nickel (Nickeline), 101, 102.
Copper Pyrites, 105, 108.
Copper Uranite, 166, 170.
Copper Vitriol, 156.
Copperas (Green Vitriol), 156.
Coquimbite, 157.
Coracite, 190.
Cordierite, 200, 211.
Corneous Lead Ore, 176.
Cornwallite (a copper arseniate).
Corundum, 196, 297.
Corynite, 103.
Cosalite, 107.
Cotunnite, 176.
Couseranite, 241.
Covelline, 130.
Crednerite, 126.
Crichtonite (Ilmenite), 118.
Crocidolite, 259.
Crocoisite, 182, 184.
Cronstedite, 256.
Crookesite, 106.
Cryolite, 178.
Cryophyllite, 235.
Cryptolite, 164.
Cryptomorphite, 172.
Cubanite, 105.
Cube Ore, 145.
Cuboite, 265.
Culsageeite, 362.
Cummingtonite, 234.
Cuprite, 116, 122, 123, 189, 193.
Cuproplumbite, 107.
Cyanite, 197, 210.
Cymophane, 197.
Cyprine, 245.

Damourite, 260.
Danaite, 103.
Danalite, 229.
Danburite, 236.

Dark Red Silver Ore, 110, 112, 121.
Datolite, 270, 275.
Daubreite, 176.
Davyne, 239.
Davidsonite (Beryl).
Dawsonite, 139.
Dechenite, 182.
Delessite, 214, 257.
Delvauxite (a lime-iron phosphate).
Demidowite, 217.
Descloizite, 182.
Desmine, 269.
Deweylite, 221.
Diadochite, 159.
Diallage, 242.
Diallogite, 137.
Diamagnetite, 124.
Diamond, 196, 206.
Dianite, 196.
Diaphorite, 111.
Diaspore, 196.
Dichroite, 200, 211.
Dihydrite, 167.
Diopside, 243.
Dioptase, 205, 216.
Diphanite, 215.
Dipyre, 241.
Disterrite (Brandisite), 216.
Disthene, 197, 210.
Dolomite, 138, 140.
Domeykite, 101.
Donacargyrite, 111.
Dopplerite, 132.
Dufrenite (Green Iron Ore, an iron phosphate).
Dufrenoysite, 103, 104.
Durangite, 147.
Dyscrasite, 113.
Dyslnite (Gahnite), 197.

Edingtonite, 264.
Egerane, 244.
Ehlite, 166.
Ekebergite (Warnerite), 240.
Elaeolite, 239.
Elastic Bitumen, 132.
Elaterite, 232.
Electrum (Amalgam), 115.
Eliasite, 190.
Embolite, 175.
Emerald, 200, 209.
Emerald-Nickel (Zaratite), 139.
Emery, 196, 207.
Emerylite, 215.
Emplectite, 107.
Enargite, 103.

INDEX. 283

Enstatite, 201.
Epichlorite, 214.
Epidote, 231, 249.
Epigenite, 103.
Epistilbite, 269.
Epsomite, 154.
Erdmannite, 228.
Eremite, 765.
Erinite, 144.
Erubescite (Bornite), 105.
Erythrine, 145.
Esmarkite, 220.
Essonite (Garnet),
Ettringite, 156.
Euchroite, 144.
Euclase, 200.
Eucolite, 237.
Eudialyte, 237.
Eudnophite, 265.
Eukairite, 106.
Eulytine, 237.
Euphyllite, 215.
Eupychroite, 163.
Eusynchite, 182.
Euxenite, 127.
Euzeolite, 269.
Evansite, 167.

Fahlerz, 110.
Fahlunite, 220.
Fargite, 266.
Farcolite, 272.
Fassaite, 232.
Faujasite, 267.
Fauserite, 158.
Fayalite, 227.
Feather Alum, 157.
Feldspar (lime), 241.
Feldspar (potash), 245.
Feldspar (soda), 245.
Feldspar Group, 253.
Felsobanyite, 155.
Fergusonite, 127.
Fibro-Ferrite, 158.
Fibrolite, 200.
Fichtelite, 133.
Figure Stone, 219.
Fire Blende, 149.
Fire Opal, 202.
Fischerite, 167.
Flint, 208
Flos Ferri, 138.

Fluellite, 178.
Fluocerite, 179.
Fluor-Apatite, 163, 169.
Fluorite, 178.
Fluor Spar, 178, 179.
Foresite (near Stilbite), 269.
Forsterite, 204.
Fowlerite, 233.
Francolite, 163.
Franklinite, 118, 124, 128, 186, 192.
Freislebenite, 111.
Frenzelite (Guanajuatite), 106.
Frugardlite, 245.
Fuchsite, 213.

Gadolinite, 204.
Gahnite, 197, 208.
Galactite, 266.
Galena, 107, 109.
Galmei (Calamine), 205.
Garnet, 230, 141.
Garnet Group, 248.
Gaylussite, 138.
Gehlenite, 204.
Geierite, 103.
Genthite, 217.
Geocronite, 133.
Geocronite, 112.
Gersdorffite, 103.
Gibbsite* (see Note, below).
Giesseckite, 220.
Gigantolite, 266.
Gilbertite, 215.
Gillingite (Hisingerite), 218.
Giobertite, 138.
Girasol, 202.
Gismondine, 271.
Glagerite, 220.
Glaserite, 153.
Glauberite, 153.
Glauber's Salt, 153.
Glaucodot, 103.
Glauconite, 259.
Glaucophane, 244.
Glingite, 204.
Glockerite, 158.
Gmelinite, 271.
Goethite, 125.
Gold, 116.
Gold-Amalgam, 115.
Goschenite (v. Beryl).

---

* Accidentally omitted from foot of page 220, where it should follow Kollyrite:
GIBBSITE (Hydrargillite) :—Al^2O^3 65·5, H^2O 34·5. In small hexagonal crystals with basal cleavage, or in mamillary or stalactitic examples of a white, greenish-yellow, or other light colour. H 2·5-3 0 ; G 2·3-2·4. BB, infusible, but commonly exfoliates. In powder, dissolved by caustic potash ; also by sulphuric acid.

Goslarite, 155.
Grahamite, 122.
Gramenite, 218.
Grammatite, 243.
Graphic Tellurium, 113.
Graphite, 117.
Green Carb. Copper, 135.
Green Earth, 258.
Green Vitriol, 156.
Grey Antimony Ore, 110, 112.
Grey Copper Ore (Tetrahedrite), 110.
Greenockite, 151.
Greenovite Sphene), 229.
Grengesite (near Delessite), 257.
Grophite, 215.
Groroilite, 188.
Grossular, 241.
Guadalcazarite, 106, 107.
Guanajuatite, 106.
Guarinite, 239.
Gummite, 190.
Gurholian, 138.
Gymnite, 221
Gypsum, 155, 161.
Gyrolite (Apophyllite?), 267.

Haarkies (Millerite), 105.
Hæmatite, 118, 120, 114, 128.
Haidingerite, 146.
Halite (Rock Salt), 174, 177.
Hallite, 262.
Halloysite, 220.
Halotrichite, 157.
Harmatite, 179.
Harmatome, 264, 274.
Harringtonite, 266.
Hartite, 133.
Hatchettine, 133.
Hauerite, 108, 124, 152.
Hausmannite, 125.
Hauyne, 246.
Haydenite, 268.
Haytorite (Quartz in pseudomorphs after Datolite).
Heavy Spar, 152, 160.
Hebronite, 164.
Hedenbergite, 232.
Hedyphane, 144.
Heliotrope (Bloodstone), 208.
Helminthite, 214.
Helvine, 229.
Helvetane, 261.
Hematite, 118, 120, 124, 128.
Hercynite, 198.
Herderite, 164.
Herrerite, 137.
Herschelite, 268.

Hessite, 113.
Hessonite (Garnet).
Heterogenite, 189.
Heterosite, 166.
Heulandite, 269, 278.
Hjelmite, 127.
Hisingerite, 218, 258.
Hœrnisite, 146.
Homichline, 105.
Hopeite, 168.
Horbachite, 105.
Hornblende, 233.
Horn Silver Ore, 175, 177.
Horseflesh Ore, 108.
Hortonolite, 284.
Hovite, 130.
Hübnerite (Wolfram).
Humboldtilite, 238.
Humboldtine, 187.
Humboldtite, 270.
Humite, 204
Hunterite, 219.
Hureaulite, 166.
Hyacinth, 198.
Hyalite, 209.
Hyalophane, 236, 246.
Hyalosiderite, 228.
Hydrargillite (see Foot-note to Gibbsite).
Hydroboracite, 172.
Hydrocuprite, 189.
Hydrodolomite, 139.
Hydroflnocerite, 139.
Hydrohematite, 125.
Hydromagnesite, 138.
Hydrophane (var. Opal).
Hydrotachylite, 273.
Hydrozincite, 139.
Hypersthene, 232.

Iberite, 220.
Ice-spar (var. Orthoclase).
Iceland Spar (var. Calcite).
Idocrase, 244, 249.
Idrialine, 130.
Iglesite, 135.
Ilmenite, 118, 120, 125, 128, 186, 192.
Ilvaite, 228, 250.
Indicolite, 199.
Iodargyrite, 175.
Iolite, 200, 211.
Iridium, 117.
Iridosmine, 117.
Iridosmium, 117.
Iron, 117.
Iron Alum, 157.
Iron Chrysolites, 250.

INDEX.

Iron Glance, 118, 120.
Iron Pyrites, 105, 108.
Ironstone, 140.
Isoclase, 168.
Ittnerite, 270.
Ixolyte, 132.

Jacobsite, 118, 186.
Jamesonite, 111, 112.
Jargon (Zircon), 198.
Jarosite, 159.
Jasper, 208.
Jefferisite, 362.
Jeffersonite, 233.
Jenite (Lievrite), 228, 250.
Jet, 132.
Johannite, 157.
Jordanite, 104.

Kammercrite, 214.
Kalaite (Turquoise).
Koinite, 154.
Kakoxene, 165.
Kalinite, 154.
Kampylite, 144.
Kaolin, 219, 225.
Karminspath, 144.
Karstenite (Anhydrite), 152.
Kastor, 215.
Keilhauite, 230.
Kenngottite (var. Miargyrite).
Keragyrite, 175, 177.
Kerasine, 135, 176.
Kermesite, 121, 123, 149.
Kerolite, 221.
Kibdelophane (Ilmenite), 118.
Kieserite, 155.
Kilbrickenite, 112.
Killinite, 262.
Kirwanite, 260.
Kjerulline, 163.
Klaprothine (Lazulite), 168.
Klipsteinite, 264.
Knebelite, 229.
Kobellite, 112.
Kœttigite, 145.
Kollyrite, 220.
Konite, 130.
Kottigite, 145.
Kongsbergite, 115.
Könleinite, 133.
Korundophyllite, 214.
Koupholite, 267.
Krantzite, 132.
Kraurite (Green Iron Ore, an iron phosphate).
Kreittonite, 197.
Kremersite, 175.

Krisuvigite, 158.
Krokidolite, 259.
Kuhnite, 146.
Kyanite, 197.

Labradorite, 241.
Labrador Feldspar, 241.
Lagonite, 172.
Lampadite, 189.
Lanarkite, 151.
Lancasterite, 139.
Langite, 158.
Lanthanite, 139.
Lapis Lazuli, 237.
Larderellite, 172.
Latrobite (var. Anorthite).
Laumontite, 272, 277.
Laxmannite, 182.
Lazulite, 168.
Lead, 115.
Lead Binnite, 104.
Lead Glance, 107, 109.
Leadhillite, 135.
Leafy Tellurium Ore, 111.
Lehrbachite, 106.
Lehuntite, 266.
Lenzinite, 220.
Leonhardite, 272.
Lepidochrocite, 125, 187.
Lepidokrokite, 125, 187.
Lepidolite, 230, 234, 247.
Lepidomelane, 227.
Lettsomite, 158.
Leuchtenbergite (Ripidolite), 214.
Leucite, 203, 212.
Leucophane, 178, 240.
Leucopyrite, 101.
Levyne, 268.
Libethenite, 167, 170.
Liebenerite, 220.
Liebigite, 139.
Lievrite, 228, 250.
Lignite, 132.
Ligurite (Sphene), 229.
Lillite, 257.
Lime Uranite, 166.
Limonite, 125, 128.
Linarite, 158.
Lincolnite, 269.
Lindakerite, 139.
Linnœite, 105.
Liroconite, 144, 147.
Litharge, 119.
Lithia Mica, 234.
Liver Ore, 131.
Lobolite, 245.
Löllingite, 101.

Loewite, 154.
Loxoclase (var. Orthoclase).
Ludlamite, 165.
Ludwigite, 171.
Luneberite, 168.
Lunnite (Phosphorchalcite), 167.
Luzonite, 103.

Magnesia Alum, 154.
Magnesite, 138, 140.
Magnetic Iron Ore, 118, 120, 124, 128, 186.
Magnetic Pyrites, 105, 108.
Magnetite, 118, 120, 124, 128, 186.
Magnoferrite, 186.
Malachite, 135, 142.
Malacolite, 243.
Malakon, 198.
Maldonite (=Bismuthic Gold).
Mangan Blende (Alabandine, 124).
Manganite, 119, 120, 125.
Manganese Alum, 158.
Manganese Spar, 137.
Manganese Vitriol, 158.
Marble, 140.
Marcasite, 105, 108.
Margarite, 215.
Margarodite, 261.
Marmatite, 151.
Marmolite, 221.
Martite, 118.
Mascagnine, 153.
Maskelynite (= Meteoric Labradorite).
Masonite, 259.
Massicot, 187.
Matlockite, 176.
Maxite, 135.
Medjidite, 157.
Meerschaum, 221.
Megabromite, 175.
Meionite, 240.
Melaconite, 189.
Melanglanee, 110.
Melanite (Black Garnet).
Melanolite, 258.
Melanochroite (Phœnicite), 182.
Melanterite, 156.

Melilite, 238.
Melinophane, 178, 240.
Meliphanite, 178, 240.
Melonite, 114.
Melopsite (near Deweylite), 221.
Menaccanite, 118.
Mendipite, 176.
Meneghnite, 111.
Mengite, 126.
Menilite, 202.
Mennige, 187.
Mercury, 115.
Mesitine, 136.
Mesolite, 266.
Mesotype, 265.
Metabrushite, 168.
Metachlorite, 214, 256.
Metacinnabarite, 107.
Metaxite, 221.
Meteoric Iron, 117.
Miargyrite, 110, 121.
Micas, 213, 224.
Microbromite, 175.
Microcline, 246.
Microlite (Pyrochlore), 126.
Microsommite, 237.
Millerite, 105.
Miloschin, 218.
Mimetesite, 144, 148.
Minium, 187.
Mirabilite, 153.
Mispickel, 103, 104.
Mizzonite, 241.
Molybdenite, 107, 109.
Molybdic Ochre, 183.
Monazite, 165.
Monrolite, 200.
Montebrasite, 164.
Monticellite, 204.
Moonstone (=Opalescent Feldspar).
Morenosite, 157.
Moroxite (var. Apatite), 163.
Mosandrite *
Mottramite, 184.
Mountain Cork, 258.
Mountain Wood, 258.
Müllerine, 113.
Mundic, 105.

---

* Accidentally omitted from Table XXVII., immediately following Heulandite, as annexed:
(H 4·0; streak distinctly yellow or brownish; decomposed by hydrochloric acid with production of chlorine fumes).

MOSANDRITE :—Essential components: Na_2O, CaO, MnO, Ce^2O^3 with Di^2O^3 and La^2O^3, TiO^2, SiO^2, H^2O. Rhombic (?), but mostly in broad-fibrous or lamellar masses of a reddish-brown colour; G 2·9–3·0. BB intumesces and fuses readily into a yellowish-brown bead. The hydrochloric-acid solution is reddish-yellow, but becomes paler, and evolves chlorine, on heating. A very rare species.

INDEX. 287

Muromontite, 228.
Muscovite, 213, 224.

Nacrite, 219.
Nadorite, 150.
Nagyagite, 111.
Nantokite, 175.
Nasturane, 127.
Native Antimony, 113.
N. Arsenic, 101, 102.
N. Bismuth, 101, 115.
N. Copper, 116.
N. Gold, 116.
N. Iridium, 117.
N. Iron, 117.
N. Lead, 115.
N. Mercury, 115.
N. Palladium, 117.
N. Platinum, 117.
N. Silver, 116.
N. Sulphur, 130.
N. Tellurium, 113.
Natrolite, 255, 274.
Natron, 138.
Naumannite, 106.
Needle Ore, 107.
Neftgil, 133.
Nemalite, 190.
Nepheline, 239.
Nephrite, 243.
Newjanskite, 117.
Nickel Glance, 110.
Nickel Green, 145.
Nickel Gymnite, 217.
Nickel Vitriol, 157.
Nickeline, 101, 102.
Nigrescite, 260.
Nigrine, 127.
Niobite (Columbite), 126.
Nipholite, 178.
Nitratine, 181.
Nitre, 181.
Nitrocalcite, 181.
Nitromagnesite, 181.
Nontronite, 218.
Noscan, 236.
Nosine, 236.
Nuttalite, 241.

Obsidian, 247.
Ochres:
   Bismuth O., 188.
   Manganese O. (Wad), 188.
   Molybdic, O., 183.
   Red O., 186, 192.
   Tungstic O., 183.
   Uran O., 190.
   Yellow O., 187.

Octahedrite, 127, 199.
Œllacherite, 261.
Oerstedite, 198.
Okenite, 268.
Olafite, 246.
Oligoclase, 246.
Oligon Spar, 136.
Olivenite, 143, 147.
Olivine, 204, 210.
Omphazite (Var. Pyroxene).
Onkosine (compact magnesian mica? Related to Phlogopite or Biotite, as Steatite to Talc).
Onofrite, 106.
Onyx (Agate), 208.
Opal, 202, 208.
Ophiolite, 225.
Orangite, 218.
Orpiment, 130, 131.
Orthite, 228.
Orthoclase, 245, 253.
Osmelite (Pectolite?), 265.
Osmium-Iridium, 117.
Osteolite, 163.
Ostranite, 198.
Ottrelite, 259.
Ouvarovite, 198.
Owenite, 256.
Oxalite, 187.
Oxhaverite. 268.
Ozarkite, 272.
Ozokerite, 133.

Pachnolite, 178.
Paisbergite, } v. Rhodonite,
Pajsbergite, }
Palagonite, 257.
Palladium, 117.
Paper Coal, 132.
Paraffine, 133.
Paragonite, 261.
Paranthine, 240.
Pargasite, 233.
Parisite, 179.
Passauite, 241.
Patrinite (Aikinite), 107.
Paulite (v. Hypersthene), 232.
Pearl Mica, 215.
Pearl Spar (Dolomite).
Pearlstone, 247.
Pectolite, 265.
Peganite, 167.
Pegmatolite, 245.
Pelicanite, 219.
Pelokonite, 189.
Pennine, 214.
Percylite, 176.

Periclase, 190.
Pericline, 246, 255.
Peridot, 204.
Peristerite, 246.
Perowskite, 126.
Perthite, 245.
Petalite, 235.
Petroleum, 132.
Petzite, 113.
Phacolite, 268, 278.
Phæstine (Altered Bronzite).
Pharmacolite, 146, 148.
Pharmacosiderite, 145, 148.
Phenakite, 200.
Phengite (Muscovite).
Phillipsite, 270, 276.
Phlogopite, 213, 224.
Phœnicite, 182.
Pholerite, 219.
Phosgenite, 135, 176.
Phosphocerite, 164.
Phosphorchalcite, 167, 170.
Phosphorite, 163
Phyllite (Ottrelite?), 259.
Physalite, 197.
Piauzite, 132.
Pickeringite, 154.
Picotite (Chromiferous Spinel).
Picrolite, 221.
Picrophyll, 215.
Picrosmine, 221.
Piedmontite, 231.
Pilinite (asbestiform amphibole?)
Pimelite, 217.
Pinguite, 218.
Pinite, 220.
Pinite group, 226.
Pisanite, 157.
Pissophane, 158.
Pistacite, 231.
Pistomesite, 136.
Pitchblende, 127, 190.
Pitchstone, 247.
Pittizite, 146.
Plagionite, 111.
Planerite, 167.
Plasma (Green Calcedony).
Platinum, 117.
Platinum-Iridium, 117.
Platinum-Iron, 117.
Plattnerite, 122.
Pleonaste, 197.
Plinian, 103.
Plumbago, 117.
Plumbo-Calcite, 135.
Plumosite (Jamesonite).
Polianite, 119.
Pollux, 203.

Polyadelphite (var. Garnet).
Polyargyrite, 110.
Polybasite, 104, 110, 121, 143.
Polycrase, 126.
Polydymite, 105.
Polyhalite, 156, 161.
Polymignite, 126.
Polyxene (N. Platinum).
Poonahlite, 272.
Porcelain Earth (Kaolin), 219.
Prase, 208.
Prascolite, 220.
Prasine (Ehlite), 166.
Pregrattite, 261.
Prehnite, 267.
Prochlorite, 214.
Prosopite, 179.
Proustite, 121, 123, 143. 147.
Przibramite, 129.
Pseudomalachite (Phosphor-chalcite), 167.
Pseudophite (Compact Chlorite).
Psilomelane, 119, 125, 193.
Psittacinite, 184.
Pucherite, 182.
Purple Copper Pyrites, 105, 108.
Puschkinite (Epidote).
Pycnite, 197.
Pycnotrope, 263.
Pyrallolite, 222.
Pyrargillite, 220.
Pyrargyrite, 110, 112, 121, 123.
Pyreneite (Black Garnet).
Pyrgom, 232.
Pyrites :
  Arsenical Pyrites.
  Capillary Pyrites (Millerite), 105.
  Cockscomb Pyrites, 108.
  Copper Pyrites, 105.
  Iron Pyrites, 105, 108.
  Magnetic Pyrites, 105, 108.
  Purple Copper Pyrites, 105.
  Radiated Pyrites (Marca-site), 105.
  Spear Pyrites, 108.
  White Iron Pyrites (Mar-casite), 105, 108.
Pyrochlore, 126.
Pyrochroite, 189, 191.
Pyrolusite, 119, 120, 125, 193.
Pyromorphite, 163, 169.
Pyrope, 231.
Pyrophyllite, 214.
Pyrophysalite (Pycnite), 197.
Pyropissite, 132.
Pyrosclerite, 264.
Pyrosmalite, 257.

Pyrostibite, 149.
Pyrostilpnite, 149.
Pyroxene, 232, 243, 250.
Pyrrhite (Pyrochlor ?).
Pyrrhosiderite (Gœthite), 187.
Pyrrhotine, 105, 108.

Quartz, 202, 208.
Quicksilver, 115.

Rabdionite, 189.
Radiated Pyrites (Marcasite), 105.
Radiolite, 266.
Rammelsbergite, 101.
Randanite, 202.
Raphilite (var. Amphibole).
Ratofkite, 178.
Realgar, 130, 131.
Red Antimony Ore, 121, 123, 149.
Red Copper Ore, 116, 189.
Red Hematite, 186, 192.
Red Iron Ore, 186, 192.
Red Lead, 187.
Red Ochre, 186, 192.
Red Silver Ores, 121, 123, 147.
Red Zinc Ore (Zincite), 188, 193.
Reddle, 192.
Redruthrite (Copper Glance), 106.
Remingtonite, 139.
Rensselaerite (Pseudomorphous Steatite).
Retinalite, 221.
Retinite, 132.
Reussin, 154.
Rhœtizite (Kyanite), 197.
Rhagite, 145.
Rhodizite, 171.
Rhodochrosite. 137, 141.
Rhodonite, 233.
Richmondite, 167.
Rionite, 112.
Ripidolite, 214.
Rittingerite, 102, 104, 121, 143.
Rivotite, 150.
Rock Crystal, 202, 208.
Rock Salt, 174, 177.
Romerite, 157.
Ropperite, 137.
Rottistite, 217.
Romanzovite (var. Garnet).
Romeite, 150.
Roscoelite, 213.
Rose Quartz, 208.
Roselite, 145.
Rubellane, 261.
20

Rubellite, 199.
Ruby, 196, 207.
Ruby Blende (Red Silver Ores), 147.
Ruby Copper (Red Copper Ore), 189.
Ruby Silver (Red Silver Ores), 121, 123, 147.
Rutile, 127, 129, 198, 206.
Ryacolite, 245.

Sagenite (var. Rutile).
Sahlite, 232, 243.
Salammoniac, 174.
Salamstone (var. Corundum).
Salmiac, 174.
Saltpetre, 181.
Salt, 174.
Samarskite, 126, 196.
Sanidine, 245.
Sapphire, 196, 297.
Sapphirine, 197.
Sarcolite, 239.
Sardianite, 151.
Sartorite, 104.
Sassoline, 171.
Saynite, 105.
Scapolite, 240, 252.
Scheelite, 183, 185.
Scheererite, 133.
Schiller-spar, 215.
Schorl, 231, 248.
Schorlomite.
Schrœtterite, 262.
Schreibersite (Meteoric Iron Phosphide).
Schwartzembergite, 176.
Scleroclase, 104.
Scolecite, 272, 277.
Scorodite, 146.
Scoulerite, 272.
Seebachite, 268.
Seladonite, 258.
Selenite, 155, 161.
Sellaite, 178.
Senarmontite, 159.
Sepiolite, 221.
Serbian, 218.
Sericite, 261.
Serpentine, 221, 225.
Serpentine Group, 225.
Seybertite (Clintonite) 219.
Siderite, 136, 140,
Sideromelane, 228.
Sideroplesite, 136.
Sideroschisolite (Cronstedtite), 256.
Sieburgite, 132.

Siegenite, 105.
Silaonite, 106.
Sillimanite, 200.
Silver, 116.
Simonyite (Blodite), 154.
Sismondine, 259.
Sisserskite, 117.
Skutterudite, 101.
Smaltine, 101, 102.
Smaragdite, 243.
Smithsonite, 137, 141.
Soapstone, 242.
Soda Alum, 154.
Soda Nitre, 181.
Sodalite, 237.
Somnite (Nepheline), 239.
Sommervillite,
Sordawalite, 260.
Spartalite, 188.
Spathic Iron Ore, 136, 140.
Spear Pyrites (Marcasite), 108.
Specular Iron Ore, 118, 120.
Sphærocobaltite, 137.
Sphærosiderite, 136.
Sphalerite, 108, 109, 124, 151, 160.
Sphene, 229.
Spinel, 197, 208.
Spodumene, 235.
Stannine, 105, 106.
Stassfurtite, 171, 173.
Staurolite, 198, 210.
Steatite, 222, 225, 242.
Stellarite, 132.
Stellite (Pectolite?).
Steppanite, 110.
Sterlingite (Damourite).
Sternbergite*.
Stiblite, 149.
Stibnite, 110, 112.
Stilbite, 269, 278.
Stilpnomelane, 250.
Stilpnosiderite, 187.
Stolzite, 183, 185.
Strengite, 165.
Striegisan, 167.
Strigovite, 257.
Strogonowite (Altered Scapolite).
Stromeyerine, 106.
Strontianite, 137, 142.
Struvite, 168.

Stypticite, 158.
Succinite, 132.
Sulphur, 130.
Susannite, 135.
Sussexite. 172.
Svanbergite, 155.
Sylvanite, 113.
Sylvine, 174.
Symplesite, 146.
Syngenite, 156.
Sysserskite, 117.
Szaibelyite, 172.

Tabergite, 214.
Tabular Spar, 238.
Tachydrite, 174.
Tachyaphalite, 198.
Tachylite, 228.
Tagilite, 166.
Talc, 214, 225.
Tolcosite, 214.
Talc-Apatite, 163.
Tallingite, 176.
Tantalite, 127.
Tapiolite, 127.
Tarmowitzite, 135.
Tauriscite, 157.
Tekoretine, 133.
Tellurium, 113.
Tengerite, 139.
Tennantite, 103, 104.
Tenorite, 116, 122.
Tephroite, 229.
Tetartin (Albite), 246.
Tetradymite. 113.
Tetrahedrite, 110, 112.
Texasite, 139.
Thallite, 231.
Thenardite, 153.
Thermonatrite, 138.
Thomsenolite, 179.
Thomsonite, 271, 276.
Thorite, 218.
Thraulite, 218, 258.
Thulite, 244.
Thuringite, 256.
Tiemannite, 106.
Tile Ore, 122, 189.
Tinkal, 171.
Tinkalzite, 172.
Tinstone, 127, 129.
Tirolite, 144.

---

* Omitted from Table III., page 105 :

(*Ag and Fe reactions*).

STERNBERGITE: Ag, Fe, S, in somewhat variable proportions. Hex., but commonly in leafy, flexible examples; tombac-brown with blue tarnish; H 1·0-1·5; G 4·2. BB, fusible into a silver-coated magnetic globule.

INDEX.

Titaniferous Iron Ore, 118, 120, 125, 128, 186, 192.
Titanite, 229.
Tecornalite, 175.
Topaz, 197, 209.
Topazolite (Yellow Garnet).
Torbarnite, 132.
Torbernite, 166.
Tourmaline, 199, 211, 231, 244.
Traversellite (var. Asbestus).
Tremolite, 243.
Tridymite, 201.
Triphane, 235.
Triphylline, 164, 170.
Triplite, 163, 169.
Tripoli, 202.
Trœgerite, 146.
Troilite, 105.
Trolleite, 168.
Trona, 138.
Troostite, 203, 222.
Tschermigite, 154.
Tschewkinite, 238.
Tungstic Ochre, 183.
Turgite, 125, 187.
Turnerite, 165.
Turquoise, 167, 170.
Tyrite, 127.
Tyrolite, 144, 147.

Ulexite, 172.
Ullmannite, 103, 110.
Unghwarite, 218.
Unionite, 244.
Uraconise (Uran Ochre), 190.
Uraninite (Pitch Blende), 127, 190.
Uranite, 166.
Uran Mica (Copper-Uranite), 166.
Uran Ochre, 190.
Uranospinnite, 146.
Uran Pitch Ore, 127.
Urao (Thermonatrite), 138.
Urpethite, 133.
Uwarowite, 198.

Valentinite, 149.
Vanadinite, 183.
Variscite, 167.
Varvicite (Warwickite), 126.
Vauquelinite, 182.
Velvet Copper Ore, 158.
Verde Antique Marble, 225.
Vermiculite, 262.
Vesuvian, 244, 249.
Villarsite, 221.

Vitreous Copper Ore (Copper Glance), 106.
Vitreous Silver Ore (Silver Glance), 107.
Vitriol, blue, 156.
Vitriol, cobalt, 157.
Vitriol, green, 156.
Vitriol, manganese, 158.
Vitriol, nickel, 157.
Vitriol-Ochre, 158.
Vivianite, 165, 170.
Voglianite, 157.
Vœlknerite, 191.
Voglite, 139.
Voigtite, 256.
Volborthite, 184.
Voltaite, 157.
Voltzine, 151.
Vulpinite (Heavy Spar), 152.

Wad, 188, 193.
Wagnerite, 163.
Walpurginite, 145.
Wapplerite, 146.
Warringtonite, 158.
Warwickite, 203, 222.
Wavellite, 167, 170.
Websterite, 155, 162.
Wehrlite, 112.
Weissite, 220.
Wernerite, 240, 252.
Whewellite, 191.
Whitneyite, 101.
Wichtisite, 234.
Wichtyne, 234.
Willemite, 203, 222.
Williamsite (var. Serpentine).
Wilsonite, 266.
Wiluite, 245.
Wiserite, 139.
Wthamite, 231.
Witherite, 137, 142.
Wittichenite, 107.
Wœhlerite, 240.
Wœlchite (var. Bournonite).
Wœrthite, 200.
Wolfachite, 103.
Wolfram, } 116, 122, 123, 183.
Wolframite, }
Wolfsbergite, 111.
Wolchonskoite, 218.
Wollastonite, 238.
Wolnyn (Heavy Spar), 152.
Wood Opal, 202, 209.
Wood Tin, 206.
Woodwardite, 158.
Wulfenite, 183, 185.
Wurtzite (Hexagonal Blende).

Xanthite, 245.
Xanthacone, 143.
Xanthophyllite, 216.
Xanthosiderite (var. Brown Iron Ore).
Xenolite, 200.
Xenotime, 164.
Xylite, 260.
Xylotile, 258.

Yellow-Ochre, 187.
Yttrocerite, 179.
Yttrotantalite, 127.
Yttrotitanite, 230.

Zalpaite, 107.
Zaratite, 139.

Zeagonite, 271.
Zeolites, 273 to 278.
Zepharovichite, 167.
Zeuncrite, 144.
Ziegenite, 105.
Zinc Blende, 108, 109, 124, 129, 151, 159.
Zinc Bloom, 139.
Zincite, 188, 193.
Zinc Spar, 137.
Zinc Vitriol (Goslarite), 155.
Zinkenite, 111.
Zippeite, 157.
Zircon, 198, 207.
Zoisite, 244, 249.
Zorgite, 106.
Zwieselite, 164.
Zygadite, 246.

www.ingramcontent.com/pod-product-compliance
Lightning Source LLC
Chambersburg PA
CBHW021956220426
43663CB00007B/834